FACTORS AFFECTING CALF CROP

Biotechnology of Reproduction

FACTORS AFFECTING CALF CROP

Biotechnology of Reproduction

Edited by

Michael J. Fields
Robert S. Sand • Joel V. Yelich

CRC Press is an imprint of the
Taylor & Francis Group, an **informa** business

CRC Press
Taylor & Francis Group
6000 Broken Sound Parkway NW, Suite 300
Boca Raton, FL 33487-2742

First issued in paperback 2019

ISBN-13: 978-0-8493-1117-8 (hbk)
ISBN-13: 978-0-367-39670-1 (pbk)

Library of Congress Cataloging-in-Publication Data

Factors affecting calf crop : biotechnology of reproduction / edited by Michael J. Fields, Robert S. Sand, Joel V. Yelich.
p. cm.
Includes bibliographical references and index.
ISBN 0-8493-1117-9 (alk. paper)
1. Beef cattle—Breeding. 2. Beef cattle—Reproduction. 3. Beef cattle—Biotechnology. 4. Calves. I. Fields, Michael J. II. Sand, Robert S. (Robert Sherman), 1941- III. Yelich, Joel V.

SF201 .F33 2001
636.2′13—dc21 2001043175

Visit the CRC Press Web site at www.crcpress.com

Library of Congress Card Number 2001043175

Visit the Taylor & Francis Web site at
http://www.taylorandfrancis.com

and the CRC Press Web site at
http://www.crcpress.com

Preface

Factors Affecting Calf Crop, published in 1994, included the following in its preface:

> "Evolution of knowledge and understanding of the biological processes involved in the reproduction of beef cattle has contributed to the growth and profitability of the industry in the Southeast as well as the rest of the nation. The integration of knowledge of genetic and physiological processes in a difficult environment so that improvement is made both in productivity and product quality has been a tremendous achievement that led to the development of a large dynamic industry producing a product that is in great demand. This book is dedicated to two of the leaders in this effort who utilized their skills as scientists to further our understanding and knowledge of the basic processes involved and at the same time assisted in conveying the information to students, producers, and industry leaders in a way they could understand and apply."

Our goal for this book is to continue the tradition established with *Factors Affecting Calf Crop* and to summarize the current knowledge of the biotechnologies impacting the beef calf crop, particularly in the subtropics. Not all of the technologies discussed here are applicable today, but we feel it is important for both the scientific community and today's well-trained ranch managers to be aware, not only of what can be used today, but also of what is being developed for future application so they can anticipate how this technology can best be applied to produce a better product at less cost.

This project has depended on the cooperation of a large number of people for its completion. We appreciate the chapter authors who have agreed to donate the royalties from this book to a scholarship fund. We are deeply indebted to Glenda Tucker, Pam Gross, and Dr. Shou-Mei Chang for their help in preparing the manuscripts for publication and to Hilary Binta, Nicole Nichols, and Sarah Balaguer for editorial assistance.

About the Editors

Michael J. Fields, Ph.D., is a Professor of Animal Physiology, Department of Animal Sciences, Institutes of Food and Agricultural Sciences, University of Florida, Gainesville. Dr. Fields graduated in 1966 from the University of Florida with a B.S. degree in animal science and obtained his M.S. degree in reproductive physiology from the same institution. Following service in the U.S. Marine Corps, he received his Ph.D. in reproductive physiology from Texas A&M University.

Dr. Fields is a member of the American Society for the Study of Reproduction, American Society of Animal Science, and the American Registry of Professional Animal Scientists. He has served as president of the Florida Chapter of Gamma Sigma Delta, Phi Kappa Phi, and Sigma Xi.

He has published over 200 scientific and popular articles of his research and has been awarded grants from the National Science Foundation, National Institutes of Health, and U.S. Department of Agriculture for his internationally recognized work on hormones secreted by the corpus luteum and their function in the cow. For these and other activities, he received the Florida Chapter of Gamma Sigma Delta International Award for Distinguished Service to Agriculture and the chapter's Senior Faculty Award, the University of Florida Professorial Award of Excellence, as well as recognition as the Outstanding Teacher in the College of Agriculture.

Robert S. Sand, Ph.D., is an Associate Professor and Extension Livestock Specialist, Department of Animal Sciences, Institute of Food and Agricultural Sciences, University of Florida, Gainesville. Dr. Sand graduated from Colorado State University with a B.S. in animal production in 1963. Following a tour in the U.S. Army, he earned an M.S. degree in 1969 and a Ph.D. in reproductive physiology in 1971 from the University of Kentucky. Dr. Sand was a Ralston–Purina and N.D.E.A. Fellow during his graduate program.

Dr. Sand provides leadership for the statewide beef cattle extension program, with special emphasis on reproductive management and cow–calf production. He is responsible for development of educational materials, and demonstrations for county extension personnel, producers, and youth. Additional responsibilities include serving as liaison with the Florida Beef Cattle Improvement Association, and as supervisor of the Florida Bull Test. Another responsibility is evaluating and determining research needs in the area of applied beef cattle management.

Dr. Sand is a member of the American Society of Animal Science, the American Registry of Professional Animal Scientists, Alpha Zeta, Gamma Sigma Delta, Sigma Xi, Phi Kappa Phi, and Epsilon Sigma Phi, and has served over 20 years as secretary of the Florida Beef Cattle Improvement Association.

Joel V. Yelich, Ph.D., is an Assistant Professor in reproductive physiology, Department of Animal Sciences, Institute of Food and Agricultural Sciences, University of

Florida, Gainesville. Dr. Yelich graduated from Montana State University in 1986 with a B.S. degree in animal science. He earned his M.S. in animal science from Colorado State University in 1989 and went on to receive his Ph.D. in animal breeding and reproduction from Oklahoma State University in 1994.

Dr. Yelich is a member of the American Society of Animal Science, Society for Reproduction, Society of Range Management, and the National Cattleman's Beef Association. Dr. Yelich is actively involved in undergraduate teaching as well as maintaining a production-oriented research program investigating factors that affect reproductive efficiency in cattle of *Bos indicus* breeding.

Contributors

R. L. Ax
Department of Animal Sciences
University of Arizona
Tucson, Arizona

W. E. Beal
Department of Animal Science
Virginia Tech
Blacksburg, Virginia

M. E. Bellin
Department of Animal Sciences
University of Arizona
Tucson, Arizona

L. Benvenisti
Department of Hormone Research
Kimron Veterinary Research
Bet Dagan, Israel

C. R. Burke
The Ohio State University
Columbus, Ohio

Chad C. Chase, Jr.
USDA/ARS
Subtropical Agricultural Research Station
Brooksville, Florida

Robert J. Collier
Department of Animal Sciences
University of Arizona
Tucson, Arizona

M. L. Day
Department of Animal Science
The Ohio State University
Columbus, Ohio

S. K. DeNise
Department of Animal Sciences
University of Arizona
Tucson, Arizona

M. Gurevich
Department of Hormone Research
Kimron Veterinary Research
Bet Dagan, Israel

E. Harel-Markowitz
Department of Hormone Research
Kimron Veterinary Research
Bet Dagan, Israel

John F. Hasler
Em Tran, Inc.
Elizabethtown, Pennsylvania

H. E. Hawkins
King Ranch, Inc.
Kingsville, Texas

T. R. Holm
Celera Ag Gen
Davis, California

E. Keith Inskeep
Division of Animal Science
West Virginia University
Morgantown, West Virginia

F. N. Kojima
Department of Animal Science
University of Missouri
Columbia, Missouri

G. Cliff Lamb
North Central Research and Outreach Center
University of Minnesota
Grand Rapids, Minnesota

J. W. Lauderdale
Emeritus
Worldwide Animal Health
Pharmacia Upjohn
Kalamazoo, Michigan

J. F. Medrano
Department of Animal Sciences
University of California–Davis
Davis, California

Karen Moore
Department of Animal Sciences
University of Florida
Gainesville, Florida

J. N. Oyarzo
Department of Animal Sciences
University of Arizona
Tucson, Arizona

D. J. Patterson
Department of Animal Science
University of Missouri
Columbia, Missouri

D. Owen Rae
Department of Large Animal Clinical Sciences
College of Veterinary Medicine
University of Florida
Gainesville, Florida

R. Michael Roberts
Department of Animal Science
University of Missouri
Columbia, Missouri

James M. Robl
Paige Laboratory
University of Massachusetts
Amherst, Massachusetts

J. J. Rutledge
Department of Animal Science
University of Wisconsin
Madison, Wisconsin

George E. Seidel, Jr.
Animal Reproduction/Biotechnology Laboratory
Colorado State University
Fort Collins, Colorado

M. Shemesh
Department of Hormone Research
Kimron Veterinary Research
Bet Dagan, Israel

L. S. Shore
Department of Hormone Research
Kimron Veterinary Research
Bet Dagan, Israel

M. F. Smith
Department of Animal Science
University of Missouri
Columbia, Missouri

Audy Spell
Cyagra of Kansas
Manhattan, Kansas

Jeffrey S. Stevenson
Department of Animal Sciences/Industry
Kansas State University
Manhattan, Kansas

Y. Stram
Department of Molecular Virology
Kimron Veterinary Institute
Bet Dagan, Israel

S. L. Wood
Pharmacia Animal Health
Kalamazoo, Michigan

Joel V. Yelich
Department of Animal Sciences
University of Florida
Gainesville, Florida

H. M. Zhang
Department of Animal Sciences
University of Arizona
Tucson, Arizona

Table of Contents

1 Developments in Reproductive Biotechnology that Will Improve the Weaned Calf Crop

Robert J. Collier

CONTENTS

The animal breeding industry worldwide is restructuring, with consolidation occurring both vertically and horizontally. This process is occurring during a time of simultaneous breakthroughs in reproductive technologies, genomics, bioinformatics, and molecular biology. The result will be a much smaller group of companies that will have evolved from traditional artificial insemination organizations into breeding companies with nucleus herds. These companies will offer a variety of products including marker-assisted selection (MAS), embryos preselected for certain traits, clones, gender-specified semen and embryos, and reproductive technologies which will improve fertility. The overall impact of these technologies will be to add value to a pregnancy and improve the uniformity and quality of beef.

The rate of genetic progress is generally described by the equation shown in Figure 1.1. The objective of a breeding program is to increase the accuracy and intensity of selection and genetic variety while decreasing the generation interval. The aim of this chapter is to review new technologies and their potential impact on the components of a breeding program.

0-8493-1117-9/02/$0.00+$1.50

$$\textbf{Rate of Genetic Progress} = \frac{\text{Accuracy} \times \text{Intensity} \times \text{Genetic Variation}}{\text{Generation Interval}}$$

Goal = Increase

accuracy of selection

intensity of selection

genetic variation

Decrease

generation interval

FIGURE 1.1 Genetic progress.

IMPROVED REPRODUCTIVE PERFORMANCE

Adding value to a pregnancy includes reducing the cost of obtaining the pregnancy. New technologies in estrous detection, synchronization, and insemination are becoming available which will reduce the cost of establishing a pregnancy and thereby justify added investment in genetics. Although only 5% of beef producers utilized artificial insemination in 1999, this is expected to increase with improved synchronization techniques. Currently, there are three commonly used synchronization methods: prostaglandin injection, Syncro-Mate-B®, and feeding melengestrol acetate (MGA). These products can be used alone or in combination to synchronize estrous (Larson and Ball, 1992). Cattle have follicular waves every 7 to 10 days under most physiological conditions. Each wave is associated with an increase in follicle stimulating hormone (FSH), and the frequency of luteinizing hormone (LH) secretory pulses determines the fate of the dominant follicle (Thatcher et al., 1993; Twagiramungu et al., 1995). Present approved methods for estrous synchronization do not regulate the follicular wave sufficiently to maximize fertility or to permit a single timed insemination by artificial insemination. A better understanding of the hormonal control of follicular waves has already led to development of improved hormonal regimens such as Ov-Synch® (Pursley et al., 1995). Future synchronization techniques will take advantage of improved understanding of the hormonal control of follicle waves to provide estrous synchronization, with sufficiently high pregnancy rates to a single artificial insemination to greatly increase the use of artificial insemination in the beef industry. Thus, improved synchronization is essential to the use of the genetic technologies available.

GENDER SELECTION

The ability to determine gender would add considerable value to a pregnancy. The only proven sexing technology is the Beltsville Sperm Sexing Technology (Johnson et al., 1987), now being commercialized by XY Inc. This technology takes advantage of the fact that there are differences in the DNA content of X- and Y-bearing sperm which lead to slight but detectable differences in the size and shape of the sperm head. Use of high-speed laser flow cytometry has brought the technology close to commercial capability. It is now possible to produce sexed embryos *in vitro* for embryo transfer (Rath et al., 1999). Combining production of sexed embryos with marker-assisted selection of those embryos has the potential to accelerate rate of genetic gain while increasing the availability of market animals. It is likely that additional breakthroughs in sorting speed and insemination techniques will occur that will allow full commercial use of the technology across domestic animal species. Value of gender selection to producers is generally estimated at two to three times present semen values. This number would be increased if genetic markers were utilized to maximize specific traits.

MARKER-ASSISTED SELECTION

The genome of cattle contains about 50,000 genes which regulate all aspects of growth, development, and function. However, very few of these genes have actually been identified. In the past, we have relied on production records and use of quantitative genetics to evaluate differences between animals. If the actual genes that regulate traits could be identified, the rate of genetic progress could be greatly increased. The rapid progress of the Human Genome Project has had a positive impact on the progress in mapping domestic animal genomes. Since there is great conservation of genetic information across species, much of the information obtained in the Human Genome Project can be utilized to help locate genes in domestic animal species.

In the pioneer days of genetics, markers were phenotypic traits, such as coat color, that could be followed in animals to identify their hereditary background. Today, we use unique genetic sequences to identify segments of a genome. Mapping programs in cattle and swine are well under way. Genetic linkages for cattle have been constructed. The general approach for a mapping program is shown in Figure 1.2. The process of mapping a genome is complex and requires several different stages. Utilization of markers on chromosomes has accelerated our discovery of quantitative trait loci (QTL) that are associated with production traits. The utilization of marker-assisted selection (MAS) for genetic selection is most informative within sire families rather than across sire families (Soller and Beckman, 1990). Also, utilization of a marker is not the same as identifying the specific genes for a trait, but it has value for the beef industry. In time, we will know the specific genes regulating production traits in beef cattle and will be able to be more accurate than present technology allows. The overall impact of the use of markers will be to increase accuracy of selection. It is likely that marker-assisted selection will find its greatest use within nucleus herds.

NUCLEUS HERDS

The objective of the use of nucleus herds is to increase the intensity of selection by combining MAS with a multiple ovulation embryo transfer (MOET) scheme. This permits maximization of best genetics across less desirable animals by using them as hosts for embryos from superior crosses. The cost of establishing a nucleus herd has proven to be a barrier to all but a few artificial insemination organizations. Use of a nucleus herd accelerates the rate of genetic progress by increasing the intensity of selection. Nucleus herds can be centralized (one location) or dispersed (several locations).

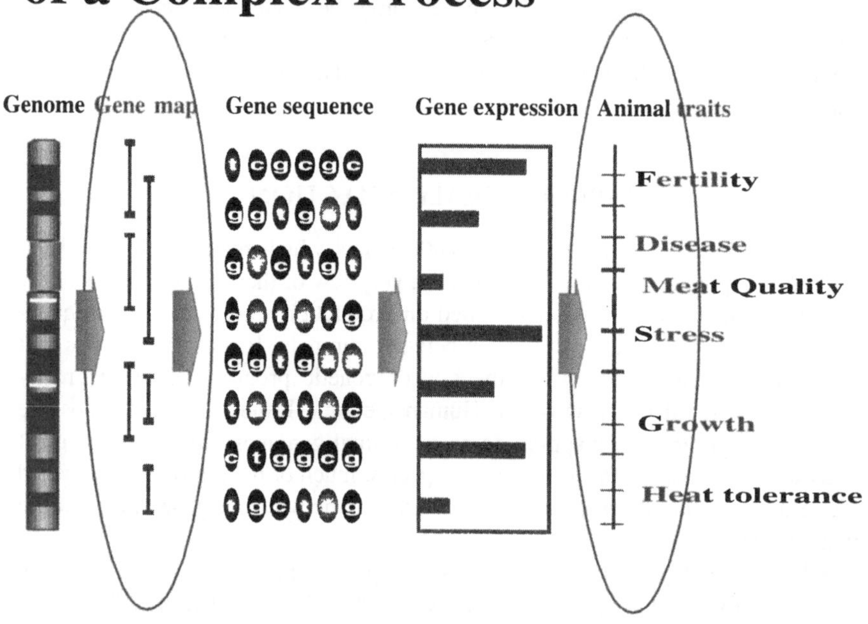

FIGURE 1.2 Strategies in genome mapping.

They can also be open or closed depending on whether or not they are willing to utilize genetics outside the herd. A challenge to owners of nucleus herds will be to control access to the genetics while obtaining maximum value. Nucleus herds will be an essential component of evolution of the artificial insemination industry into a breeding industry which offers a wide variety of value-added germ plasm.

CLONING

Cloning can be utilized to maximize the number of embryos from a superior cross via embryo cloning or to maximize the presence of a specific animal via adult cloning. The advantage of adult cloning is that you already have a record on a given animal. Recent breakthroughs in the production of clones from stem cells offer great promise for reducing the costs of cloning. However, overall success of cloning must be improved to make this technology a commercial reality. Only 0.5% of clones survive to postnatal age. Access to cloned animals does not mean that spreading copies of a given animal across an industry will result in the same production records. Instead, we should expect a bell-shaped curve as herd–year–season effects are expressed. The shape of this curve should be steeper, however, than in the general population since the genetics of the animals will be identical. Copies of cloned embryos can be frozen while records are obtained on a subset of the clones, permitting identification of the superior embryo(s) to be utilized. It is thus likely that embryo and adult cloning will each occupy a niche to accelerate genetic progress in the beef industry.

A great value of cloning will be the first opportunity to address traits of low heritability, such as fertility and disease resistance. In the beef industry, it is likely that the first animals to be cloned will be bulls to be used for breeding purposes.

GENE INSERTION

Identification of genes of value to the beef industry is not limited to production traits. It is now possible to consider improving the value of domestic animal food products by removing compounds that are harmful to human health and maximizing those that are beneficial to human health. Strategies utilizing gene insertion techniques in cattle have been identified (Hoeschele, 1990). It is clear that the two major trends of the next decade, and possibly the century, will be consumerism and environmentalism. Use of gene insertion techniques offers the potential to address both of these issues. It is also clear that consumers are wary of moving genes from plants to animals or across animals and are especially concerned about placing human genes in animals. We should remember that there are plenty of opportunities to stick to bovine genes to make use of unique differences within cattle to maximize the value of our products and the health of our animals, and minimize the impact on the environment.

REFERENCES

Hoeschele, I. 1990. Potential gain from insertion of major genes in dairy cattle. *J. Dairy Sci.* 73:2601.

Johnson, L.A., J.P. Flook, and M.V. Look. 1987. Flow cytometry of X and Y chromosome-bearing sperm for DNA using an improved preparation method and staining with Hoechst 33342. *Gamete Res.* 17:203.

Larson, L.L. and P.J.H. Ball. 1992. Regulation of estrous cycles in dairy cattle: a review. *Theriogenology* 38:255.

Pursley, J.R., M.O. Mee, and M.C. Wiltbank. 1995. Synchronization of ovulation in dairy cows using $PGF_{2\alpha}$ and GnRH. *Theriogenology* 44:915.

Rath, D.C., R. Long, J.R. Dobrinsky, G.R. Welch, L.L. Schrier, and L.A. Johnson. 1999. *In vitro* production of sexed embryos for gender preselection: high speed sorting of X-chromosome bearing sperm to produce pigs after embryo transfer. *J. Anim. Sci.* 77:3346.

Soller, M. and J.S. Beckman. 1990. Marker-based mapping of quantitative trait loci using replicated progenies. *Theor. Appl. Genet.* 80:205.

Thatcher, W.W., M. Drost, J.D. Savio, K.L. Macmillan, K.W. Entwistle, E.J. Schmitt, R.L. De la Sota, and G.R. Morris. 1993. New clinical uses of GnRH and its analogues in cattle. *Anim. Reprod. Sci.* 33:27.

Twagiramungu, H., L.A. Guilbault, and J.J. Dufour. 1995. Synchronization of ovarian follicular waves with a gonadotropin-releasing hormone agonist to increase the precision of estrous in cattle: a review. *J. Anim. Sci.* 73:3141.

2 Bovine Estrus: Tools for Detection and Understanding

D. Owen Rae

CONTENTS

Bovine estrus has been described as a short period (approximately 15 to 18 hours) of sexual receptivity that is manifested every 18 to 24 days, with ovulation occurring 10 to 14 hours after the cessation of behavioral signs of estrus (Foote, 1975; Esslemont et al., 1980; Smith, 1986; Allrich, 1993; Table 2.1).

Signs of estrus include standing to receive mounting or mounting other cows. A skilled observer will notice more subtle signs, such as following, standing with, head resting, sniffing, nuzzling, licking, and grouping with other cows in or near estrus. These animals are active, nervous, restless, bawling, walking, and searching (Williamson et al., 1972a,b; Allrich, 1993). As a result of this behavior other signs may be evident, such as a roughened tail head and clear vaginal mucus hanging from the vulva or smeared at the thigh. The vulva may appear swollen, moist, or reddened. As estrus and ovulation occur, evidence of these events may be noted by a bloody mucus discharge from the vagina 2 to 4 days following estrus. It is then too late for insemination.

ESTROUS DETECTION

Estrous detection is the process by which those behaviors representing physiological changes leading to ovulation and receptivity to insemination are detected. Indeed, the sole purpose for estrous detection is to identify an appropriate time to inseminate

0-8493-1117-9/02/$0.00+$1.50

TABLE 2.1
Signs Associated with Estrus during an 18-Hour Period

Mounting	Behavior	External Genitalia	Mucus	Bloody Discharge	Tail Head Hair
		Early			
Mounts other cows	Bawling, fence walking, butting, following, nervousness	Lips of vulva red and slightly swollen	Very little, watery	None	Not ruffled or matted
		Middle			
Stands to be ridden, also will mount	Complacent, friendly, following, licking, not eating, restless	Lips of vulva red and slightly swollen, walls of vagina moist and glistening	Abundant, clear, copious	Seldom	Slight to observable roughening and matting
		Late			
Will not stand to be mounted, will mount other cows	All signs of nervousness disappear	Swelling decreased	Decreased, very sticky, rubbery	1–3 days following signs of standing	Pronounced roughening and matting

From Herrick (1978). With permission.

the female in order to impregnate her. Standing to be mounted or standing estrus is considered to be the most important sign (Foote, 1975; French et al., 1989; Allrich, 1993). Of course, the bull is our most efficient detector of estrus. Human observation to detect estrus is of practical importance in exclusively female herds where artificial insemination is performed or where genetic improvement by use of artificial insemination is elected. However, undetected and falsely detected estrus results in missed and untimely inseminations, with consequent loss of income.

Studies conducted over the years have evaluated these preovulatory mounting behaviors and associated estrous behaviors. The objective of many of these studies has not only been to gain a better understanding of the estrous cycle and estrous behaviors, but also to improve the detection of estrus and, more important, the time of ovulation to optimize timing of insemination, thus improving the likelihood of pregnancy. Detection of estrus has been accomplished in many ways with varying levels of success. Detection of standing estrus by human observation has been the method of choice in identifying a cow in estrus and for establishing an insemination time (Williamson et al., 1972a; Foote, 1975; Lehrer et al., 1992; Allrich, 1993; Senger, 1994). The success of estrous detection, regardless of the method, is measured by two important indices: detection efficiency and detection accuracy. These are defined and calculated as shown in Table 2.2 (Lehrer et al., 1992).

TABLE 2.2
Estrous Detection Measures

Efficiency = Number of correct detections/total number of estrous periods (or progesterone-determined nonluteal periods) × 100.
How well are cows recognized in estrus during the interval of the estrous period?

Accuracy = Number of correct detections/number of correct and false-positive detections × 100.
Of cows presented for AI, how many are truly in estrus?

The ideal system for detection of estrus has been described by Senger (1994) as providing: 1) continuous (24 hours/day) surveillance of the cow; 2) accurate and automatic identification of the cow in estrus; 3) operation for the productive lifetime of the cow; 4) minimized or no labor requirements; and 5) high accuracy (95%) in identifying the appropriate physiologic or behavioral events that correlate highly with ovulation. The methods employed to detect estrus have been many and divergent in their approach. Estrous detection aids have been developed to assist human observation, including nonautomated, automated, and telemetric methods. Some of these include mount detectors, tail chalk, teaser bulls or androgenized cows, and video recordings, which have been well reviewed in several manuscripts (Williamson et al., 1972b; Foote, 1975; Lehrer et al., 1992; Senger, 1994). More recent advances in detection methods include automated methods such as activity monitoring by pedometer and rump-mounted, pressure-sensitive electronic mount detection devices (Pennington et al., 1986; Senger, 1994; Stevenson et al., 1996). This chapter will review these methods and discuss the advantages and disadvantages of each (Table 2.3) and how they might fit into a management plan. Also, using a pressure-sensitive electronic mount detection system, University of Florida researchers, as well as others, have now reported new and interesting findings relevant to the expression of estrus. These are briefly reported to illustrate the use of these tools.

ESTROUS DETECTION METHODS

OBSERVATION

The standard established for the observed detection of estrus is a twice daily, 30-minute observation for signs consistent with estrus in the morning and evening. The observer detects overt and more subtle signs representing estrus in the observed cows. An astute observer becomes very familiar with the cows within the herd and identifies those behaviors representing estrus. Observation is made easier by the patterned gathering of cows for observation. This may be accomplished by feeding a supplement at the time of observation. The factor, however, that is most critical to the success of detection by observation is the motivation of the human observer.

TABLE 2.3
Summary of Detection Methods and Aids are Presented, Briefly. Noting Advantages and Disadvantages of Each, the Reported Range of Efficiency and Accuracy for Each Method, and Its Estimated Cost

Detection Method	Advantage	Disadvantage	Range Reported (%)		
			Efficiency	Accuracy	Cost
Observation	Actual observation of animals.	Time commitment is required.	51–94	50–98	Time
Painted/ chalked	Simple and inexpensive.	Repeated chalk application is required.	66+		$1/cow
Kamar® detection	Relatively simple and inexpensive.	Detectors can be lost and erroneously triggered.	80+		$1–2/cow head
Chin ball marker on bulls and androgenized cows	Bulls and androgenized cows are very effective detectors.	Animal and equipment must be maintained.	79+		$150 plus ink
Continuous video recording	A continuous visually recorded record of activity is produced.	Placement and maintenance of video equipment. Video evaluation time.	56–94	~50	$400–800 plus videotape.
Hormonal assay—progesterone	It is best able to confirm estrous detection errors.	Poor accuracy of detection.	60–100	<50	$4–10/cow
Electrical impedance	A quick, simple procedure in gathered cows.	Large variations in values require continual monitoring.	65–82	57–82	$50–300/probe
Elevation in body or milk temperature	A quick, simple procedure in gathered cows.	Large variations in values require continual monitoring.	50–74	55+	

TABLE 2.3 (Continued)
Summary of Detection Methods and Aids are Presented, Briefly. Noting Advantages and Disadvantages of Each, the Reported Range of Efficiency and Accuracy for Each Method, and Its Estimated Cost

Detection Method	Advantage	Disadvantage	Range Reported (%) Efficiency	Accuracy	Cost
Radiotelemetric activity monitor	Continuous monitoring of activity.	Cost and maintenance demands of equipment and pedometers.	62–81	22–100	Individual cow and system costs.
Radiotelemetric mount detector	Continuous mount detection.	Cost and maintenance demands of equipment and transmitters.	89–92	88–100	See Table 2.5.

NONAUTOMATED MEANS OF DETECTION

To aid the observer, several commercial and noncommercial detection aids have been employed. One method is the simple application of chalk or paint to the rump of the female, followed by watching for the smearing or rubbing-off of the chalk. Observation is required and continual renewing of the chalk is necessary. A slightly higher application of technology is the rump-mounted mount detector (i.e., Kamar®) which provides a signal giving evidence of a mounting event. The Kamar patch, for example, changes from white to bright red when the contents of the tube are compressed under the pressure of a mounting herd mate.

A recently introduced technology builds upon the basic principle of these methods. It is an electronic heat detection device (MountCount™) designed to bridge the gap between low-technology products such as Kamar patches and high-technology products such as Heat-Watch® (to be described later). It provides accuracy similar to the radiotelemetry devices at a price nearer that of the Kamar. It has a pressure-sensitive button that detects mounting events and alerts the observer that the cow or heifer is in estrus and ready to breed.

These commercial aids are designed to augment the observer's effectiveness in detecting cows in estrus. The cost of these aids is relatively low, and they are generally easy to implement. The major drawback is the continued monitoring and care required to keep these aids attached to the cow. And certainly, the chance triggering of detectors reduces the accuracy of detection.

An alternate approach to detection is to permit an animal observer to detect estrus. The use of bulls or androgenized females as teaser animals, with or without a chin ball marking device, provides a continuous method of observation within the group of sexually active females. Bulls not suitable for use as herd sires, yet healthy and sound, may be used as "teaser bulls" when subjected to vasectomy, epididymectomy, penile deviation, penile block, or a combination of these. These procedures are best performed by a veterinarian and are well described by Walker (1990). It is usually preferable that the bull not be capable of intromission and be sterile, while maintaining a high level of sex drive. These criteria provide an aggressive detector of estrus and one which will not cause an unwanted pregnancy or venereal disease (Table 2.4).

TABLE 2.4
Preparation of Teaser Animals for Detection of Estrus, Including Advantages and Disadvantages of Each

Procedure	Advantages	Disadvantages	
Bull penile block	A penile blocking device is installed in the prepuce.	1. Allows for normal mounting. 2. Prevents extension of penis, thus insemination. 3. Helps prevent the spread of venereal diseases.	1. Effective for only 1 year, bulls tend to lose sex drive.
Bull vasectomy	The vas deferens is surgically severed, causing sterility.	1. Normal libido. 2. Biostimulation from the sterile bull increases conception.	1. Can spread venereal disease.
Bull penile deviation	The prepuce and penis are moved, being redirected to one side, to make intromission unlikely.	1. Provides better detection than vasectomy or penile block. 2. No venereal disease spread. 3. No loss of libido, thus longer work life.	1. None noted.
Bull caudal epididymectomy	The tail of the epididymis is removed, preventing sperm from reaching the penis.	1. Relatively simple to perform. 2. Economical method of preparation.	1. Can spread venereal disease.
Androgenized female	Testosterone propionate or enanthate injections or Synovex-H implants are administered prior to the breeding season.	1. Longer work life. 2. Safer than bulls. 3. Treatments are less expensive than surgical bull procedures. 4. The and rogenized cow can be one that lost her calf, thus no extra maintenance costs.	1. None noted.

Modified from *Beef* (February 2000).

Androgenized cows are also effective as detectors of estrus. Cows to be androgenized can be selected from those that have lost their calves during the calving season or those awaiting culling. Androgenized cows are those treated with testosterone. Numerous protocols with differing products and treatment intervals have been proposed and used effectively (Anonymous, 1979; Britt, 1980; Mortimer et al., 1990). Treated cows function as estrous detectors and are then marketable following an appropriate drug withdrawal period. The application of more than a single detection method improves the chances of identifying females in estrus (Lehrer et al., 1992). Gwazdauskas et al. (1990) demonstrated that androgenized cows in conjunction with a rump-mounted device (Kamar) improved estrous detection by 1 to 6% above either method alone. It has also been reported that dogs and other mammals may be trained to detect odors characteristically associated with estrus (Allrich, 1993).

One or more video cameras placed so as to observe the area where females are maintained provides a continuous record of their activity. Cameras are placed in a fixed or panning motion. Cameras can be purchased which permit a time-delay recording (i.e., snap shots are taken at timed intervals), thus reducing the amount of tape that must be reviewed. This technique has obvious limitations, which include: the need to potentially review large quantities of videotape, the need for adequate camera coverage of the area in which cows are maintained, and the potential difficulty of identifying animals within the group demonstrating the estrous behavior. However, this technique offers advantages over those mentioned before in that mounting activity may be quantified and the duration of behaviors associated with estrus can be determined.

Hormonal assay of plasma progesterone, which has been used widely in research and applied to a limited degree in the dairy industry, has only been infrequently applied in the beef industry. Most popular are cow-side tests for progesterone in milk, serum, or urine. Hormone assay is best used to detect periestrous nonluteal periods and to indicate estrous detection errors (Lehrer et al., 1992). The efficiency of these tests may vary from 56 to 94%, but usually exceeds that of a human observer. However, the accuracy is usually less than 50%, which is lower than visual observation. This technology is best utilized as a method of evaluating cows that are suspect, that perhaps have triggered a detection device but show no signs of estrus. The progesterone assay will accurately identify those that are not in estrus (Moore and Spahr, 1991).

During the periestrous period, one of the characteristic signs of estrus is a swollen vulva due to an increased hydration of the genitalia. As a result, there is a change in the electrical impedance or resistance of the vulva, the vagina, and its secretions. The tissue hydration is inversely related to tissue electrical impedance. Thus, as a cow enters the follicular phase of the estrous cycle, the electrical resistance of vaginal tissues and secretions decreases, reaching a nadir during estrus. By measuring the electrical resistance (ohms) manually by means of a vaginal probe over a course of days or by use of a radio-telemeterable indwelling ohm meter, the cycle can be monitored (Senger, 1994). The efficiency of this procedure is 56 to 82%, and the accuracy is reported to be in a similar range, i.e., 57 to 82%. This technique, however, has an unpredictable pattern, and a marked variation exists among animals and even within an animal measured at different times (Lehrer et al., 1992).

Elevation in intravaginal temperatures from 0.3 to 1.1°C or milk temperature from 0.2 to 0.4°C may aid in estrous detection but is not reliable as a sole means of detection (Lehrer et al., 1992). Certainly, in pastured cattle in warmer climates this method is of limited value, but may have potential, especially in the dairy industry, as an automated or telemeterable application in combination with other detection methods.

AUTOMATED AND TELEMETERABLE METHODS

It is important to develop and apply newer technologies that provide efficient, accurate methods of identifying females in estrus, that permit elimination of the reliance on visual observation and the less effective tools of detection. It should be noted that several of the methods previously described have potential for automated and/or telemeterable applications. This section will focus on the applications which detect and monitor activity and mounting.

It has been recognized for many years that cows are much more active during estrus than during other times of the reproductive cycle. It has been observed that activity during estrus is two to four times that of cows in diestrus (Senger, 1994). This observation led to the first application of an activity monitoring tool in the form of a pedometer. This method is particularly popular in larger dairies, where the pedometer also provides a means of identification.

The pedometer (i.e., Afikim®) is a leg band that electronically transmits signals to a receiver, which is often mounted in the milking parlor, providing a twice- or three-times-per-day source of information on cow activity and distance traveled. The efficiency of this method has been reported (Lehrer et al., 1992) as better than a twice-per-day visual observation and equal to a four-times-per-day visual observation. The efficiency is from 60 to 100%. The method detects 63 to 79% of nonluteal periods and ovulations unaccompanied by behavioral estrus (Lehrer et al., 1992). The accuracy is from 22 to 100%. This wide variation is attributable to limitations of the pedometer, i.e., unstable cattle handling or management and erratic environmental conditions. Under stable management and environmental conditions, the accuracy is reported at 91 to 92%, depending on the threshold established for significantly increased activity (Pennington et al., 1986; Moore and Spahr, 1991). Efficiency and accuracy of the pedometer are inversely related.

The electronically transmitted mount detector system (i.e., HeatWatch®) records mounting activity in cows fitted with a transmitter, thus providing a 24-hours-a-day mounting record. This system includes a transmitter placed within a disposable pouch glued to the rump of the cow, a receiver, and buffer. The transmitter has a pressure-activated trigger, which sends a transmitter-specific signal to the receiver unit located within the transmission reception area of the unit (about one quarter of a mile). The receiver is hardwired to a buffer unit located adjacent to a personal computer. There, mount data are stored until called up by the proprietary software programmed to collate and organize mount data. The raw data transmitted include the transmitter number, date, time, duration of signal transmission, and strength of signal. From this raw data, information useful to the producer is generated. Data generated include cow lists of animals in estrus, animals coming in or leaving estrus (suspects), numbers of mounts, when mounting activity began, and when mounting last occurred. Xu et al. (1998) reported its efficiency at 92% and the accuracy of this method at 100%.

Although radiotelemetric systems provide continuous "observation" and an automation of the tedious chore of observation, including quantification of activities (movement or mounting), they are still not without their own set of challenges. As with any such mechanized system, a level of maintenance is required. The receiver, buffer, computer, wiring, and software require care and upkeep. The patches that house the transmitter must be regularly inspected to ensure that they remain firmly attached to the cow. Losses from a group of cows may reach 50% over a 21-day period. This can be especially challenging when the cow has an especially hot, active estrous period with many mounts. If the patch is rubbed free of the rump, the transmitter may be lost—a potentially expensive loss. The transmitter, too, requires regular assessment for functionality, including battery replacement. The estimated initial and recurring costs associated with the mount detection system are shown in Table 2.5.

Recently introduced to the market place is an adaptation of the HeatWatch® system, the HeatWatch Xpress®. This system is designed for operations that do not have (or want) a personal computer, but still desire a higher level of accuracy in mount detection. The system continually monitors mounting activity via the radio transmitter affixed to the tailhead of the cow, sending mount data to a radio receiver. The receiver relays the mount data to an electronic mail box, a buffer, located in the breeder's office. This system does not use a computer or software to collect and display the data. The breeder extracts standing estrous information by selecting a button on the buffer and having that data printed on a thermal printer. Two other buttons provide additional status information. This system is priced lower than its more sophisticated counterpart.

TABLE 2.5
Costs Associated with an Electronic Mount Detection Device

Costs	Per Cow	100 Cows
Initial		
Transmitter	$60.00 each	$6000
Receiver, buffer, wiring	$2200.00	$2200
Computer software	$2000.00	$2000
Total initial cost	$4260.00	$10,200
Recurring		
Replacement batteries	~$5 every 6 to 9 mo	$500
Patches, glue	~$5/cow/21days	$500
Labor	~$5/cow/21days	$500
Total recurring costs	$15	$1500

APPLICATION OF A MOUNT DETECTION SYSTEM

Electronic mount detection systems have been used for several years at the University of Florida and many other research institutes to provide information on the estrous behavior of cows and heifers subjected to estrous detection and artificial insemination programs. Preliminary applications have asked questions such as:

- When do cows show estrus?
- Are there breed-type differences?
- What is the frequency of mounting activity by time of day?
- What is the duration of estrus?
- How are mounting activity and duration of estrus related?
- How does time of breeding affect pregnancy outcome?
- Is pregnancy percentage improved by application of estrous detection aids?
- When does ovulation occur relative to detected estrus?

Some of the findings reported relative to these questions are briefly presented in this section. There has been a suspicion that cows may show signs of estrus more commonly and more actively during the night than during daylight hours. Table 2.6 reports the findings of several studies which suggest that there is not a great difference between activity and expression of estrus throughout the 24 hours of the day.

The length of estrous expression can be a critical factor if that expression occurs outside a scheduled observation time, such as the nighttime hours. Estrous length is quite variable, with an average of about 8 to 10 hours, as measured by the radio-telemetric mount detection devices (Table 2.7). It is interesting to note that the duration is reported to be considerably longer when measured by continuous observation (Stevenson et al., 1996) and activity monitoring by pedometer (Pennington et al., 1986; ~15 h, Table 2.7). Esslemont et al. (1980) addresses the range of behaviors indicative of estrus and their association with mounting activities as measured by the mount detection devices; there are definite signs of impending estrus, including increased activity prior to receiving the first mount.

Some animals and breed types are thought to express a much shorter estrous length. An example is the Brahman cow, which appears to have a much shorter duration of estrus than *Bos taurus* cattle, but has mounting activity which may not be that much different than other cattle. For all cattle, the number of mounts demonstrated during estrus is quite variable; it may be anywhere between only a few to several hundred. Most will receive 20 to 40 mounts. This number seems to vary considerably, and the factors associated with this variability have been well discussed in the literature (Esslemont et al., 1980; Allrich, 1993).

Probably the most critical question, once an accurate detection of estrus is made, is when should insemination be performed. Timing of insemination has traditionally been based on the a.m./p.m. rule (i.e., a cow that is observed in estrus in the morning is inseminated in the evening, and a cow observed in estrus in the evening is inseminated the following morning). This rule has worked well when the primary method of estrous detection has been visual observation, alone or in conjunction with other nontelemetric methods (Table 2.8). However, telemetric detection methods require a reappraisal of the rule when they are utilized.

TABLE 2.6
A Summary of Several Studies Utilizing Cows or Heifers Fitted with Radiotelemetric Mount Detection Devices to Answer the Questions, "When Do Cows Show Estrus?" and "What is the Frequency of Mounting Activity by Time of Day?"

		Time (24-hour clock)				
	No.	**0.00–6.00**	**6.00–12.00**	**12.00–18.00**	**18.00–24.00**	**Mean**
Rae unpublished (1995). Beef Research Unit, University of Florida.						
Angus cows						
Onset of standing event (%)	85	22.4	25.9	29.3	22.4	
Brahman cows						
Onset of standing event (%)	39	20.5	23.1	25.6	30.8	
Mean mounts (range 3 to 153)		41	40	40	30	38
Mean estrus interval (hours)		8.3	8.0	9.3	7.5	8.3
Dransfield et al. (1998). Dairy cows located in 17 different herds.						
Onset of first standing event (%)	2055	24.5	28.4	19.8	27.3	
Onset of last standing event (%)	2055	24.8	27.8	23.4	24.0	
Stevenson et al. (1996). University Agriculture Research Center- Hays, Kansas State. Angus × Hereford × Brahman yearling heifers.						
Onset of first standing event (%)	40	35.0	12.5	27.5	25.0	
Onset of last standing event (%)	40	22.5	30.0	27.5	20.0	

Stevenson et al. (1996) observed that the radiotelemetric detection device effectively identified the start of estrus earlier than visual observation. Cavalieri and Fitzpatrick (1995) reported a difference in pregnancy rate for heifers subjected to estrous synchronization and bred at either a fixed-time insemination or at 12 hours following the onset of detected estrus (34.5 and 57.1%, respectively), where insemination occurred at a mean of 3.6 and 12.6 hours, respectively, following the onset of estrus. Dransfield et al. (1998, Table 2.9) found a significant association between

TABLE 2.7
A Summary of Several Studies Utilizing Cows and Heifers Fitted with Radiotelemetric Mount Detection (RTD) or Activity Monitoring Devices to Answer the Questions, "Are There Breed-Type Differences?", "What is the Duration of Estrus?", and "How Are Mounting Activity and Duration of Estrus Related?"

	Parameters of Estrus Measured by Radiotelemetric Mount Detection Device (Mean ± sem)			
	No.	Time to Estrus[a] (hours)	Estrous Length[b] (hours)	Mounts[c] (No.)
Rae et al. 1999. Heifers subjected to estrous synchronization with Synchro-Mate-B[d].				
Angus (An)	29	37 ± 3	8.5 ± 1.2*	19 ± 4*
Brahman (Bh)	13	32 ± 3	6.7 ± 1.2	25 ± 5
Crossbred (An × Bh)	24	40 ± 2	11.9 ± 1.2	37 ± 6
Stevenson et al. (1996). University Agriculture Research Center—Hays, Kansas State. Angus × Hereford × Brahman yearling heifers subjected to estrous synchronization with MGA and $PGF_{2\alpha}$.				
Observation + RTD	30	58 ± 6	15.6 ± 1.3*	61 ± 10*
RTD	11	69 ± 6	8.4 ± 1.3	19 ± 10
Dransfield et al. (1998). Dairy cows located in 17 different herds, observed for spontaneous estrus.				
	2055		7.1 ± 5.4	9 ± 7
Xu et al. (1998). Dairy cows in two herds observed for spontaneous estrus in pastured cattle.				
Herd 1	48		9.7 ± 0.1*	14 ± 2*
Herd 2	41		7.3 ± 0.6	9 ± 1
Both	89		8.6 ± 0.5	11 ± 1
Pennington et al. (1986). Holstein cows subjected to $PGF_{2\alpha}$ during the luteal phase. Observation + activity. Cows showing signs of estrus had a 677% increase in pedometer activity.				
	26		15.7 ± 2.6	25 ± 5

[a] Time from an estrous synchronization event to the first standing estrous event (hours).
[b] Estrous length (hours).
[c] Number of mounts recorded during the synchronized estrus.
[d] Synchro-Mate-B, Rhone Merieux, Inc., Athens, GA.
* Difference exists at $P < 0.05$ compared with corresponding values in the same column and in the same study.

conception and the time interval from the first mount to artificial insemination; the greatest conception rates in these mature dairy cows occurred at 4 to 12 hours after the onset of standing to receive a mount.

The appropriate timing of insemination ultimately is that time which permits the bull's semen to encounter the cow's released ova at a time and place that permits conception to occur. Ovulation has been reported to occur 23 to 33 hours following the onset of behavioral estrus. Vaca et al. (1985) reported that ovulation in a group of 20 nonlactating Zebu cows occurred at a mean of 28.2 ± 5.0 hours following the

TABLE 2.8
"When Should Insemination Be Performed?" The Following Represents Current Training Materials in the Artificial Insemination Industry

Preheat 6 to10 hours	Standing Heat 18 hours		Ovulation 10 to14 hours	Life of Ova 6 to10 hours	Bleeding
Too early	Possible	Best time	Possible	Too late	

Modified from American Breeders Service (1991).

TABLE 2.9
Summary of a Study of Dairy Cows Fitted with a Radiotelemetric Mount Detection Monitoring Device to Answer the Question, "When Should Cows Be Inseminated Relative to Estrous Mounting Events?"

	Interval from Onset of Estrus to AI (hours)						
	0–4	4–8	8–12	12–16	16–20	20–24	24–26
Number	327	735	677	459	317	139	7
Conception rate (%)	43.1	50.9	51.1	46.2	28.1	31.7	14.3
Odds of conception	1.00	1.35	1.33	1.12	0.51	0.57	0.18
95% C.I.*		1.03–1.77	1.01–1.75	0.83–1.50	0.36–0.71	0.37–0.87	0.02–1.56

* Confidence Interval

Modified from Dransfield et al. (1998).

start of sexual receptivity. Walker et al. (1996) found that the mean time of ovulation (27.6 ± 5.4 hours) did not differ between a $PGF_{2\alpha}$-induced estrus and a naturally occurring estrus in lactating Holstein cows. In their study, 78% of the cows ovulated within 40 hours of the onset of estrus. In a study of Nelore cattle with either a natural or induced estrus, Pinheiro et al. (1998) reported an estrus-to-ovulation interval of 26.6 ± 0.4 hours and an estrous length of 10.9 ± 1.4 hours.

It appears, then, that when an estrous detection method can identify onset of mounting activity and insemination is performed on a morning-and-evening schedule, the greatest conception rate is achieved when artificial insemination is performed such that the time interval to insemination does not exceed 12 hours from the onset of mounting nor is less than 4 hours from onset.

CONCLUSIONS

The use of estrous detection aids has generally been shown to be more efficient in estrous detection than human visual observation, but it is not necessarily more accurate. It is the differences in managerial conditions and the commitment to visual observation for estrus that makes the difference in its success as a sole or major component in a detection program. Certainly, estrous detection efficiency can be improved by consecutive or simultaneous use of more than one detection method. Automated and computerized detection is possible now by radio transmission from pedometer and mount detection aids that are available commercially.

Again, the purpose for detection of estrus is singularly to identify an appropriate time to breed/inseminate the female. Much effort has been made to identify estrus and deliver semen at a targeted time relative to estrus. We have noted the technological advancement of aids in the detection of estrus, but the real advancement in bovine breeding will be the elimination of any need to detect estrus, that is, performing insemination at a fixed time relative to a known and planned ovulation. This aim must be accomplished with a high level of reproductive efficiency and at a cost in labor and materials that is cost effective.

REFERENCES

Allrich, R.D. 1993. Estrous behavior and detection in cattle. *Vet. Clin. N. Am. Food Anim. Pract.* 9:249.

American Breeders Service. 1991. Heat detection and synchronization. *A.I. Manual,* third edition. pp. 28.

Anonymous. 1979. Androgenized cows. *American Breeders Service Division.* 11:1.

Beef. 2000. The heat is on: accurate heat detection is the key to a successful AI program. *Beef* 36:30.

Britt, J.H. 1980. Testosterone treatment of cows for detection of estrus. In: D.A. Morrow (Ed.) *Current Therapy in Theriogenology.* pp. 174. W.B. Saunders, Philadelphia, PA.

Cavalieri, J. and L.A. Fitzpatrick. 1995. Oestrus detection techniques and insemination strategies in *Bos indicus* heifers synchronised with norgestomet-oestradiol. *Aust. Vet. J.* 72:177.

Dransfield, M.B.G., R.L. Nebel, R.E. Pearson, and L.D. Warnick. 1998. Timing of insemination for dairy cows identified in estrus by a radiotelemetric estrus detection system. *J. Dairy Sci.* 81:1874.

Esslemont, R.J., R.G. Glencross, M.J. Bryant, and G.S. Pope. 1980. A quantitative study of pre-ovulatory behaviour in cattle (British Friesian heifers). *Applied Anim. Ethol.* 6:1.

Foote, R.H. 1975. Estrus detection and estrus detection aids. *J. Dairy Sci.* 58:248.

French, J.M., G.F. Moore, C.P. Graham, and S.E. Long. 1989. Behavioral predictors of oestrus in domestic cattle, *Bos taurus. Anim. Behav.* 38:913.

Gwazdauskas, F.C., R.L. Nebel, D.J. Sprecher, W.D. Whittier, and M.L. McGilliard. 1990. Effectiveness of rump-mounted devices and androgenized females for detection of estrus in dairy cattle. *J. Dairy Sci.* 73:2965.

Herrick, J. 1978. Breeding time requires good eyesight and cow knowledge. *Adv. Anim. Breeder,* pp. 15. May 1978.

Lehrer, A.R., G.S. Lewis, and E. Aizinbud. 1992. Estrus detection in cattle: recent developments. *Anim. Reprod. Sci.* 28:355.

Moore, A.S. and S.L. Spahr. 1991. Activity monitoring and an enzyme immunoassay for milk progesterone to aid in the detection of estrus. *J. Dairy Sci.* 74:3857.

Mortimer, R.G., M.D. Salman, M. Gutierrez, and J.D. Olson. 1990. Effects of androgenizing dairy heifers with ear implants containing testosterone and estrogen on detection of estrus. *J. Dairy Sci.* 73:1773.

Pennington, J.A., J.L. Albright, and C.J. Callahan. 1986. Relationships of sexual activities in estrous cows to different frequencies of observation and pedometer measurements. *J. Dairy Sci.* 69:2925.

Pinheiro O.L., C.M. Barros, R.A. Figueiredo, E.R. do Valle, R.O. Encarnação, and C.R. Padovani. 1998. Estrous behavior and the estrus-to-ovulation interval in Nelore cattle (*Bos indicus*) with natural estrus or estrus induced with prostaglandin F_2 or norgestomet and estradiol valerate. *Theriogenology* 49:667.

Rae, D.O., P.J. Chenoweth, M.A. Giangreco, P.W. Dixon, and F.L. Bennett. 1999. Assessment of estrus detection by visual observation and electronic detection methods and characterization of factors associated with estrus and pregnancy in beef heifers. *Theriogenology* 51:1121.

Senger, P.L. 1994. The estrus detection problem: new concepts, technologies, and possibilities. *J. Dairy Sci.* 77:2745.

Smith, R.D. 1986. Estrus detection. In: D.A. Morrow (Ed.), *Current Therapy in Theriogenology*. pp. 153. W.B. Saunders, Philadelphia, PA.

Stevenson, J.S., M.W. Smith, J.R. Jaeger, L.R. Corah, and D.G. LeFever. 1996. Detection of estrus by visual observation and radiotelemetry, estrus-synchronized beef heifers. *J. Anim. Sci.* 74:729.

Vaca, L.A., C.S. Galina, S. Fernandez-Baca, F.J. Escobar, and B. Ramirez. 1985. Oestrous cycles, oestrus and ovulation of the Zebu in the Mexican tropics. *Vet. Rec.* 117:434.

Walker, D.F. 1990. Genital surgery of the bull. In: D.A. Morrow (Ed.) *Current Therapy in Theriogenology*. pp. 396. W.B. Saunders, Philadelphia, PA.

Walker, W.L., R.L. Nebel, and M.L. McGilliard. 1996. Time of ovulation relative to mounting activity in dairy cattle. *J. Dairy Sci.* 79:1555.

Williamson, N.B., R.S. Morris, D.C. Blood, and C.M. Cannon. 1972a. A study of oestrus behavior and oestrus detection methods in a large commercial dairy herd. I. The relative efficiency of methods of oestrus detection. *Vet. Rec.* 91:50.

Williamson, N.B., R.S. Morris, D.C. Blood, C.M. Cannon, and P.J. Wright. 1972b. A study of oestrus behavior and oestrus detection methods in a large commercial dairy herd. II. Oestrus signs and behavior patterns. *Vet. Rec.* 91:58.

Xu, Z.Z., D.J. McKnight, R. Vishwanath, C.J. Pitt, and L.J. Burton. 1998. Estrus detection using radiotelemetry or visual observation and tail painting for dairy cows on pasture. *J. Dairy Sci.* 81:2890.

3 Use of Prostaglandin $F_{2\alpha}$ ($PGF_{2\alpha}$) in Cattle Breeding

J. W. Lauderdale

CONTENTS

From a historical perspective, in the 1930s, ejaculates of both ram and man were demonstrated to possess *in vitro* uterine strip contractility. The biologic activity could not be accounted for by known substances. Sheep prostate glands became the source of material in the search for the biologic activity and chemical structure of the unknown substance(s). In 1937, the material was named "prostaglandin" by Von Euler. Of some interest is that "prostaglandin" and "progesterone" appeared as new names on the same page of a German journal in 1937. Not until 1965, when biosynthesis rather than extraction was achieved to produce prostaglandins, were gram quantities of prostaglandins available for research. Prostaglandins are, for the most part, 20 carbon fatty acids. Prostaglandin $F_{2\alpha}$ ($PGF_{2\alpha}$) is the most often discussed prostaglandin relative to domestic animal research and practical utility. Prostaglandins have been detected in every mammalian tissue studied. Generally, prostaglandins are produced in and elicit their biological activity within a tissue. Thus, the uterine production of $PGF_{2\alpha}$ but luteolytic biologic activity at the ovary is unusual for the prostaglandins in general. Research over the past 25 years has established that $PGF_{2\alpha}$

0-8493-1117-9/02/$0.00+$1.50

of uterine origin is a key component of the natural luteolytic mechanism that contributes to the natural estrous cycle control for the cow. Although $PGF_{2\alpha}$ has been developed to commercial applicability primarily as a luteolytic agent, $PGF_{2\alpha}$ has biologic activity associated with the normal parturient process, is involved in release of various pituitary hormones, is associated with the normal process of ovulation, and is associated with normal ova and sperm transport in the female reproductive tract.

During the 1970s, extensive research led to the identification of various $PGF_{2\alpha}$ analogs with luteolytic activity, and the luteolytic dose–response curves were characterized for various PGFs. Published data support the conclusion that injection of cattle with a luteolytic dose of $PGF_{2\alpha}$ prior to day 5 after ovulation will not result in luteolysis as consistently as injection on or after day 6 (Lauderdale, 1972). The double injection of $PGF_{2\alpha}$ at an 11-day interval was the cattle management system studied most extensively in an attempt to manage the failure of $PGF_{2\alpha}$ to regress the corpus luteum (CL) during the first 4 to 5 days after estrus. Two additional cattle management systems with $PGF_{2\alpha}$ were investigated; in one system, the nonresponders of days 1 to 4 or 5 after estrus were addressed, but in the second, these cattle were accepted as nonresponders.

MATERIALS AND METHODS

For this chapter, the data will be cited primarily from studies using 5 mL Lutalyse® sterile solution (25 mg $PGF_{2\alpha}$/33.5 mg dinoprost tromethamine) and the intramuscular (IM) route of administration (Lauderdale et al., 1981), since data demonstrated luteolytic and estrous grouping effectiveness of that dose and route of administration (Lauderdale et al., 1977; Moody and Lauderdale, 1977).

Cattle were assigned to experimental groups in an attempt to balance age, postpartum interval, and semen source across groups. Cattle were not palpated per rectum for the presence of a CL prior to assignment to the study, and no attempt was made to assign only estrous cycling cattle to the study. The design was a randomized block, with each herd providing a control group and one or two groups of cattle administered Lutalyse. All cattle were maintained in the same pasture or pen. The data were analyzed statistically using the model $Y = T + TH + \text{Error}$ where T was the fixed Treatment effect and H was the random Herd effect. The TH interaction term was used to test for the Treatment effect significance.

DOUBLE INJECTION SYSTEM

Cattle were assigned randomly in replicates to either a control group or to Lutalyse groups. Cattle assigned to the Lutalyse groups were injected IM in the hip gluteus maximus muscle region twice at an 11- (10 to 12) day interval with 5 mL Lutalyse. Cattle injected with Lutalyse were artificially inseminated (AI) either at detected estrus (LLAIE) or at about 80 hours (LLAI80) after the second injection (Figure 3.1). Generally, cattle of the control and LLAIE groups were observed for estrus twice daily and artificially inseminated about 6 to 13 hours after first observation of estrus. Cattle of the LLAI80 were artificially inseminated at about 77 to 80 hours after the second injection of Lutalyse and were rebred at any estrus detected 5 days or more

Program Designation			Breeding Method			
LLAIE	↓	↓	AIE	AIE or Bull	AIE or Bull	
LLAI80	↓	↓	TAI	AIE or Bull	AIE or Bull	
LaIE		↓	AIE	AIE or Bull	AIE or Bull	
AILAI			AIE	AIE	AIE or Bull	AIE or Bull
	−14 to −12	−1	5	9	22	27
Days before Breeding Season			Days of Breeding Season			

FIGURE 3.1 Cattle breeding management with 5 mL Lutalyse sterile solution (25 mg $PGF_{2\alpha}$/33.5 mg dinoprost tromethamine; IM). AIE: inseminated at detected estrus. TAI: inseminated at about 77 to 80 h after the second injection of Lutalyse.

after the 80-hour artificial insemination. Dates of injections of Lutalyse were established such that the second injection would be administered the day before initiation of the normal breeding season within herd.

SINGLE INJECTION SYSTEMS

Cattle were assigned randomly in replicates to either control group or to a Lutalyse group. The AILAI cattle management system was observation of cattle for estrus and artificial insemination for the first 4 days, inject cattle not detected in estrus with 5 mL Lutalyse IM on the morning of day 5 and continue to observe cattle for estrus and artificial insemination accordingly on days 5 through 9, i.e., a 9-day artificial insemination season (Figure 3.1). Breeding for the remainder of the breeding season was by artificial insemination, bulls, or some combination of artificial insemination and bulls. The LAIE cattle management system was IM injection of cattle with 5 mL Lutalyse on the day before initiation of the breeding season, followed by observation of cattle for estrus and artificial insemination for 5 days (Figure 3.1). Breeding for the remainder of the breeding season was by artificial insemination, bulls, or some combination of artificial insemination and bulls. Within-herd comparisons were made between control and LAIE cattle and between control and AILAI cattle. In three additional herds, within-herd comparisons were made among control, LLAIE, and LAIE cattle.

The pregnancy status of cattle was evaluated between 60 to 100 days after start of the artificial insemination season by rectal palpation of the uterus for products of conception. Comparisons were made for percent detected in first estrus, first service conception rate, and pregnancy rate. On a within-experimental group basis:

$$\text{Estrous Detection (ED)} = \frac{\text{Number Detected in Estrus} \times 100}{\text{No. Assigned}}$$

Estrous percent was calculated for each interval of interest.

$$\text{Conception Rate (CR)} = \frac{\text{Number Pregnant} \times 100}{\text{No. Detected in Estrus and AI}}$$

Conception rate was calculated for first service only.

$$\text{Pregnancy Rate (PR)} = \frac{\text{Number Pregnant} \times 100}{\text{Number Assigned}}$$

Pregnancy rate was calculated for each interval of interest.

RESULTS AND DISCUSSION

Double-Injection Cattle Breeding Management

Significantly ($P < 0.05$) greater percentages of cattle were detected in estrus during the first 5 days of the artificial insemination season for the LLAIE cattle compared to controls (Table 3.1). Similar percentages of control and LLAIE cattle were detected in estrus at least once during the first 24 days (one estrous cycle) of the artificial insemination season (Table 3.1).

First service conception rates were similar between control and LLAIE cattle for both the first 5 days and days 1 to 24 of artificial insemination (Table 3.1). These data

TABLE 3.1
Least Square Means for Data Derived from Field Investigations for the Double Injection of Lutalyse at an 11-Day Interval Followed by Either AI at Detected Estrus or AI at 80 Hours

Type of Cattle	No. Herds (No. Cattle)	Experimental Group Designation	% Estrus Days		First Service Conception Rate Days		Pregnancy Rate Days		
			1–5	1–24	1–5	1–24	1–5	1–24	1–28
Suckled	43	Control	17[a]	64[a]	60[a]	61[a]	13[a]	50[a]	55[a]
beef cow	(2958)	LLAIE	52[b]	72[a]	57[a]	62[a]	37[b]	57[b]	62[b]
Beef	36	Control	17[a]	80[a]	49[a]	56[a]	9[a]	50[a]	54[a]
heifer	(3798)	LLAIE	64[b]	84[a]	52[a]	53[a]	34[b]	54[a]	57[a]
Dairy	6	Control	12[a]	74[a]	69[a]	66[a]	24[a]	51[a]	52[a]
heifer	(786)	LLAIE	73[b]	82[b]	64[a]	56[a]	56[b]	66[a]	68[a]
Suckled	32	Control	nm	nm	nm	nm	11[a]	48[a]	52[a]
beef cow	(2291)	LLAI80	nm	nm	nm	nm	37[b]	51[a]	58[a]
Beef	27	Control	nm	nm	nm	nm	11[a]	51[a]	54[a]
heifer	(1764)	LLAI80	nm	nm	nm	nm	35[b]	49[a]	49[a]
Dairy	6	Control	nm	nm	nm	nm	24[a]	51[a]	52[a]
heifer	(810)	LLAI80	nm	nm	nm	nm	43[a]	53[a]	57[a]

[a,b] Means in the same column within type of cattle within comparison without a common superscript differed significantly ($P < 0.05$). nm = not measured.

See the Materials and Methods Section for a description of each group. Some herds had the three-way comparison of C vs. LLAIE vs. LLAI80. Therefore, control cattle of those herds contributed to the numbers of cattle in both the LLAIE and LLAI80 comparisons.

From Lauderdale, J.W. et al., 1981. With permission.

reinforce previously reported data that conception rate was not altered significantly following use of $PGF_{2\alpha}$ (Hafs et al., 1975; Inskeep, 1973; Lauderdale et al., 1974).

Pregnancy rate reflects both estrous synchronization and conception rate. During the synchronized interval, the first 5 days after the second injection of Lutalyse, pregnancy rates were greater for both LLAIE and LLAI80 cattle compared to controls (Table 3.1). Although not specifically reflected in Table 3.1, these investigations did not identify a significant difference between cattle of LLAIE and LLAI80 groups based on within-herd comparisons (can be inferred from Table 3.1, 10). Pregnancy rates generally were similar between control and either LLAIE or LLAI80 cattle for days 1 to 24 and 1 to 28 (Table 3.1). About 0.7 more services per conception were required for the LLAI80 than LLAIE program for pregnancies during the 28-day artificial insemination interval. This increase is due to the cattle that are not estrous cycling, but are inseminated at 77 to 80 hours after $PGF_{2\alpha}$ in the LLAI80 group.

The 80-hour timed artificial insemination reported herein was very successful. Success was measured by similar pregnancy rates between cattle artificially inseminated at 80 hours after the second Lutalyse injection compared to contemporary control cattle observed daily for estrus and artificial insemination at detected estrus over 24 days. However, the success of timed artificial insemination has been highly variable among herds and within herds over time since we completed the studies reported herein. The basis for this variation in response is the variation, primarily, in "control" of follicular waves but also the percent of cattle anestrus. In those groups of cattle where timed artificial insemination has worked well, the incidence of anestrus or prepuberty was very low, and the cattle were in the stage of the estrous cycle where follicular waves were "similar" among the cohort of cattle treated. We now know, based on an understanding of follicle waves, that, to achieve consistently high pregnancy rates using timed artificial insemination, both follicular waves must be synchronized/managed and the lifespan of the CL must be managed. Follicle waves can be managed through the use of GnRH, and the CL lifespan can be managed by use of $PGF_{2\alpha}$.

Single-Injection Breeding Management

The percentage of cattle detected in estrus the first time for days 1 through 5 was similar between AILAI (25%) and control (24%) beef heifers, but was different ($P < 0.05$) for AILAI (17%) compared to control (21%) cows. The percentage of cattle detected in estrus the first time during days 1 through 9 was greater ($P < 0.01$) for AILAI than for controls for both cows (54% vs. 38%) and heifers (64% vs. 38%). First estrous detection rates for the first 24 days of breeding were similar between AILAI and control cattle for both cows (70% vs. 73%) and heifers (77% vs. 78%), respectively (Table 3.2).

First service conception rates were concluded to be not different between cattle assigned to AILAI and control groups for days 1 through 5, 1 through 9, and 1 through 24, respectively (Table 3.2).

Pregnancy rates for days 1 through 5 were similar between AILAI and control heifers (16% vs. 15%), but tended to be different ($P < 0.09$) for cows (12% vs. 14%), respectively (Table 3.2). Pregnancy rates were greater ($P < 0.01$) for AILAI than for

TABLE 3.2
Least Squares Means for Control vs. AILAI for Percent of Beef Cows and Beef Heifers Observed in First Estrus, Conception Rate at First Estrus, and Pregnancy Rate

Measurement	Type of Cattle	Experimental Group Designation	Days of Breeding Season			
			1–5	1–9	1–24	1–28
Percent in estrus	Beef cows	Control	21^{a}	38^{a}	73^{a}	
		AILAI	17^{b}	54^{b}	70^{a}	
	Beef heifers	Control	24^{a}	38^{a}	78^{a}	
		AILAI	25^{a}	64^{b}	77^{a}	
First service conception rate	Beef cows	Control	59^{a}	64^{a}	63^{a}	
		AILAI	64^{a}	58^{a}	59^{a}	
	Beef heifers	Control	62^{a}	56^{a}	59^{a}	
		AILAI	62^{a}	53^{a}	57^{a}	
Pregnancy rate	Beef cows	Control	14^{c}	26^{a}	54^{c}	59^{c}
		AILAI	12^{d}	39^{b}	56^{d}	63^{d}
	Beef heifers	Control	15^{a}	24^{a}	55^{a}	59^{a}
		AILAI	16^{a}	45^{b}	56^{a}	63^{a}

[a,b] Means without a common superscript differed ($P < 0.05$).
[c,d] Means without a common superscript differed ($P < 0.10$).
See the Materials and Methods Section for a description for each group. Thirty-eight herds with a total of about 5400 cows and 19 herds with a total of about 2761 heifers were included in these studies.

From Lauderdale, J.W. et al., 1981. With permission.

control cattle for days 1 through 9 for cows (39% vs. 26%) and heifers (45% vs. 24%) and for days 1 through 24 for cows (56% vs. 54%, $P < 0.08$), respectively. Pregnancy rates were not significantly different between control (55%) and AILAI (56%) heifers for days 1 through 24. Pregnancy rates for days 1 through 28 were 63% and 59% for AILAI and control cows ($P < 0.06$) and were 63% and 59% for AILAI and control ($P < 0.16$) heifers, respectively (Table 3.2).

The percentages of cattle detected in estrus the first time, first service conception rates, and pregnancy rates should be similar between controls and cattle assigned to the AILAI group for days 1 through 5 since the AILAI cattle would not have been injected with $PGF_{2\alpha}$. That was the case for beef heifers. In contrast, both percentage of cows detected in estrus and pregnancy rate were elevated for control cows for days 1 through 5 (Table 3.2). The basis for that difference is unknown since the cows were assigned randomly in replicates to the experimental groups; these differences are assumed to be chance observations.

The data on enhanced pregnancy rates after 9 days of artificial insemination with the AILAI management system are consistent with data published previously (Greene et al., 1977; Lambert et al., 1976; Lambert et al., 1975). The greater pregnancy rate in the AILAI group for days 1 through 9 demonstrated the effectiveness of the use of $PGF_{2\alpha}$ in that system of cattle management. The trend for more pregnancies in the

TABLE 3.3
Least Squares Means for Control vs. LAIE for Percent of Beef Cows and Beef Heifers Observed in First Estrus, Conception Rate at First Estrus, and Pregnancy Rate

Measurement	Type of Cattle	Experimental Group Designation	Days of Breeding Season 1–5	1–24	1–28
Percent in Estrus	Beef cows	Control	31[a]	68[a]	
		LAIE	57[b]	76[a]	
	Beef heifers	Control	28[a]	82[a]	
		LAIE	52[b]	83[a]	
First service conception rate	Beef cows	Control	49[a]	53[a]	
		LAIE	54[a]	63[b]	
	Beef heifers	Control	47[a]	53[a]	
		LAIE	52[a]	57[a]	
Pregnancy rate	Beef cows	Control	14[a]	56[a]	60[a]
		LAIE	30[b]	60[a]	66[a]
	Beef heifers	Control	12[a]	49[a]	52[a]
		LAIE	28[b]	55[a]	57[a]

[a,b] Means without a common superscript differed ($P < 0.05$).

See the Materials and Methods Section for a precise description for each group. Twelve herds with a total of about 1592 cows and 5 herds with a total of about 727 heifers were included in these studies.

From Lauderdale, J.W. et al., 1981. With permission.

AILAI group after 28 days of artificial insemination reinforces the conclusion that the AILAI management system was effective as measured by percent of the herd pregnant.

The percentage of cattle detected in estrus the first time during days 1 through 5 was greater for LAIE than for control for both cows (57% vs. 31%, $P < 0.01$) and heifers (52% vs. 28%, $P < 0.05$, Table 3.3). The percentage of cattle detected in estrus the first time during days 1 through 24 was similar between LAIE and controls for cows (76% vs. 68%) and heifers (83% vs. 82%, Table 3.3). The percentages of cattle detected in estrus during the first 24 days of artificial insemination were 68 and 82 for control cows and heifers. These values should be an overestimate of the percent of the herd having estrous cycles on the day of Lutalyse injection, since the cattle had 24 more days to initiate estrous cycles. Since $PGF_{2\alpha}$ has been shown to be ineffective during days 1 through 4 or 5 after estrus and cattle have an 18- to 24- ($x = 21$) day estrus cycle, a single injection of $PGF_{2\alpha}$ would be expected to regress the CL and synchronize about 75 to 80% of a group of estrous cycling cattle. Calculation of the predicted estrus detection rates for cattle of this study would be as follows for the single injection: 75% effective of 68% of estrous cycling cows equals 51% expected (actual was 57% for LAIE cows, Table 3.3), and 75% effective of 82% of estrous cycling heifers equals 62% expected (actual was 52% for LAIE heifers). Thus, the predicted and observed estrus detection rates of 51% and 57%

for cows and 62% and 52% for heifers appeared to be similar, which reinforces the conclusion that a single injection of $PGF_{2\alpha}$ yielded the predicted response.

First service conception rates were similar for cattle of the control and LAIE groups, as would be expected (Table 3.3).

Pregnancy rates for days 1 through 5 for LAIE and control cattle were 30 and 14% ($P < 0.01$) for cows and 28 and 12% ($P < 0.04$) for heifers (Table 3.3). Pregnancy rates for days 1 through 24 for LAIE and control cattle were 60 and 56% for cows and 55 and 49% for heifers. Pregnancy rates for days 1 through 28 for LAIE and control cattle were 66 and 60% ($P < 0.13$) for cows and 57 and 52% for heifers (Table 3.3).

These data are similar to those reported previously relative to use of the LAIE management system (Inskeep, 1973; Lauderdale et al., 1974; Moody, 1979; Turman et al., 1975). The pregnancy rates for 5 days of breeding in the LAIE management system demonstrated that system to be effective.

Comparison of LAIE and LLAIE

Data on percent cattle detected in estrus, first service conception rates, and pregnancy rates are presented in Table 3.4. Cattle of the LLAIE system compared to cattle of the LAIE system should have about a 20 to 25% greater first estrus detection rate and pregnancy rate for breeding the first 5 days after $PGF_{2\alpha}$ since $PGF_{2\alpha}$ is ineffective or less effective as a luteolytic agent when injected during the first 5 days after ovulation (Lauderdale, 1972). The observed percentage differences between LAIE and LLAIE cattle for first estrus were 10% (cows) and 23% (heifers), and for pregnancy rate were 22% (cows) and 23% (heifers) (Table 3.4). Thus, the expected percentage differences of about 20 to 25% and the observed percentage differences of 10, 23, 22, and 23% were similar in this limited study.

POTENTIAL IMPACT OF TREATED CATTLE ON NONTREATED CONTROL CATTLE IN THE SAME PEN/PASTURE

Inspection of the data from the studies cited herein identified, in some herds, an unusually high rate of pregnancy in the control cattle during the first 21 days of the breeding season. This observation was made most frequently in herds with a "high" pregnancy rate during the first 21 days of the breeding season. In an attempt to measure if treated cattle were effecting a response in control cattle in the same pen/pasture, we assumed "normal" for control cattle to be 100% of the cattle estrus cycling (thus, 5% in estrus each day) and a 75% first service conception rate (FSCR). Inspection of the FSCR data in Tables 3.1 through 3.3 identified FSCR to be in the 50 and 60% range; thus, use of 75% is conservative in the estimate (underestimate) of the impact of treated cattle on control cattle. Since $PR = ED \times CR$, and since we set $CR = 75\%$ (a value most likely not to be exceeded), the contrast between the observed pregnancy rate (OPR) and the theoretical pregnancy rate (TPR) yields an estimate of whether or not the control cattle were becoming pregnant (due primarily to more of them

TABLE 3.4
Least Squares Means for Control vs. LAIE vs. LLAIE for Percent of Beef Cows and Beef Heifers Observed in First Estrus, Conception Rate at First Estrus, and Pregnancy Rate

Measurement	Type of Cattle	Experimental Group Designation	Days of Breeding Season			Percent Different
			1–5	1–24	1–28	
Percent in estrus	Beef cow	Control	32^a	64^a		
		LAIE	67^b	76^a		
		LLAIE	74^b	67^a		10
	Beef heifer	Control	27^a	68^a		
		LAIE	40^a	68^a		23
		LLAIE	52^a	70^a		
First service conception rate	Beef cow	Control	51^a	53^a		
		LAIE	60^a	63^a		
		LLAIE	64^a	63^a		
	Beef heifer	Control	34^a	42^a		
		LAIE	40^a	42^a		
		LLAIE	44^a	41^a		
Pregnancy rate	Beef cow	Control	16^a	48^a	49^a	
		LAIE	36^b	58^b	61^a	
		LLAIE	46^b	59^b	62^a	22
	Beef heifer	Control	7^a	37^a	41^a	
		LAIE	17^b	40^a	43^a	
		LLAIE	22^b	38^a	42^a	23

^{a,b} Means without a common superscript differed ($P < 0.05$).

See the Materials and Methods Section for a description for each group. Three herds each with a total of about 527 cows and 881 heifers were included in these studies. Percent difference between LAIE and LLAIE for estrus and pregnancy rate for days 1–5.

From Lauderdale, J.W. et al., 1981. With permission.

detected in estrus) at an unusually high rate. For the herd used in this example, the PR for the treated cattle was 58% pregnant by day 2 of the breeding season as a result of the synchronized breeding following the second Lutalyse injection. For the control cattle in this herd, the TPR and OPR for days of the breeding season were 7.5 and 10% by day 3, 11 and 25% by day 4, 15 and 30% by day 5, 19 and 45% by day 6, 34 and 50% by day 10, 53 and 70% by day 15, and 80 and 80% by day 28. This observation identifies that when "large numbers" of estrous synchronized cattle are in estrus over a few days they may influence expression of fertile estrus in other cattle in the same pen/pasture that were not estrous synchronized. Also, this observation allows for an interpretation that, in estrus synchronization studies where the control cattle are in the same pen/pasture as the treated cattle, the treated cattle can increase the reproductive efficiency of the control cattle, thus making more challenging the interpretation of the effectiveness of the treatment.

SUMMARY

This chapter is a summary of research completed during the 1970s and 1980s. The research summarized is that which was completed in order to secure the approval of the Center for Veterinary Medicine (CVM) of the Food and Drug Administration (FDA) to market Lutalyse sterile solution in the United States. Ovarian follicular waves were not recognized at that time. Thus, the papers of Stevenson and Patterson (herein) are very important to an understanding of the value (biologically and economically) of the use of GnRH and progestogens to "control" follicular waves relative to the biological effectiveness of $PGF_{2\alpha}$. Additionally, with an understanding of follicular waves, the current interval between $PGF_{2\alpha}$ injections, for the double-injection program, is recommended to be 14 days for a greater degree of and precision of estrus synchrony and increased pregnancy rate.

For the studies discussed in the chapter, cattle were assigned randomly in replicates to experimental groups within each herd. Cattle were injected intramuscularly (IM) twice with 5 mL Lutalyse sterile solution (25 mg $PGF_{2\alpha}$/33.5 mg dinoprost tromethamine) at a 10- to 12-day interval and were artificially inseminated (AI) either at each detected estrus following the second injection (LLAIE) or at about 80 hours (usually between 77 and 80 hours) after the second injection and at subsequent detected estruses (LLAI80). Cattle injected once were assigned to either 1) a group in which cattle were artificially inseminated according to detected estrus for 4 days followed by 5 mL Lutalyse IM on the morning of day 5 for all cattle not detected in estrus through day 4, followed by detection of estrus and by artificial insemination on days 5 through 9 (AILAI); or 2) injected once with 5 mL Lutalyse the day before the start of the breeding season followed by artificial insemination at detected estruses (LAIE). Percent cattle pregnant (number pregnant × 100/number in herd) is reflective of both degree of estrus control and conception rate (number pregnant × 100/number AI). First service conception rates were not altered significantly for Lutalyse-injected cattle. Pregnancy rates for days 1 to 5 of the breeding season were greater ($P < 0.05$) for cattle of both the LLAIE and LLAI80 groups than for the control cattle. Pregnancy rate was similar between cattle of the LLAIE and LLAI80 groups for the first 5 days of the artificial insemination season. Pregnancy rates were either greater than or similar to controls for LLAIE and LLAI80 cattle for breeding for 28 days. A greater percent of cattle ($P < 0.05$) were pregnant after 9 days of artificial insemination for cattle of the AILAI group than controls. Pregnancy rates tended to be greater for AILAI cattle than for controls when bred for 28 days. A greater percent of cattle ($P < 0.05$) were pregnant after 5 days of artificial insemination for cattle of the LAIE group than for controls. Pregnancy rates were similar between cattle of LAIE and control groups for breeding for 28 days. Cattle of the LAIE group had a pregnancy rate 74% as great as cattle of the LLAIE group; thus, the observed difference between the two treatment regimens approximated the theoretical difference of 75 to 80%.

CONCLUSIONS

Lutalyse sterile solution, and other approved $PGF_{2\alpha}$, are effective as used for regression of the corpus luteum and resultant estrus synchrony. The versatility of the $PGF_{2\alpha}$ make them attractive for use under numerous beef breeding management systems.

Although not addressed specifically in this chapter, bull breeding can be used in place of artificial insemination, as long as the bulls are properly managed. With the understanding of follicular waves and the role of progestogens to help initiate estrus cyclically, as discussed especially by Drs. Day, Beal, Stevenson, and Patterson in this book, the modern beef reproduction manager has $PGF_{2\alpha}$, GnRH, and MGA (as a progestogen) as products to use in beef cattle estrus control. Thus, the reader is encouraged to read those chapters for both updated use programs for beef reproduction management and for current references to that research and practical utility. Expected in the future will be additional progestogens, such as PRID and/or CIDR. Through management of both the corpus luteum and follicular waves, the modern beef reproduction manager can achieve "acceptable" pregnancy rates with timed artificial insemination under many circumstances.

REFERENCES

Greene, W.M., D.K Han, P.W. Lambert, and E.L. Moody. 1977. Effect of two consecutive years in a $PGF_{2\alpha}$ or conventional AI breeding system. *J. Anim. Sci. (Suppl. 1)* 45:355.

Hafs, H.D., J.G. Manns, and G.E. Lamming. 1975. Fertility of cattle from AI after $PGF_{2\alpha}$. *J. Anim. Sci.,* 41:355.

Inskeep, E.K. 1973. Potential uses of prostaglandins in control of reproductive cycles of domestic animals. *J. Anim. Sci.* 36:1149–1157.

Lambert, P.W., D.R. Griswold, V.A. LaVoie, and E.L. Moody. 1975. Artificial insemination beef management with prostaglandin $F_{2\alpha}$ controlled estrus. *J. Anim. Sci.* 41:364.

Lambert, P.W., W.M. Greene, J.D. Strickland, D.K. Han, and E.L. Moody. 1976. $PGF_{2\alpha}$ controlled estrus in beef cattle. *J. Anim. Sci.* 42:1565.

Lauderdale, J.W. 1972. Effects of $PGF_{2\alpha}$ on pregnancy and estrous cycles of cattle. *J. Anim. Sci.* 35: 246.

Lauderdale, J.W., B.E. Seguin, J.N. Stellfiug, J.R. Chenault, W.W. Thatcher, C.K. Vincent, and A.F. Loyancano. 1974. Fertility of cattle following $PGF_{2\alpha}$ injection. *J. Anim. Sci.* 38:964–967. 8.

Lauderdale, J.W., E.L. Moody, and C.W. Kasson. 1977. Dose effect of $PGF_{2\alpha}$ on return to estrus and pregnancy in cattle. *J. Anim. Sci. (Suppl. 1)* 45:181. 9.

Lauderdale, J.W., J.F. McAllister, D.D. Kratzer, and E.L. Moody. 1981. Use of prostaglandin $F_{2\alpha}$ ($PGF_{2\alpha}$) in cattle breeding. *Acta Vet. Scand. Suppl.* 77: 181–191.

Moody, E.L. 1979. Studies on Lutalyse use programs for estrus control. *Proceedings of the Lutalyse Symposium,* Brook Lodge, Augusta, MI, August 6-8, 33–41.10.

Moody, E.L. and J.W. Lauderdale. 1977. Fertility of cattle following $PGF_{2\alpha}$ controlled ovulation. *J. Anim. Sci. (Suppl. 1)* 45:189.11.

Turman, E.J., R.P. Wettemann, T.D. Rich, and R. Totusek. 1975. Estrous synchronization of range cows with $PGF_{2\alpha}$. *J. Anim. Sci.* 41:382–383.

4 Estrous Synchronization of Cyclic and Anestrous Cows with Syncro-Mate-B®

W. E. Beal

CONTENTS

The history of estrous cycle synchronization and the use of artificial insemination in cattle is a testament to how discoveries in basic science can be applied to advance the techniques used for livestock breeding and management. Several authors described the experiments that have been conducted since the discovery of ovarian steroids and which have led to the effective control of the length of the bovine estrous cycle and the timing of estrus and ovulation (Hansel and Schechter, 1972; Hansel and Beal, 1979; Patterson et al., 1989; Odde, 1990; Larson and Ball, 1992).

Research and development of estrous synchronization products has left us with two approaches to controlling cycle length: 1) to regress the corpus luteum (CL) of the animal before the time of natural luteolysis, and thereby shorten the cycle, or 2) to administer exogenous progestins to delay the time of estrus following natural or induced luteolysis, which may extend the length of the estrous cycle. In either case, the emphasis is placed on controlling or mimicking luteal function to control the time of estrus. The two approaches to cycle control are the basis for commercially available products that successfully synchronize estrus. This chapter describes the results and mechanisms of estrous synchronization in cyclic and anestrous cows treated with Syncro-Mate-B (SMB)®.

0-8493-1117-9/02/$0.00+$1.50

ESTROUS CYCLE CONTROL

Exogenous Progestins

Isolation and synthesis of estrogen (Allen and Doisy, 1923) and progesterone (Corner and Allen, 1929) were followed by studies that revealed estrus could be delayed and thereby synchronized by exogenous administration of progesterone to cattle or sheep (Christian and Casida, 1948; Dutt and Casida, 1948). This led to a flurry of activity in which progesterone or synthetic progestins were injected, released intravaginally, or fed for a period up to and exceeding the length of the estrous cycle to synchronize estrus following the cessation of progestin administration (see Hansel and Beal, 1979). In general, the longer the progestin was administered to cattle, the higher the rate of estrous synchronization, but the lower the fertility of the synchronized animals.

Twenty-five years after long-term progestin feeding to control estrus was abandoned due to low fertility, several laboratories have been able to use ultrasonography to demonstrate that progestin administration at "subluteal" levels inhibits estrus and ovulation and synchronizes estrus in cattle, but that a persistent, estrogen-secreting follicle develops when progestin treatment extends the estrous cycle (Lucy et al., 1990; Sirois and Fortune, 1990; Cupp et al., 1992). Development of the persistent follicle is caused by increased pulsatile secretion of gonadotropins during the period when the exogenous progestin is inhibiting estrus, but the corpus luteum has regressed (Kojima et al., 1992; Savio et al., 1993; Stock and Fortune, 1993; Custer et al., 1994). The low fertility of cows bred at the synchronized estrus following long-term administration of progestin is due to premature resumption of meiosis of ova or abnormal development of embryos derived from ova of persistent follicles (Wishart and Young, 1974; Mihm et al., 1994; Ahmad et al., 1995; Revah and Butler, 1996).

Removing the persistent follicle that develops during long-term progestin administration or causing the regression of the dominant persistent follicle and development of a "new" follicular wave during progestin administration results in normal ovulatory follicular development (Roberson et al., 1989; Savio et al., 1993). It is clear today that shortening the period of progestin administration or applying a consistent method of "turning over" a persistent follicle and initiating a new wave of follicular development improves the fertility of cattle treated with exogenous progestin to synchronize estrus (Anderson and Day, 1994; Schmitt et al., 1996).

Combining Progestins and Estradiol

Although at the time they lacked the knowledge of why progestin-treated cattle exhibited lower fertility, Wiltbank et al. (1961) recognized that reducing the period of progestin administration improved conception rates at the synchronized estrus. Simultaneously, Kaltenbach et al. (1964), Loy et al. (1960), and Wiltbank (1966) demonstrated that estradiol was luteolytic when administered early in the bovine estrous cycle. Hence, combining progestin treatment with the administration of estradiol at the initiation of treatment enabled the period of progestin treatment to be shortened (9 to 14 days) without reducing the percentage of animals exhibiting a synchronized estrus. What researchers at the time did not know, but what has been demonstrated conclusively since (Bo et al., 1995; Ginther et al., 1996), is that an injection of estradiol regresses the existing dominant follicle and initiates a new wave of follicular development. Hence, in treatments combining estradiol and short-term progestin administration, the estradiol acts as a luteolysin for cattle treated beginning early in the estrous cycle, and it also acts to initiate growth of a new preovulatory follicle in all the treated animals. This treatment regimen is the basis for the commercial product, SMB (Syncro-Mate-B®, Merial), marketed in the United States, as well as the PRID® (progesterone-releasing intravaginal device, Sanofi Animal Health Inc., Paris, France) and CIDR®, EAZI-BREED™ (controlled intravaginal drug release device, InterAg, Hamilton, New Zealand), marketed in Canada, Europe, Australia, Mexico, and New Zealand.

RESPONSE TO SYNCRO-MATE-B

CYCLIC COWS AND HEIFERS

The treatment of cyclic cows or heifers with exogenous progestin preceded by an injection of estradiol is usually followed by a high incidence (>90%) of estrus during the 5 days following progestin removal. Odde (1990) reviewed 15 trials conducted with 1032 puberal heifers that were observed for signs of estrus following SMB treatment. Of those heifers, 92.5% were observed in estrus within 5 days after treatment. The failure to achieve synchronization rates of 100% in cyclic heifers or cows treated with SMB was related to the response of animals treated at different stages of the estrous cycle. Reports by Miksch et al. (1978) indicated that the SMB treatment was effective in only 80 to 86% of the heifers that began treatment on days 1 through 8 of the cycle. Pratt et al. (1991) went on to report that estrus was synchronized in only 48% of the cows treated on day 3, but that synchronization was 100% when treatment began on day 9 of the estrous cycle.

The distribution of estrus following SMB treatment is highly synchronized (Figure 4.1). In 15 separate trials in which the standard SMB treatment was used to synchronize estrus in 736 cows or heifers, a majority (65%) of the animals were observed in estrus between 24 and 48 hours after implant removal (Miksch et al., 1978; Spitzer et al., 1978). The "tight" synchrony of estrus that occurs following either SMB treatment of heifers or SMB treatment and 48-hour calf removal in postpartum beef cows makes these treatments logical for use with timed insemination. Mares et al. (1977) reported that pregnancy rates following timed breeding 48 to 54 hours after implant removal were actually higher (51%) than when SMB-treated heifers were inseminated 12 hours after estrous detection (39%) in herds in which the majority of heifers were cycling prior to SMB treatment. Hence, one of the principal advantages of SMB treatment is the tight synchrony of estrus, which makes this treatment compatible with timed insemination, especially in situations where careful heat detection is difficult.

Conception rates of cattle treated with SMB and bred 12 hours after estrous detection have been reported to be not significantly different from those of untreated controls in the same trials (see Odde, 1990). However, upon closer inspection of the fertility of cattle treated with SMB, it became apparent that while the reduction in conception rates of all the animals treated may not have been statistically significant,

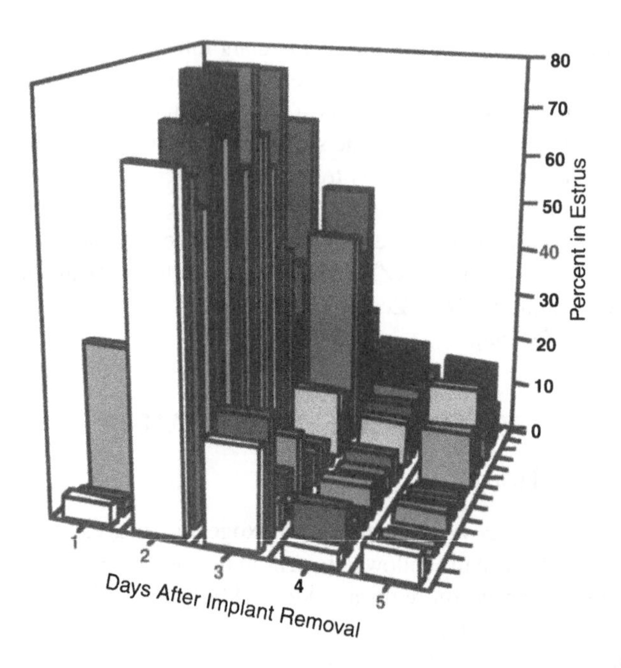

FIGURE 4.1 The distribution of estrus following Syncro-Mate-B treatment in 15 trials (Adapted from Miksch et al., 1978; Spitzer et al., 1978).

the conception rates of those cattle that began SMB treatment late in the estrous cycle (>day 14) were significantly lower (Brink and Kiracofe, 1988; Beal, 1995).

Anestrous Postpartum Cows

The ovaries of well-nourished postpartum, suckled beef cows are capable of ovulating and initiating estrous cycles within 2 weeks after calving (Short et al., 1990). The suckling stimulus of the calf, however, usually inhibits the cow from initiating estrous cycles for 45 (mature cows) to 65 days (first-calf heifers) after calving. Often this prevents a cow from cycling or becoming pregnant at the beginning of the breeding season. Estrus synchronization treatments involving progestins can be used to "induce" estrus in some noncyclic postpartum cows. This effect is enhanced if the calves are temporarily weaned from the cows.

The most dramatic success of inducing estrus in noncyclic cows was reported by Smith et al. (1979). They evaluated the effects of treatment with SMB and 48-hour calf removal alone or in combination on the estrous response and pregnancy rate of noncyclic cows in three herds (Table 4.1). Syncro-Mate-B was able to induce estrus sooner and in a larger proportion of cows than was calf removal. However, the most effective treatment for inducing a synchronized estrus was the combination of SMB treatment and 48-hour calf removal beginning at the time of implant removal. The combined effect of SMB and calf removal resulted in synchronized pregnancy rates of 44, 46, and 35% in three herds in which fewer than 25% of the cows were cycling prior to treatment. Kiser et al. (1980) demonstrated that it was necessary to remove

TABLE 4.1
Estrous Response and Fertility of Postpartum Cows Treated with Syncro-Mate-B (SMB) and 48-hour Calf Removal (CR)

		Estrous Response (%)[a]		Pregnancy Rate (%)[b]	
Treatment	**No.**	**4 day**	**21 day**	**4 day**	**21 day**
Trial I					
SMB	18	61	61	17	—
SMB + CR[c]	18	78	78	44	—
Trial II					
CR[d]	22	32	73	14	27
SMB + CR[c]	24	96	96	46	58
Trial III					
Control	52	11	31	8	17
CR[d]	52	19	62	18	44
SMB	53	60	68	27	40
SMB + CR[c]	53	85	88	35	58

[a] Detected in estrus within 120 hours after treatment.
[b] Pregnancy rate = number pregnant/number in group × 100.
[c] 48-hour calf removal at SMB implant removal.
[d] 48-hour calf removal at beginning of breeding season.

Adapted from Smith et al. (1979).

calves for 48 hours, rather than 24 hours, in conjunction with SMB to induce estrus in noncyclic cows. They also noted that the beneficial effects of SMB and calf removal were not evident in herds where cows were in marginal (BCS 4) body condition.

Fertility of noncyclic cows induced to exhibit estrus in response to SMB or SMB plus calf removal should not be expected to be equal to that of animals exhibiting estrous cycles prior to the breeding season. We compared conception rates of cycling animals synchronized with SMB or PGF2 to that of noncyclic cows after estrus was induced with SMB. Conception rates of cyclic cows synchronized with SMB (35/58; 60%) or PGF2 (54/84; 64%) were both greater than the conception rate of noncyclic cows bred after an induced estrus (24/55; 44%). Hence, if SMB and/or calf removal is used to induce estrus in some noncyclic cows, lower fertility should be expected following the induced estrus.

SUMMARY

Syncro-Mate-B and other products combining estradiol and progestin are effective in synchronizing estrus in cycling cows and in inducing estrus in some anestrous cows. The efficacy of SMB is limited by the failure of estradiol to consistently regress the corpus luteum in cyclic animals that begin treatment during the first 5 days of the estrous cycle. The fertility of cows and heifers that begin treatment late (>day 14) in the estrous cycle may be lower than that of animals synchronized at other times.

The advantage of SMB is the high incidence of estrus following treatment. Synchronization of estrus following SMB treatment is also confined to a very short period in the majority of animals, thereby facilitating the use of timed insemination without estrous detection. The ability to induce estrus in many noncyclic postpartum cows enhances the response to SMB in mixed groups of cyclic and anestrous animals and the induction of estrus is enhanced by 48-hour calf removal. When used to synchronize cyclic cows and heifers or groups of postpartum cows in which the majority (>60%) of the animals are cyclic, SMB is likely to result in greater than 80% estrous response and, following artificial insemination breeding, the pregnancy rate is usually greater than 40%.

REFERENCES

Ahmad, N., F. N. Schrick, R. L. Butcher, and E. K. Inskeep. 1995. Effect of persistent follicles on early embryonic losses in beef cows. *Biol. Reprod.* 52:1129.

Allen, E. and E. A. Doisy. 1923. An ovarian hormone. Preliminary report on its localization, extraction and partial purification, and action in test animals. *J. Am. Med. Assoc.* 81:819.

Anderson, L. H. and M. L. Day. 1994. Acute progesterone administration regresses persistent dominant follicles and improves fertility of cattle in which estrus was synchronized with melengestrol acetate. *J. Anim. Sci.* 72:2955.

Beal, W. E. 1995. Estrus synchronization programs what works and what doesn't. In: *Proc. Beef Improv. Fed. Twenty-Seventh Res. Symp. Ann. Mt.* pp. 13, Colby, KS.

Bo, G. A., G. P. Adams, R. A. Pierson, and R. J. Mapletoft. 1995. Exogenous control of follicular wave emergence in cattle. *Theriogenology* 43:31.

Brink, J. T. and G. H. Kirakofe. 1988. Effect of estrous cycle stage at SyncroMate B treatment on conception and time to estrus in cattle. *Theriogenology* 29:513.

Christian, R. E. and L. E. Casida. 1948. The effect of progesterone in altering the estrus cycle of the cow. *J. Anim. Sci.* 7:540 (Abstr.).

Corner, G. W. and W. M. Allen. 1929. Physiology of the corpus luteum II. Production of a special uterine reaction (progestational proliferation) by extracts of the corpus luteum. *Am. J. Physiol.* 88:326.

Cupp, A., M. Garcia-Winder, A. Zamudio, V. Mariscal, M. Wehrman, N. Kojima, K. Peters, E. Bergfeld, P. Hernandez, T. Sanchez, R. Kittock, and J. Kinder. 1992. Two concentrations of progesterone (P4) in circulation have a differential effect on the pattern of ovarian follicular development in the cow. *Biol. Reprod.* 45 (Suppl. 1):106.

Custer, E. E., W. E. Beal, S. J. Hall, A. W. Meadows, J. G. Berardinelli, and R. Adair. 1994. Effect of melengestrol acetate (MGA) or progesterone releasing intravaginal devices (PRID) on follicular development, estradiol-17β and progesterone concentrations and luteinizing hormone release during an artificially lengthened bovine estrous cycle. *J. Anim. Sci.* 72:1282.

Dutt, R. H. and L. E. Casida. 1948. Alteration of the estrual cycle in sheep by use of progesterone and its effect upon subsequent ovulation and fertility. *Endocrinology* 43:208.

Ginther, O. J., M. C. Wiltbank, F. M. Fricke, J. R. Gibbons, and K. Kot. 1996. Minireview. Selection of the dominant follicle in cattle. *Biol. Reprod.* 55:1187.

Hansel, W. and R. J. Schechter. 1972. Biotechnical procedures concerning the control of the estrous cycle in domestic animals. *VIIth Inter. Cong. Anim. Reprod. A.I. Munich.* 1:78.

Hansel, W. and W. E. Beal. 1979. Ovulation control in cattle. In: *BARC Symposia III. Anim. Reprod.* pp. 91. Allanheld, Osmun and Co., Montclair, NJ.

Kaltenbach, C. C., G. D. Niswender, D. R. Zimmerman, and J. N. Wiltbank. 1964. Alterations of ovarian activity in cycling, pregnant and hysterectomized heifers with exogenous estrogens. *J. Anim. Sci.* 23:995.

Kiser, T. E., S. E. Dunlap, L. L. Benyshek, and S. E. Mares. 1980. The effect of calf removal on estrous response and pregnancy rate of beef cows after Syncro-Mate-B treatment. *Theriogenology* 13:381.

Kojima, N., T. T. Stumpf, A. S. Cupp, L. A. Worth, M. S. Roberson, M. W. Wolfe, R. J. Kittock, and R. J. Kinder. 1992. Exogenous progesterone and progestins as used in estrous synchrony regimens do not mimic the corpus luteum in regulation of LH and 17β-estradiol in circulation of cows. *Biol. Reprod.* 47:1009.

Larson, L. L. and P. J. H. Ball. 1992. Regulation of estrous cycles in dairy cattle: a review. *Theriogenology* 38:255.

Loy, R. G., R. G. Zimbelman, and L. E. Casida. 1960. Effects of injected ovarian hormone on the corpus luteum of the estrual cycle in cattle. *J. Anim. Sci.* 19:175.

Lucy, M. C., W. W. Thatcher, and K. L. Macmillan. 1990. Ultrasonic identification of follicular populations and return to estrus in early postpartum dairy cows given intravaginal progesterone for 15 days. *Theriogenology* 34:325.

Mares, S. E., L. A. Peterson, E. A. Henderson, and M. E. Davenport. 1977. Fertility of beef herds inseminated by estrus or by time following Syncro-Mate-B (SMB) treatment. *J. Anim. Sci.* 45 (Suppl. 1):185.

Mihm, M., N. Curran, P. Hyttel, M. P. Boland, and J. F. Roche. 1994. Resumption of meiosis in cattle oocytes from preovulatory follicles with a short and a long duration of dominance. *J. Reprod. Fertil.* 13:14.

Miksch, E. D., D. G. LeFever, G. Mukembo, J. C. Spitzer, and J. N. Wiltbank. 1978. Synchronization of estrus in beef cattle. II. Effect of an injection of norgestomet and an estrogen in conjunction with a norgestomet implant in heifers and cows. *Theriogenology* 10:201.

Odde, K. J. 1990. A review of synchronization of estrus in postpartum cattle. *J. Anim. Sci.* 68:817.

Patterson, D. J., G. H. Kiracofe, J. S. Stevenson, and L. R. Corah. 1989. Control of the bovine estrous cycle with melengestrol acetate (MGA): a review. *J. Anim. Sci.* 67:1895.

Pratt, S. L., J. C. Spitzer, G. L. Burns, and B. B. Plyler. 1991. Luteal function, estrous response, and pregnancy rate after treatment with norgestomet and various dosages of estradiol valerate in suckled cows. *J. Anim. Sci.* 69:2721.

Revah, I. and W. R. Butler. 1996. Prolonged dominance of follicles and reduced viability of bovine oocytes. *J. Reprod. Fertil.* 106:39.

Roberson, M. S., M. W. Wolfe, T. T. Stumpf, R. J. Kittok, and J. E. Kinder. 1989. Luteinizing hormone secretion and corpus luteum function in cows receiving two levels of progesterone. *Biol. Reprod.* 41:997.

Savio, J. D., W. W. Thatcher, L. Badinga, R. L. de la Sota, and D. Wolfenson. 1993. Regulation of dominant follicle turnover during the oestrus cycle in cows. *J. Reprod. Fertil.* 97:197.

Schmitt, É. J.-P., T. Diaz, M. Drost, and W. W. Thatcher. 1996. Use of gonadotropin-releasing hormone agonist or human chorionic gonadotropin for timed insemination in cattle. *J. Anim. Sci.* 74:1084.

Short, R. E., R. A. Bellows, R. B. Staigmiller, J. G. Berardinelli, and E. E. Custer. 1990. Physiological mechanisms controlling anestrus and infertility in postpartum beef cattle. *J. Anim. Sci.* 68:799.

Sirois, J. and J. E. Fortune. 1990. Lengthening the bovine estrous cycle with low concentrations of exogenous progesterone: a model for studying ovarian follicular dominance. *Endocrinology* 127:916.

Smith, M. F., W. C. Burrell, L. D. Shipp, L. R. Sprott, W. H. Songster, and J. N. Wiltbank. 1979. Hormone treatments and use of calf removal in postpartum beef cows. *J. Anim. Sci.* 48:1285.

Spitzer, J. C., D. L. Jones, E. D. Miksch, and J. N. Wiltbank. 1978. Synchronization of estrus in beef cattle. III. Field trials in heifers using a norgestomet implant and injections of norgestomet and estradiol valerate. *Theriogenology* 10:223.

Stock, A. E. and J. E. Fortune. 1993. Ovarian follicular dominance in cattle: relationship between prolonged growth of the ovulatory follicle and endocrine parameters. *Endocrinology* 132:1108.

Wiltbank, J. N., J. E. Ingalls, and W. W. Rowden. 1961. Effects of various forms and concentrations of estrogens alone or in combination with gonadotropins on the estrous cycle of beef heifers. *J. Anim. Sci.* 20:341.

Wiltbank, J. N. 1966. Modification of ovarian activity in the bovine following injection of oestrogen and gonadotropin. *J. Reprod. Fertil.* (Suppl. 1):1.

Wishart, D. F. and I. M. Young. 1974. Artificial insemination of progestin (SC21009) treated cattle at predetermined times. *Vet. Rec.* 95:503.

5 Use of GnRH to Synchronize Estrus and/or Ovulation in Beef Cows with or without Timed Insemination

Jeffrey S. Stevenson

CONTENTS

Traditional estrous-synchronization programs were developed to control the estrous cycle by either inhibiting the occurrence of estrus by providing an exogenous source of progestin (e.g., feeding, drenching, injecting, or implanting) or managing the corpus luteum by administering an exogenous luteolysin (e.g., prostaglandin $F_{2\alpha}$ or $PGF_{2\alpha}$). The challenge faced by the cow–calf producer desiring to use artificial insemination at the beginning of the breeding season is to not only synchronize successfully the estrous cycle but to induce a fertile estrus or ovulation in cows that are yet anestrus when the breeding season begins.

0-8493-1117-9/02/$0.00+$1.50

LACTATIONAL ANESTRUS

Anestrus is a major contributor and the most limiting factor in achieving high pregnancy rates (Short et al., 1990) because a significant proportion of cows are not cycling at the start of the breeding season when an artificial insemination program is implemented. Continual presence of a suckling calf prolongs anestrus and delays the reinitiation of estrous cycles (Williams, 1990). Although insufficient energy and protein intake and insufficient body condition at calving are limiting factors, temporary or permanent weaning of the calf usually initiates estrus within a few days (Williams, 1990). Younger cows with nursing calves generally have more prolonged anestrus because of their additional growth requirement (Randel, 1990).

Nutrients are used by cows according to an established priority (Short and Adams, 1988). The first priority is maintenance of essential body functions to preserve life. Once that maintenance requirement is met, remaining nutrients accommodate growth. Finally, lactation and then initiation of estrous cycles are supported. Because older cows have no growth requirement, nutrients are more likely to be available for milk synthesis (first) and estrous cycle initiation (second). Because of this priority system, younger growing cows generally produce less milk and are anestrus longer after calving.

INCIDENCE OF ANESTRUS

During the past 6 years, we have treated more than 2200 beef cows with various hormonal treatments to synchronize estrus, ovulation, or both, in an attempt to achieve conception during the first week of the breeding season and maximize the proportion of cows pregnant to genetically superior artificial insemination sires (Stevenson et al., 1997; Thompson et al., 1999; Stevenson et al., 2001). As part of these studies, we measured the incidence of cyclicity at the beginning of the breeding season, both prior to hormonal treatments and in response to these treatments. The major risk factors that limit a high rate of cyclicity at the beginning of the breeding season include age of cow, body condition, and days since calving. Despite the fact that 2-year-old cows may calve up to 3 weeks earlier than older cows, their cycling rates are still less than those of older cows (Figure 5.1). Cows with body condition scores less than 5 (1 = thin and 9 = fat) are less likely to be cycling than those with greater body condition scores (Figure 5.2). Body condition represents a repeatable visual appraisal of the nutrition program. The literature indicates that suckled cows must calve with a minimum body condition score of 5 (Short et al., 1990) to prevent prolonged postpartum intervals to estrus. Cows that have calved less than 70 days before the beginning of the breeding season also are less likely to be cycling (Figure 5.3). In summary, our studies have demonstrated that only 50 to 60% of suckled beef cows are cycling between 7 and 11 days before the onset of the breeding season. With those cycling rates, applying an estrus-synchronization system that controls only the onset of estrus in cycling cows, one could only expect pregnancy rates of 30 to 40% if conception rates are 60 to 70%. In order to maximize pregnancy rates during the first week of the breeding season, one must employ a system that will induce fertile estrus

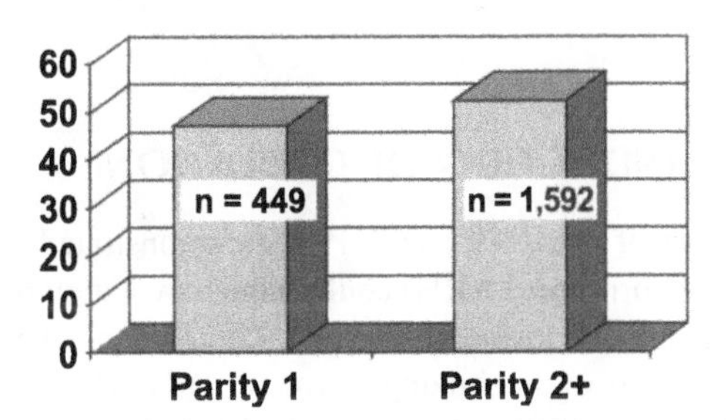

FIGURE 5.1 Percentage of suckled cows cycling (elevated concentrations of serum progesterone) at the beginning of the breeding season based on their parity. Parity 1 cows calved an average 3 weeks ahead of parity 2 cows. A composite of four studies (Stevenson et al., 1997; 2000).

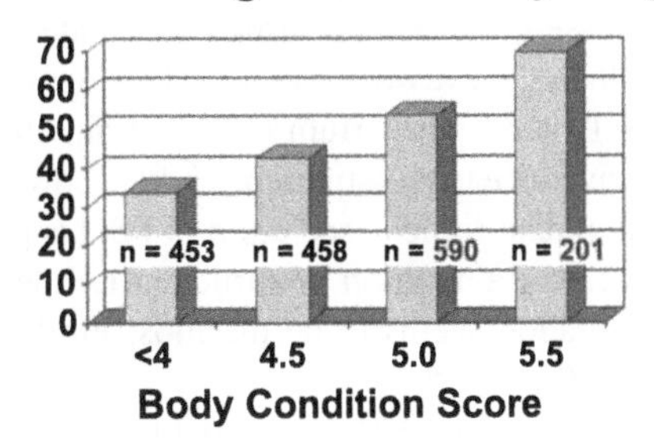

4

FIGURE 5.2 Percentage of suckled cows cycling (elevated concentrations of serum progesterone) based on body condition scores assessed at the beginning of the breeding season. A composite of four studies (Stevenson et al., 1997; 2000).

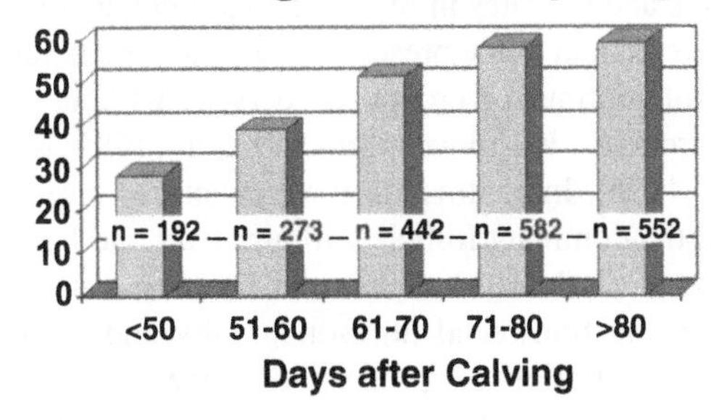

FIGURE 5.3 Percentage of suckled cows cycling (elevated concentrations of serum progesterone) based on days postpartum at the beginning of the breeding season. A composite of four studies (Stevenson et al., 1997; 2000).

or ovulation in suckled anestrous cows in addition to synchronizing the estrous cycles of cycling cows.

INDUCTION OF OVULATION

The scientific literature of the 1970s and 1980s demonstrated that injections of gonadotropin-releasing hormone (GnRH) could induce ovulation of ovarian follicles in milked (Britt et al., 1974) and suckled cows (Schams et al., 1973). Injections of GnRH induced the release of luteinizing hormone (LH) (Britt et al., 1974) and follicle-stimulating hormone (FSH) (Foster et al., 1980) from the anterior pituitary gland. Resulting elevated blood concentrations of LH and FSH either caused follicular rupture (ovulation), if a follicle(s) was present, or may have induced some new follicular development. In other studies (Entwistle and Oga, 1977; Webb et al., 1977), unless two injections were given 7 to 14 days apart (Webb et al., 1977; Fonseca et al., 1980) and cows were at least 30 days postpartum at the time of treatment, the incidence of ovulation was low. In some cases, cows continued to show regular estrous cycles after ovulation induction, and in a few cases, when the short-lived induced corpus luteum regressed, the cows returned to a state of anestrus.

Based on current knowledge, waves of follicular development occur in suckled cows, beginning as early as 14 days postpartum (Stagg et al., 1995). Because waves of follicles develop and turn over in early postpartum suckled cows, only to be replaced by a new wave and a new dominant follicle, secretion of FSH is probably not limiting the recurrence of estrous cycles. Reinitiation of sufficient LH pulses to support final follicular maturation and ovulation seems to be the most limiting factor to estrous-cycle initiation (Williams, 1990).

In recent studies, injections of GnRH initiated turnover of large follicles or induced the dominant follicle to ovulate when at a proper stage of maturity followed by emergence of a new follicular wave (Twagiramungu et al., 1995; Thompson et al., 1999). Unless a follicle is at least 10 mm in diameter, GnRH-induced LH release will not be effective in stimulating ovulation (Crowe et al., 1993), because once a dominant follicle is selected, it possesses LH receptors and becomes LH-dependent (Lucy et al., 1992). In the absence of progesterone, a dominant follicle will continue to grow until it matures and ovulates in response to a preovulatory LH surge. In the absence of a corpus luteum and in the presence of an exogenous source of progestin, a dominant follicle will continue to grow and "persist" without ovulation until the exogenous source of progestin is removed (Sanchez et al., 1995; Kojima et al., 1995; Smith and Stevenson, 1995). In contrast, in the presence of progesterone produced by the corpus luteum, the dominant follicle will turn over and be replaced by a new dominant follicle (Adams et al., 1992; Smith and Stevenson, 1995).

In recent studies using transrectal ultrasonography and daily collected blood samples to monitor concentrations of progesterone, we have demonstrated that a single GnRH injection is quite effective (>80%) in inducing ovulation and formation of the first corpus luteum in late-calving, suckled anestrous cows (34 ± 6 days postpartum), whether cows were treated with GnRH alone, received GnRH plus a norgestomet implant for 7 days (Syncro-Mate-B, Merial implant only), or the implant alone

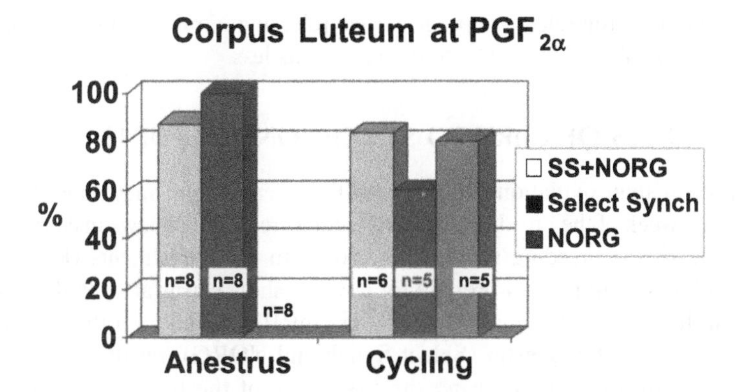

FIGURE 5.4 Percentage of cycling and noncycling suckled cows with a corpus luteum 7 days after GnRH (just prior to an injection of $PGF_{2\alpha}$). Cows averaged 34 ± 6 days postpartum at the time of GnRH injection. (From Thompson et al., 1999. With permission.)

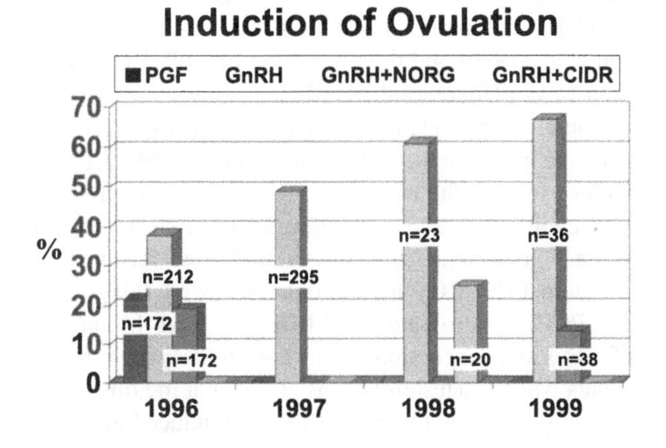

FIGURE 5.5 Percentage of cows in which ovulation was induced by GnRH. A composite of six studies (Stevenson et al., 1997; 2000; unpublished results).

(Thompson et al., 1999). Figure 5.4 illustrates the percentage of cycling and noncycling cows that had a corpus luteum 7 days after the injection of GnRH.

In further studies, based only on blood samples, the percentage of cows with elevated progesterone (induction of ovulation occurred) 7 days after GnRH is illustrated in Figure 5.5. During 1996, the treatment with GnRH alone increased the rate of induced ovulation. Thus, subtracting the control percentage ($2 \times PGF_{2\alpha}$) of induced ovulations resulting from spontaneous ovulation from the percentage of cows treated with GnRH shows that 16.5% of the noncycling cows were induced to ovulate. In a second study in 1996, rate of induced ovulation was 28% in anestrous cows treated with GnRH plus a norgestomet implant. From 1997 to 1999, anywhere from 50 to 67% of the anestrous cows treated with GnRH alone had elevated progesterone 7 days later. In those cows treated with GnRH plus either a norgestomet implant or a CIDR (controlled internal drug release; a progesterone-releasing intravaginal insert, InterAg, Hamilton, New Zealand), induction of ovulation was less.

USES OF GnRH TO SYNCHRONIZE ESTRUS

More important than ovulation induction to the overall pregnancy rates achieved during the first week of the breeding season was the percentage of cows in estrus during the first week of the breeding season in response to these treatments (Figure 5.6). Clearly, GnRH is an important prelude for cows to show estrus after $PGF_{2\alpha}$ (Select Synch), but the ovulation induction ability of GnRH coupled with the addition of a pre-estrus source of progestin (Select Synch and NORG), resulted in a greater percentage of cows in estrus during the first week of the breeding season, than administering only $PGF_{2\alpha}$.

In addition, pretreatment of the cows with one norgestomet implant promoted increased size of the dominant follicle in the absence of a corpus luteum (Garcia-Winder et al., 1987; Rajamahendran and Taylor, 1991; Smith and Stevenson, 1995) by allowing the LH pulse frequency to increase (Garcia-Winder et al., 1986). Norgestomet treatment before GnRH administration increased the amount of GnRH-induced LH release (Thompson et al., 1999) and the proportion of GnRH-induced ovulations in noncycling, suckled cows (Troxel et al., 1993).

Not only is a source of progestin important for expression of estrus, but it is critical for consistent improvement in pregnancy rates. Table 5.1 summarizes the detection of estrus for cows and their conception rates based on serum concentrations of progesterone collected 14, 7, and 0 days before $PGF_{2\alpha}$. Select Synch cows received GnRH 7 days before $PGF_{2\alpha}$, Select Synch + NORG cows received an injection of GnRH plus a norgestomet (NORG) implant only (removed 7 days later) just before $PGF_{2\alpha}$, and the 2 × $PGF_{2\alpha}$ controls received $PGF_{2\alpha}$ on days 14 and 0.

Expression of estrus and conception rates were similarly high in all cycling cows of all three treatments when progesterone was elevated at the time of $PGF_{2\alpha}$. Part of this group included those cows that were anestrus and induced to ovulate (e.g., LLH

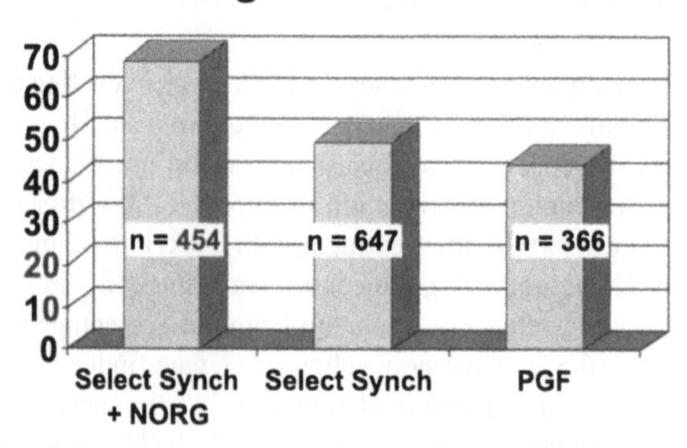

FIGURE 5.6 Percentage of cows detected in estrus during the first 7 days of the breeding season after various treatments. A composite of four studies (Stevenson et al., 1997; 2000).

TABLE 5.1
Distribution of Cows Detected in Estrus Based on Serum Concentrations of Progesterone on Days −14, −7, and 0 before the Second (2 × $PGF_{2\alpha}$ Treatment) or before the Only $PGF_{2\alpha}$

Progesterone Profile[a]	Treatment		
	Select Synch	Select Synch + NORG	2 × $PGF_{2\alpha}$
	% in Estrus (No.)		
	Cycling + elevated progesterone at $PGF_{2\alpha}$		
HHH	88.9 (18)	89.5 (19)	88.9 (36)
HLH	86.7 (30)	90.5 (21)	65.2 (23)
LHH	91.2 (36)	83.8 (37)	86.0 (43)
LLH	66.2 (65)	93.6 (31)	83.3 (36)
Overall[b]	78.5^x (149)	88.8^x (108)	83.1^x (138)
Conception rate[c]	65.8^x	65.3^x	65.5^x
	Cycling + low progesterone at $PGF_{2\alpha}$		
HHL	80.0 (25)	100 (26)	20.0 (5)
HLL	75.0 (4)	100 (6)	40.0 (5)
LHL	77.8 (9)	100 (5)	26.7 (15)
Overall[b]	78.4^x (38)	100^x (37)	25.0^y (25)
Conception rate[c]	57.1^{xy}	41.7^x	83.3^y
	Noncycling + low progesterone at $PGF_{2\alpha}$		
LLL[b]	27.9^y (111)	51.7^x (151)	16.3^z (135)
Conception rate[c]	71.4^x	59.2^x	20.0^y

[a] Blood serum progesterone was determined on day −14 (first injection of 2 ×$PGF_{2\alpha}$ treatment), day −7 (injection of GnRH for Select Synch or injection of GnRH + implant only of norgestomet for Select Synch + NORG treatment), or day 0 ($PGF_{2\alpha}$ injection for all treatments). Concentrations of progesterone were either low (L; <1 ng/mL) or high (H; ≥1 ng/mL) in each of three serum samples, making up eight permutations (HHH, HHL, HLH, HLL, LHH, LHL, LLH, or LLL). Norgestomet, progestin in Syncro-Mate-B (SMB) implant only.

[b] Treatment × progesterone profile interaction ($P < .001$).

[c] Treatment × progesterone profile interaction ($P < .05$).

[xyz] Percentages within a row lacking a common superscript letter differ ($P < .05$).

progesterone profile in Table 5.1). Among cows that were cycling but had low concentrations of progesterone on day 0, expression of estrus was less in controls but conception was greater than that in Select Synch (GnRH injection 7 days before $PGF_{2\alpha}$) + NORG cows. Among noncycling cows that were not induced to ovulate in response to GnRH on day 7, expression of estrus was greatest if they were pretreated with GnRH and norgestomet compared to controls. Conception rates were greater in both groups that received GnRH than in cows that only received $PGF_{2\alpha}$.

These results were validated by a subsequent experiment in which cows were treated with GnRH and norgestomet (Table 5.2). In this experiment, cows were treated with the Select Synch + NORG protocol and either inseminated after detected

TABLE 5.2
Effect of Using GnRH, $PGF_{2\alpha}$, and Norgestomet on Estrus, Conception Rates, and Pregnancy Rates[a]

Item	AI Estrus	Estrous Detection + Timed AI
	% (No.)	
Detected in estrus	79.5 (78)	30.2 (86)
Anestrus	80.9 (21)	20.0 (15)
Cycling	78.9 (57)	32.4 (71)
Conception rate[b]	67.7 (62)	59.3 (86)
Anestrus	64.7 (17)	26.6 (15)
Cycling	68.8 (45)	66.2 (71)
Pregnancy rate[c]	53.8 (78)	59.3 (86)
Anestrus	52.4 (21)	26.6 (15)
Cycling	54.4 (57)	66.2 (71)

[a] Cows received the Select Synch + NORG protocol. Estrus-bred cows were inseminated after detected estrus, whereas cows in the estrus detection + timed artificial insemination group were inseminated after detected estrus until 48 hours after $PGF_{2\alpha}$, then the remaining cows were given GnRH at 48 hours and inseminated 16 hours later.
[b] AI protocol × cycling status ($P = .08$).
[c] AI protocol × cycling status ($P < .05$).

estrus or inseminated after detected estrus until 48 hours after $PGF_{2\alpha}$ and then all remaining cows were administered GnRH at 48 hours and inseminated 16 hours later. Pregnancy rates exceeded 50% after both inseminating protocols, but timed insemination of noncycling cows produced a lower pregnancy rate.

A further advantage for using a progestin concurrently with GnRH is that most cows (87%) have normal luteal phases after insemination compared to 71% of cows receiving only GnRH or 43% of cows receiving only the norgestomet implant before $PGF_{2\alpha}$. Pregnancy rates in the latter three groups were 71, 31, and 15%, respectively. These results are consistent with an earlier report in which norgestomet treatment prior to calf weaning reduced the incidence of short estrous cycles (Ramirez-Godinez et al., 1981).

USES OF GnRH IN TIMED AI PROTOCOLS

OVSYNCH

The evolution of the newest breeding programs began with the Ovsynch protocol in dairy cows (Figure 5.7). This protocol uses the initial GnRH injection (100 μg) to induce ovulation of a dominant follicle (largest follicle in either ovary) that develops into a new corpus luteum in the cycling dairy cow. In cycling dairy cows, about 60 of

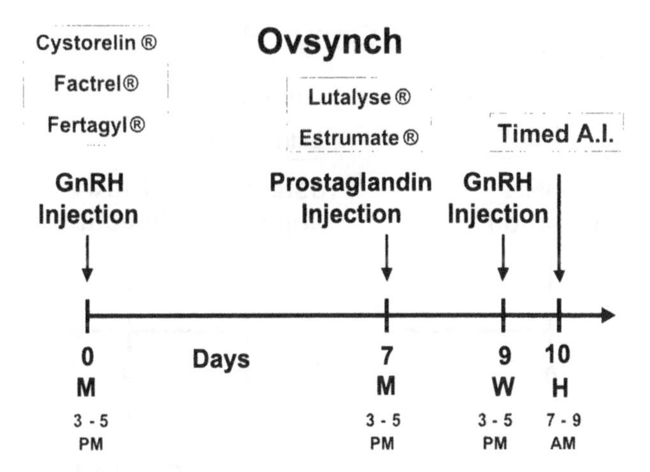

FIGURE 5.7 The Ovsynch protocol with market-available hormone products, Cystorelin (Merial, Iselin, NJ), M = Monday, W = Wednesday, and H = Thursday. Factrel (Fort Dodge, Overland Park, KS); Fertogyl (Intervet, Boxneer, The Netherlands). Lutalyse (Pharmacia & Upjohn, Kalamazoo, MI). Estrumate (Schering-Plough, Kenilworth, NJ).

the cows given the initial GnRH injection ovulate a follicle in response to the LH released by the GnRH injection, depending on the stage of their estrous cycle. Following this induced ovulation, a new wave of follicles emerges from both ovaries within 48 hours, from which a new dominant follicle develops. Seven days after the initial GnRH injection, $PGF_{2\alpha}$ is injected to lyse or kill the original corpus luteum (if one was present at the time of the initial GnRH injection) and the new corpus luteum induced by GnRH. During the next 48 hours, the new dominant follicle rapidly matures; and at 48 hours after $PGF_{2\alpha}$, a second GnRH injection (100 μg) is administered, and the cow is inseminated in the next 8 to 24 hours. A time–response study was conducted in which dairy cows were inseminated at the same time as the second GnRH injection (0 hours) or at 8, 16, 24, or 32 hours later; the best pregnancy rates occurred when the timed artificial insemination occurred at 16 hours (Pursley et al., 1998). In practice, on dairy farms, cows are inseminated anytime they are detected in estrus during this protocol, and further hormonal injections are discontinued. For dairy cows, a half dose (50 μg) of the Cystorelin product has been used successfully in the Ovsynch protocol at both GnRH injection times. Research conducted with the half dose in beef cows has produced equivocal results.

The first application of the Ovsynch protocol in beef cattle was reported recently (Geary et al., 1998). In those early studies, the Ovsynch protocol was compared to a standard Syncro-Mate-B (SMB) protocol with 48-hour calf removal. Overall, the Ovsynch protocol produced greater pregnancy rates (54 vs. 42%) than SMB, particularly among cycling cows (59 vs. 38%).

Cosynch

Based on the distributions of estrus after the Select Synch protocol and the early success with Ovsynch in beef cows, the Cosynch protocol was developed (Figure 5.8) and tested (Geary and Whittier, 1997; Stevenson et al., 2000). The logic behind this

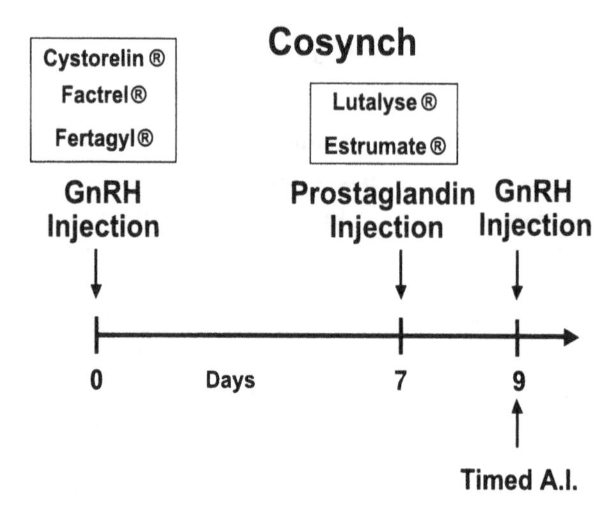

FIGURE 5.8 The Cosynch protocol with market-available hormone products cited (see Figure 5.7 for source of hormones).

protocol was to reduce the number of trips through the working facility to three. Comparisons of Cosynch to Ovsynch were made and pregnancy rates were identical at 48% (Geary and Whittier, 1997). A further study (Geary et al., 2001) also incorporated 48-hour calf removal (between the injections of $PGF_{2\alpha}$ and the second GnRH injection) after both the Ovsynch and Cosynch protocols. Calf removal produced pregnancy rates that were 9 percentage points greater than rates after each protocol without calf removal.

We conducted a similar study in Kansas on three ranches in which we treated all cows with either Select Synch or Cosynch or inseminated a third group of cows after detected estrus until 54 hours after $PGF_{2\alpha}$, when the remaining cows were time inseminated and given a second GnRH injection at artificial insemination (Figure 5.9). Pregnancy rates among noncycling cows were very similar, but among cycling cows, those inseminated at estrus after the Select Synch protocol had greater pregnancy rates than those in the other protocols (Thompson et al., 1999).

Cosynch + Progestin

The combination of Cosynch and a progestin (CIDR) or a norgestomet implant is an effective treatment (Figure 5.10). During 1998 when the CIDR insert was used as a source of progesterone, the average cycling rate exceeded 77% and both the Cosynch and Cosynch + CIDR protocols resulted in pregnancy rates that exceeded 50% (Table 5.3). The addition of the CIDR tended (P = 0.05) to further improve pregnancy rates by 15 percentage points. During 1999, the average cycling rate was only 59% in the same herd of cows, but the Cosynch protocol with the addition of norgestomet produced pregnancy rates greater than 50%, whereas those after Cosynch alone were only 31% (Table 5.3).

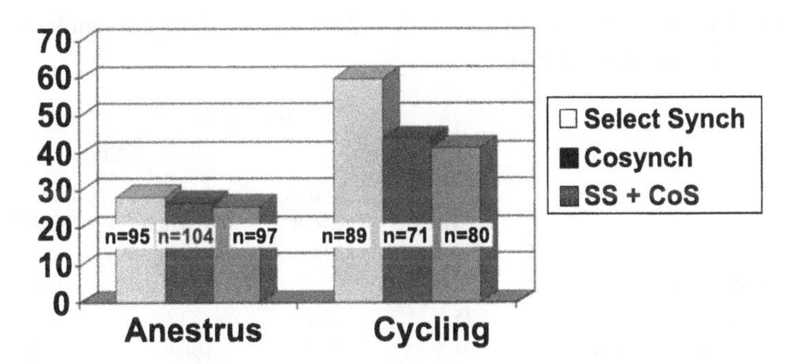

FIGURE 5.9 Pregnancy rates of noncycling and cycling suckled cows treated with the Select Synch, Cosynch, and Select Synch + Cosynch protocols. (From Stevenson et al., 2000. With permission.)

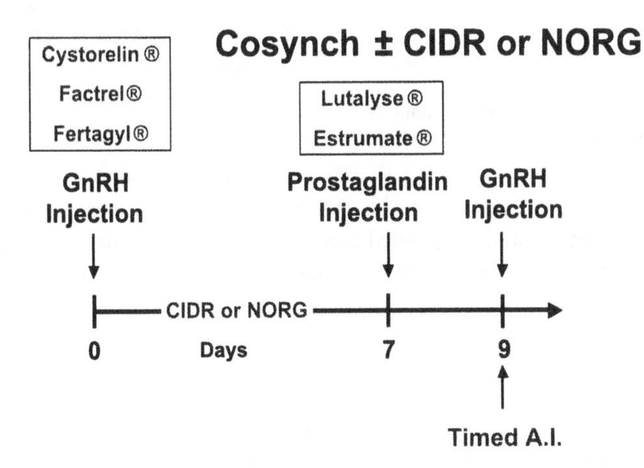

FIGURE 5.10 Cosynch protocol with or without an exogenous source of progestin administered during 7 days via a norgestomet implant (NORG) or an intravaginal progesterone-releasing insert (CIDR).

IMPLICATIONS

Treatment of suckled cows with gonadotropin-releasing hormone 7 days before an injection of $PGF_{2\alpha}$ partially resolved the problem of anestrus before the beginning of the breeding season that cannot be resolved with $PGF_{2\alpha}$ systems alone. The injection of GnRH induced ovulation of a follicle in a variable percentage of anestrous suckled cows. More cows were detected in estrus with normal fertility after GnRH. Addition of a progestin at the time of GnRH injection further improved most responses (e.g., rates of detected estrus and pregnancy). Fixed-time inseminations that followed a

TABLE 5.3
Comparison of Cosynch vs. Cosynch + Norgestomet or Cosynch + CIDR in Suckled Beef Cows

	Treatments			
	1998		1999	
Item	**Cosynch**	**Cosynch+ CIDR**	**Cosynch**	**Cosynch+ NORG**
Number of cows	92	95	91	92
Cycling by day −7, %	77.0	78.9	60.4	58.7
High progesterone on day 0, %	75.0	59.0[a]	73.6	47.8[a]
Low progesterone at +48 hours, %	94.6	98.9	98.9	94.6
Body condition score on day 0	—	—	4.8	4.6
Average days postpartum on day 0 (range)	71 (26–108)	70 (29–108)	76 (37–103)	78 (37–106)
Pregnancy rates, % (No.)	51.1	66.3[a]	30.7	51.1[b]
Breed				
Angus	58.8 (51)	62.2 (53)	37.0 (54)	56.6 (53)
Hereford	30.4 (23)	70.8 (24)	18.7 (16)	31.2 (16)
Simmental	55.6 (18)	72.2 (18)	23.8 (21)	52.1 (23)

[a] Different ($P = .05$) from Cosynch within year.
[b] Different ($P < .09$) from Cosynch within year.

second gonadotropin-releasing hormone injection after $PGF_{2\alpha}$ generally reduced fertility unless the treatment also included a progestin. Further refinement and success of these treatments should increase the convenience and appeal of applying artificial insemination in beef cows.

REFERENCES

Adams, G. P., R. L. Matteri, and O. J. Ginther. 1992. Effect of progesterone on ovarian follicles, emergence of follicular waves, and circulating follicle-stimulating hormone in heifers. *J. Reprod. Fertil.* 95:627.

Britt, J. H., R. J. Kittok, and D. S. Harrison. 1974. Ovulation, estrus and endocrine response after GnRH in early postpartum cows. *J. Anim. Sci.* 39:915.

Crowe, M. A., D. Goulding, A. Baguisi, M. P. Boland, and J. F. Roche. 1993. Induced ovulation of the first postpartum dominant follicle in beef suckler cows using a GnRH analogue. *J. Reprod. Fertil.* 99:551.

Entwistle, K. and L. A. Oga. 1977. Effect of plane of nutrition on luteinizing hormone (LH) response to luteinizing hormone releasing hormone (LHRH) in anestrus postpartum beef cows. *Theriogenology* 8:190.

Fonseca, F. A., J. H. Britt, M. Kosugiyama, H. D. Ritchie, and E. U. Dillard. 1980. Ovulation, ovarian function, and reproductive performance after treatments with gonadotropin releasing hormone in postpartum suckled cows. *Theriogenology* 13:171.

Foster, J. P., G. E. Lamming, and A. R. Peters. 1980. Short-term relationships between plasma LH, FSH and progesterone concentrations in postpartum dairy cows and the effect of GnRH injection. *J. Reprod. Fertil.* 59:321.

Garcia-Winder, M., P. E. Lewis, D. R. Deaver, V. G. Smith, G. S. Lewis, and E. K. Inskeep. 1986. Endocrine profiles associated with lifespan of induced corpora lutea in postpartum beef cows. *J. Anim. Sci.* 62:1353.

Garcia-Winder, M., P. E. Lewis, E. C. Townsend, and E. K. Inskeep. 1987. Effects of norgestomet on follicular development in postpartum beef cows. *J. Anim. Sci.* 64:1099.

Geary, T. M. and J. C. Whittier. 1997. Modifications of the Ovsynch estrous synchronization protocol for use in beef cows. *J. Anim. Sci.* 75 (Suppl. 1):236.

Geary, T. M., J. C. Whittier, E. R. Downing, D. G. LeFever, R. W. Silcox, M. D. Holland, T. M. Nett, and G. D. Niswender. 1998. Pregnancy rates of postpartum beef cows that were synchronized using Syncro-Mate-B or the Ovsynch protocol. *J. Anim. Sci.* 76:1523.

Geary, T. M., J. C. Whittier, D. M. Hallford, and M. D. MacNeil. 2001. Calf removal improves conception to the Ovsynch and Co-Synch protocols. *J. Anim. Sci.* 79:1.

Kojima, F. N., J. R. Chenault, M. E. Wehrman, E.G. Bergfeld, A. S. Cupp. L. A. Werth,. V. Mariscal, T. Sanchez, R. J. Kittok, and J. E. Kinder. 1995. Melengestrol acetate at greater doses than typically used for estrous synchrony in bovine females does not mimic endogenous progesterone in regulation of secretion of luteinizing hormone and 17β-estradiol. *Biol. Reprod.* 52:455.

Lucy, M. C., J. D. Savio, L. Badinga, R. L. de la Sota, and W. W. Thatcher. 1992. Factors that affect ovarian follicular dynamics in cattle. *J. Anim. Sci.* 70:3615.

Pursley, J. R., R. W. Silcox, and M. C. Wiltbank. 1998. Effect of time of artificial insemination on pregnancy rates, calving rates, pregnancy loss, and gender ratio after synchronization of ovulation in lactating dairy cows. *J. Dairy Sci.* 81:2139.

Rajamahendran, R. and C. Taylor. 1991. Follicular dynamics and temporal relationships among body temperature, oestrus, the surge of luteinizing hormone and ovulation in Holstein heifers treated with norgestomet. *J. Reprod. Fertil.* 92:461.

Ramirez-Godinez, J. A., G. H. Kiracofe, R. M. McKee, R. R. Schalles, and R. J. Kittok. 1981. Reducing the incidence of short estrous cycles in beef cows with norgestomet. *Theriogenology* 15:613.

Randel, R. D. 1990. Nutrition and postpartum rebreeding in cattle. *J. Anim. Sci.* 68:853.

Sanchez, T., M. E. Wehrman, F. N. Kojima, A. S. Cupp, E. G. Bergfeld, K. E. Peters, V. Mariscal, R. J. Kittok, and J. E. Kinder. 1995. Dosage of the synthetic progestin, norgestomet, influences luteinizing hormone pulse frequency and endogenous secretion of 17β-estradiol in heifers. *Biol. Reprod.* 52:464.

Schams, D. F. Hofer, B. Hoffman, M. Ender, and H. Karg. 1973. Effects of synthetic LH–RH treatment on bovine ovarian function during estrus cycle and postpartum period. *Acta Endocrinologica Suppl.* 177:273.

Short, R. E. and D. C. Adams. 1988. Nutritional and hormonal interrelationships in beef cattle reproduction. *Can. J. Anim. Sci.* 68:29.

Short, R. E., R. A. Bellows, R. B. Staigmiller, J. G. Berardinelli, and E. E. Custer. 1990. Physiological mechanisms controlling anestrus and infertility in postpartum beef cattle. *J. Anim. Sci.* 68:799.

Smith, M. W. and J. S. Stevenson. 1995. Fate of the dominant follicle, embryonal survival, and pregnancy rates in dairy cattle treated with prostaglandin $F_{2\alpha}$ ($PGF_{2\alpha}$) and progestins in the absence or presence of a functional corpus luteum. *J. Anim. Sci.* 73:3743.

Stagg, K., M. G. Diskin, J. M. Sreenan, and J. F. Roche. 1995. Follicular development in long-term anestrous suckler beef cows fed two levels of energy postpartum. *Anim. Reprod. Sci.* 38:49.

Stevenson, J. S., D. P. Hoffman, D. A. Nichols, R. M. McKee, and C. L. Krehbiel. 1997. Fertility in estrus-cycling and noncycling virgin heifers and suckled beef cows after induced ovulation. *J. Anim. Sci.* 75:1343.

Stevenson, J. S., K. E. Thompson, W. L. Forbes, G. C. Lamb, D. M. Grieger, and L. R. Corah. 2000. Synchronizing estrus and(or) ovulation in beef cows after combinations of GnRH, norgestomet, and prostaglandin $F_{2\alpha}$ with or without timed inseminations. *J. Anim. Sci.* 78:1747.

Thompson, K. E., J. S. Stevenson, G. C. Lamb, D. M. Grieger, and C. A. Löest. 1999. Follicular, hormonal, and pregnancy responses of early postpartum suckled beef cows to GnRH, norgestomet, and $PGF_{2\alpha}$. *J. Anim. Sci.* 77:1823.

Troxel, T. R., L. C. Cruz, R. S. Ott, and D. J. Kesler. 1993. Norgestomet and gonadotropin-releasing hormone enhance corpus luteum function and fertility of postpartum suckled beef cows. *J. Anim. Sci.* 71:2579.

Twagiramungu, H., L. A. Guilbault, and J. J. Dufour. 1995. Synchronization of ovarian follicular waves with a gonadotropin-releasing hormone agonist to increase the precision of estrus in cattle: a review. *J. Anim. Sci.* 73:3141.

Webb, R. G. E. Lamming, N. B. Haynes, H. D. Hafs, and J. G. Manns. 1977. Response of cyclic and postpartum suckled cows to injections of synthetic LH–RH. *J. Reprod. Fertil.* 50:203.

Williams, G. L. 1990. Suckling as a regulator of postpartum rebreeding in cattle: a review. *J. Anim. Sci.* 68:831.

6 Current and Emerging Methods to Synchronize Estrus with Melengestrol Acetate (MGA)

D. J. Patterson, S. L. Wood, F. N. Kojima, and M. F. Smith

CONTENTS

Improving traits of major economic importance in beef cattle can be accomplished most rapidly through selection of genetically superior sires and widespread use of artificial insemination (AI). Procedures that facilitate synchronization of estrus in cycling females and induction of an ovulatory estrus in peripubertal heifers and anestrous postpartum cows will increase reproductive rates and expedite genetic progress.

0-8493-1117-9/02/$0.00+$1.50

Estrous synchronization can be an effective means of increasing the proportion of females that become pregnant early in the breeding season, resulting in shorter calving seasons and more uniform calf crops (Dziuk and Bellows, 1983). Females that conceived to a synchronized estrus calved earlier in the calving season and weaned calves that were, on average, 13 days older and 21 pounds per calf heavier than calves from nonsynchronized females (Schafer et al., 1990).

Effective estrous synchronization programs offer the following advantages: 1) cows or heifers are in estrus at a predicted time, which facilitates artificial insemination, embryo transfer, or other assisted reproductive techniques; 2) the time required for detection of estrus is reduced, thus decreasing labor expense associated with estrous detection; 3) cycling females will conceive earlier during the breeding period; and 4) artificial insemination becomes more practical (Kiracofe, 1988).

Although hormonal treatment of heifers and cows to group estrous periods has been a commercial reality for years, producers have been slow to adopt this management practice. Perhaps this is because of past synchronization failures, which resulted when females that were placed on synchronization treatments failed to reach puberty or resume normal estrous cycles following calving. To avoid problems when using estrous synchronization, females should be selected for treatment when: 1) adequate time has elapsed from calving to the time synchronization treatments are imposed; 2) cows are in average or above average body condition (scores of 5 or higher on a scale from 1 = emaciated to 9 = obese); 3) minimal or preferably no calving problems are experienced; and 4) replacement heifers are developed to prebreeding target weights that represent 65% of their projected mature weight and reproductive tract scores are performed to time the initiation of specific estrous synchronization treatments (Patterson et al., 2000b).

Current research efforts are directed toward the development of a highly effective and economical estrous synchronization program(s) for postpartum beef cows and replacement beef heifers that result in excellent pregnancy rates following artificial insemination at a fixed time. The following protocols and terms will be referred to throughout this chapter.

Protocols:

PG: prostaglandin $F_{2\alpha}$ (PG; Lutalyse®, Estrumate®, Prostamate®, In Synch®).

MGA–PG: melengestrol acetate (MGA; 0.5 mg/hd/day) is fed for a period of 14 days, with PG administered 17 or 19 days after MGA withdrawal.

GnRH–PG: gonadotropin-releasing hormone injection (Cystorelin®, Factrel®, Fertagyl®), followed in 7 days with an injection of PG.

MGA–GnRH–PG: MGA is fed for 14 days, GnRH is administered 10 or 12 days after MGA withdrawal, and PG is administered 7 days after GnRH.

7–11 Synch: MGA is fed for 7 days, PG is administered on the last day MGA is fed, GnRH is administered 4 days after the cessation of MGA, and a second injection of PG is administered 11 days after MGA withdrawal.

Terms:

Estrous response: the number of females that exhibit estrus during a synchronized period.

Synchronized period: the period of time during which estrus is expressed after treatment.

Synchronized conception rate: the proportion of females that become pregnant of those exhibiting estrus and inseminated during the synchronized period.

Synchronized pregnancy rate: the proportion of females that become pregnant of the total number treated.

A REVIEW OF THE ESTROUS CYCLE

Progesterone plays a dominant role in regulation of the estrous cycle of the cow (Table 6.1). Progesterone inhibits release of luteinizing hormone (LH) through negative feedback control of the hypothalamus and anterior pituitary gland (Jainudeen and Hafez, 1987) and inhibits behavioral estrus. A drop in progesterone removes negative feedback inhibition from the hypothalamus and anterior pituitary, resulting in increased LH pulse frequency. Rapid follicle growth accompanied by increased secretion of estradiol-17β is stimulated by increased release of LH. Estradiol-17β reaches threshold concentrations 2 to 3 days after the drop in progesterone, which stimulates the preovulatory LH surge. High concentrations of estradiol-17β are necessary for the female to exhibit signs of behavioral estrus, and the preovulatory LH surge stimulates final maturation of the oocyte and subsequent ovulation.

The physiological basis for estrous synchronization followed the discovery that progesterone inhibited maturation of ovarian Graafian follicles (Nellor and Cole, 1956). Synchronizing estrous cycles of domestic cattle depends on control of the functional life span of the corpus luteum (Hansel and Convey, 1983). There are two ways to facilitate control of the corpus luteum that result subsequently in estrus and ovulation (Hansel and Convey, 1983). The first method involves long-term administration of a progestin with subsequent regression of the corpus luteum during the time the progestin is administered (Britt, 1987). Estrus and ovulation occur within 2 to 8 days after progestin withdrawal. The second method involves the administration of prostaglandin $F_{2\alpha}$ (PG) that shortens the normal life span of the corpus luteum. This is accompanied generally with estrus and ovulation within 48 to 120 hours

TABLE 6.1
Characteristics of the Estrous Cycle of the Cow

Period	Day	Characteristics
Estrus	0	Behavioral estrus (heat)
Metestrus	1 to 4	Ovulation (CL formation)
Diestrus	5 to16	Progesterone secretion
Proestrus	17 to 21	Rapid follicle growth

after injection. The corpus luteum is responsive to PG only during the period of diestrus. Estrus will be synchronized only among females that are within this responsive stage at the time of treatment (Britt, 1987).

The development of precise methods to synchronize estrus requires control of both luteal life span and follicular waves. Ovarian follicular growth in cattle occurs in waves that are divided into three stages—recruitment, selection, and dominance (Adams, 1999). A transient increase in follicle-stimulating hormone (FSH) precedes each follicular wave and likely serves as the initiator of recruitment. Following recruitment, one of the "recruited follicles" is selected from the cohort, escapes atresia, and continues toward ovulation. At some point after the selection process, a follicle becomes dominant and thereby avoids atresia. If the dominant follicle fails to ovulate, it then loses its dominance and becomes atretic. Synchronization of new follicular waves in cattle occurs following ovulation (GnRH injection) or turnover (estradiol or progesterone injection) of dominant follicles (Bo et al., 1995; McDowell et al., 1998).

DEVELOPMENT OF METHODS TO SYNCHRONIZE ESTRUS

The development of methods to control the estrous cycle of the cow has occurred in five distinct phases. The physiological basis for estrous synchronization followed the discovery that progesterone inhibited preovulatory follicular maturation and ovulation (Nellor and Cole, 1956; Hansel et al., 1961; Lamond, 1964). Regulation of estrous cycles was believed to be associated with control of the corpus luteum, whose life span and secretory activity are regulated by trophic and lytic mechanisms (Thimonier et al., 1975). Phase I included efforts to prolong the luteal phase of the estrous cycle or to establish an artificial luteal phase by administering exogenous progesterone. Later, progestational agents were combined with estrogens or gonadotropins in Phase II; whereas Phase III involved PG and its analogs as luteolytic agents. Treatments that combined progestational agents with PG characterized Phase IV.

Precise monitoring of ovarian follicles and corpora lutea over time by transrectal ultrasonography expanded our understanding of the bovine estrous cycle and particularly the change that occurs during a follicular wave. Growth of follicles in cattle occurs in distinct wave-like patterns, with new follicular waves occurring approximately every 10 days (6- to 15-day range). We now know (Phase V) that precise control of estrous cycles requires the manipulation of both follicular waves and luteal life span.

A single injection of gonadotropin-releasing hormone (GnRH) to cows at random stages of their estrous cycles causes release of luteinizing hormone, leading to synchronized ovulation or luteinization of most large dominant follicles. Consequently, a new follicular wave is initiated in all cows within 2 to 3 days of GnRH administration. Luteal tissue that forms after GnRH administration is capable of undergoing PG-induced luteolysis 6 or 7 days later (Twagiramungu et al., 1995). This method will be referred to as the GnRH–PG protocol throughout this chapter. The GnRH–PG protocol increased estrous synchronization rate in beef (Twagiramungu et al., 1992a,b) and dairy (Thatcher et al., 1993) cattle. A drawback of this method is that approximately 5 to 15% of the cows are detected in estrus on or before the day of PG injection, thus reducing the proportion of females that are detected in estrus and inseminated during the synchronized period (Kojima et al., 2000).

SYNCHRONIZATION OF ESTRUS AND OVULATION WITH THE GnRH–PG–GnRH PROTOCOL

Administration of PG alone is commonly utilized to synchronize an ovulatory estrus in cycling cows. However, this method is ineffective in anestrous females, and variation among animals in the stage of the follicular wave at the time of PG injection directly contributes to the variation in onset of estrus during the synchronized period (Macmillan and Henderson, 1984; Sirois and Fortune, 1988; Kastelic et al., 1990; Savio et al., 1990). Consequently, the GnRH–PG–GnRH protocol was developed to synchronize follicular waves and timing of ovulation. The GnRH–PG–GnRH protocol for fixed-time AI results in development of a preovulatory follicle that ovulates in response to a second GnRH-induced LH surge 48 hours after PG injection. Addition of a GnRH injection 48 hours after PG was given the trademark, Ovsynch (Pursley et al., 1995). Ovsynch was validated recently as a reliable means of synchronizing ovulation for fixed-time AI in lactating dairy cows (Pursley et al., 1995; 1997a,b; Burke et al., 1996; Schmitt et al., 1996). Time of ovulation with Ovsynch occurs 24 to 32 hours after the second GnRH injection and is synchronized in 87 to 100% of lactating dairy cows (Pursley et al., 1997a). Pregnancy rates among cows that were inseminated at a fixed time following Ovsynch ranged from 32 to 45%, rates comparable to controls (Pursley et al., 1997b; 1998). The Ovsynch protocol, however, did not effectively synchronize estrus and ovulation in dairy heifers (35% pregnancy rate compared with 74% in PG controls; Pursley et al., 1997b).

Protocols for fixed-time insemination were recently tested in postpartum beef cows. Pregnancy rates for Ovsynch-treated beef cows were compared with those of cows synchronized and inseminated at a fixed time following treatment with Syncro-Mate-B (Geary et al., 1998). Calves in both treatment groups were removed from their dams for a period of 48 hours beginning either at the time of implant removal (Syncro-Mate-B) or at the time PG was administered (Ovsynch). Pregnancy rates following fixed-time insemination after Ovsynch (54%) were higher than for Syncro-Mate-B-II (42%) treated cows. One should note that on the day following timed insemination, cows were exposed to fertile bulls of the same breed; no attempt was made to determine progeny paternity. Additionally, we do not know the incidence of short cycles among cows that were anestrous prior to treatment and that perhaps returned to estrus prematurely and became pregnant to natural service (Geary et al., 1998).

Recently, variations of the Ovsynch protocol (CO-Synch and Select Synch) were tested in postpartum beef cows (Figure 6.1). It is important to understand that treatment variations of Ovsynch currently being used in postpartum beef cows have not undergone the same validation process that Ovsynch underwent in lactating dairy cows. At this point, we do not know whether response in postpartum beef cows to

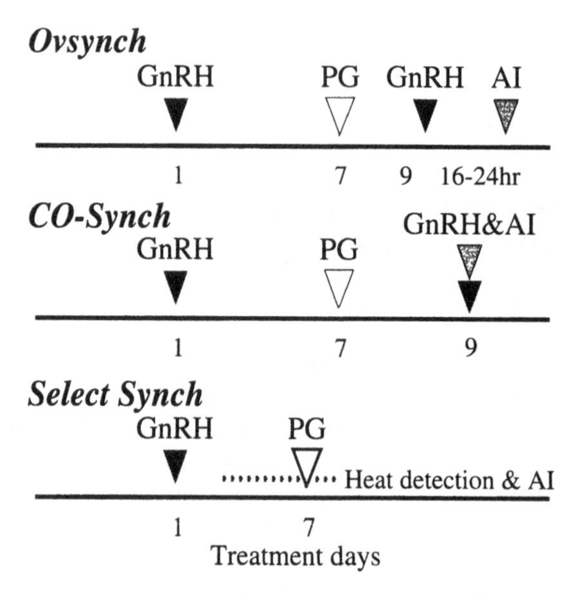

FIGURE 6.1 Methods currently being used to synchronize ovulation in postpartum beef cows: Ovsynch, CO-Synch, and Select Synch.

the protocols outlined in Figure 6.1 is the same or different from lactating dairy cows due to potential differences in follicular wave patterns. Differences in specific response variables may include: a) the relative length of time to ovulation from the second GnRH injection; b) the anticipated range in timing of ovulation; and c) the degree of ovulation synchrony that occurs.

Two variations from Ovsynch are being used most extensively in postpartum beef cows. CO-Synch (Geary et al., 2001) is similar to Ovsynch in that timing and sequence of injections are the same and all cows are inseminated at a fixed time. CO-Synch differs from Ovsynch, however, in that cows are inseminated when the second GnRH injection is administered, compared to the recommended 16 hours after GnRH for Ovsynch-treated cows. Select Synch (Downing et al., 1998) differs, too, in that cows do not receive the second injection of GnRH and are not inseminated at a fixed time. Cows synchronized with this protocol are inseminated 12 hours after detected estrus. It is currently recommended for Select Synch-treated cows that detection of estrus begin as early as 4 days after GnRH injection and continue through 5 days after PG (Kojima et al., 2000). Select Synch, similar to Ovsynch, was less effective than the melengestrol acetate (MGA)–PG protocol in synchronizing estrus in beef heifers (Stevenson et al., 1999).

BACKGROUND AND DEVELOPMENT OF METHODS TO SYNCHRONIZE ESTROUS CYCLES WITH MGA

Melengestrol acetate (MGA; Pharmacia & Animal Health, Kalamazoo, MI), an orally active progestational steroid (6-methyl-17-alpha-acetoxy-16-methylene-pregn-4,6-diene-3,20-dione) was developed in 1962. Melengestrol acetate is capable of maintaining pregnancy and delaying menstrual and estrual activity (Zimbelman and Smith, 1966); however, minimal effective doses of MGA required for complete maintenance of pregnancy (4 mg/day) are 8 to 20 times greater than those needed to inhibit ovulation (Zimbelman, 1963). Melengestrol acetate was first marketed for use in feedlot heifers to improve feed efficiency and rate of gain by allowing ovarian follicular development while inhibiting estrus and ovulation (Zimbelman and Smith, 1966). Complete suppression of estrus and ovulation occurred in most cattle when MGA was fed at a dosage of 0.42 mg.

The development of synchronization methods that utilized MGA spanned the major periods of growth in estrous cycle control that began in 1963. The cumulative results of these studies indicated that short treatment periods (10 days or less) with progestins resulted in near normal fertility at the controlled estrus (Roche, 1974; Miksch et al., 1978), but synchrony of estrus was lower compared with heifers given progesterone for 20 days (Roche, 1974). Luteolytic compounds (prostaglandins) were capable of inducing luteal regression but were incapable generally of inhibiting ovulation. Based on these observations, Heersche et al. (1979) concluded that luteolytic compounds needed to be used more than once or combined with progestins to obtain a maximal synchronization response. Combining progestins, which inhibit estrus and ovulation, with a luteolytic product circumvented that portion of the luteal phase that is insensitive to PG (Heersche et al., 1979). This treatment allowed single injections of PG to synchronize entire groups of cattle, on or after day 5 of progestin treatment.

The theory behind combined progestin–PG treatment was that animals beginning treatments with progestins early in their estrous cycles had normal luteal development, and estrus occurred after administration of PG. When treatment with a progestin began late in the estrous cycle, the corpus luteum regressed spontaneously during treatment, and females were held out of estrus until the source of the progestin was removed. In these cases, the injection of PG was not needed, even though it was administered to all animals (Patterson et al., 1989b).

Shorter feeding periods of MGA combined with PG at the end of treatment received renewed interest after available alternatives for regulation of the estrous cycle seemed too costly or labor intensive for widespread commercial acceptance. Melengestrol acetate fed to cattle for 5 days combined with an injection of PG on the last day was tested first as a means of estrous synchronization (Moody et al., 1978). The proportion of animals detected in estrus during the synchronized period was similar to or higher than that of controls, but conception rates at first service were reduced.

Treatment with MGA for 7 days combined with PG on day 7 reduced fertility at the synchronized estrus (Patterson et al., 1986; Beal et al., 1988) or showed no marked advantage over PG alone (Chenault et al., 1987). Decreased conception rates at the synchronized estrus occurred only in those heifers in which treatment with MGA began after day 12 of the estrous cycle (Table 6.2; Patterson et al., 1989a). Animals treated late in the estrous cycle had lower fertility because of extended interovulatory intervals (Patterson et al., 1989a). Conception rates were reduced also when beef cows were implanted with Syncro-Mate-B on or after day 11 of the estrous cycle and inseminated 48 hours after implant removal without regard to estrus (Table 6.2; Brink and Kiracofe, 1988). The Syncro-Mate-B treatment can

TABLE 6.2
Conception Rates of Heifers Treated with Syncro-Mate-B or MGA–PG on Different Days of the Estrous Cycle

	Aggregate Conception Rates for Day of the Estrous Cycle Treatment Began	
Day of the Estrous Cycle[a]	Syncro-Mate-B[b]	MGA–PG[c]
0 to 5	10/21 (47%)	18/22 (82%)
6 to 11	12/26 (46%)	9/13 (69%)
12 to 16	9/24 (37%)	5/12 (42%)
17 to 21	5/14 (36%)	3/14 (21%)

[a] Day of the estrous cycle on which respective treatments began (estrus = day 0).

[b] Adapted from Brink and Kiracofe (1988).

[c] 7 days of MGA with PG administered on day 7 (adapted from Patterson et al., 1989a).

Adapted from Patterson et al. (1989b).

induce estrus independent of the ovaries (McGuire et al., 1990), which may lead to asynchrony of estrus and ovulation. This may result in a reduction in conception rate and perhaps contributes to the variability in conception rate observed after treatment with Syncro-Mate-B (McGuire et al., 1990).

THE MGA–PG PROTOCOL

Longer feeding periods of MGA (14 days) with PG injected 17 days after MGA withdrawal were developed simultaneously with the 7 day program (Figure 6.2; Brown et al., 1988). Treatment with MGA for 14 days places all animals in the mid to late luteal phase of the estrous cycle (days 11 to 14) at the time PG is administered. Consequently, the variability in the interval from PG injection to estrus was reduced (Stevenson et al., 1984; Watts and Fuquay, 1985) and maximized conception rate (King et al., 1982). No reduction in fertility was observed by inseminating at the second estrus after MGA withdrawal. Brown et al. (1988) compared this system with the Syncro-Mate-B system. Despite similarities in the degree of synchrony between treatments, conception rates were significantly higher among heifers on the MGA–PG system than for heifers that were synchronized with Syncro-Mate-B (Table 6.3; Brown et al., 1988).

Melengestrol acetate is fed at a rate of 0.5 mg/head/day for 14 days. The MGA is fed generally in a grain carrier and either top-dressed onto other feeds or batch mixed with larger quantities. This aspect of the treatment is critical. Melengestrol acetate will suppress estrus during treatment, and animals that fail to consume the required amount will return to estrus during the feeding period, hence reducing the synchronization response. Animals should be observed for signs of behavioral estrus each day of the feeding period. This may be accomplished as animals approach the feeding area and prior to feed distribution. This practice will ensure that all animals receive adequate intake. Cows or heifers will exhibit estrus beginning 48 hours after

TABLE 6.3
Estrous Response, Conception, and Pregnancy Rates in Beef Heifers Comparing the MGA–PG System to Syncro-Mate-B

	Heifers	Estrous Response		Synchronized Conception Rate		Synchronized Pregnancy Rate	
Treatment	**No.**	**No.**	**%**	**No.**	**%**	**No.**	**%**
MGA–PG	157	130/157	83[a]	90/130	69[a]	90/157	57[a]
Syncro-Mate-B	153	138/153	90[a]	56/138	41[b]	56/153	37[b]

[a,b] Percentages within columns with different superscripts differ (P < .05).

Adapted from Brown et al. (1988).

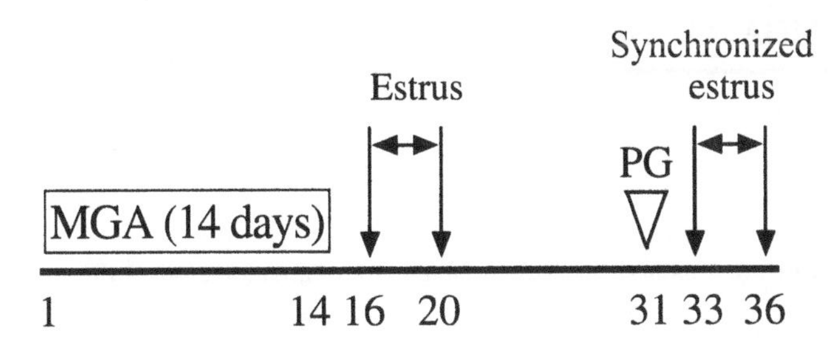

FIGURE 6.2 The MGA–PG protocol. (Adapted from Brown et al., 1988.)

MGA withdrawal, but should not be inseminated or exposed for natural service at this time. Fertility at the first estrus after withdrawal of MGA is generally low. Prostaglandin $F_{2\alpha}$ should be administered 17 days after MGA withdrawal, with insemination based on detection of estrus.

Pexton et al. (1989) proposed that the use of single-sire matings to breed synchronized groups of females could be advantageous in assisting producers to make a transition from natural service to AI. The MGA system is easily adapted for use in natural service breeding programs (Patterson et al., 1991). Cows or heifers receive the normal 14-day feeding period of MGA; however, in this scenario PG is not administered after MGA withdrawal (Patterson et al., 1991; LeFever and Odde, 1994). Fertile bulls may be exposed to treated groups of females as early as 10 to 17 days after MGA withdrawal (Figure 6.3; Patterson et al., 1991).

This system works effectively; however, careful attention to bull-to-female ratios should be observed. It is recommended that 15 to 20 synchronized females be exposed per bull. Age and breeding condition of the bull and results of breeding soundness examinations should be considered carefully. Farin et al. (1989) reported that classification of bulls by mean libido score aids in identifying bulls that service more estrous synchronized females; however, classification of bulls by results from breeding soundness examinations was more useful in identifying bulls that impregnate more females.

TABLE 6.4
Summary of Estrous Synchronization Field Trials Using MGA Prior to Natural Service or MGA–PG Prior to AI

Breeding Program	Heifers No.	Estrous Response No.	Estrous Response %	Synchronized Conception Rate No.	Synchronized Conception Rate %	Synchronized Pregnancy Rate No.	Synchronized Pregnancy Rate %
Natural service	1749	—	—	—	—	1151/1749	66
AI	4245	3354/4245	79	2414/3354	72	2414/4245	57

Patterson, unpublished data.

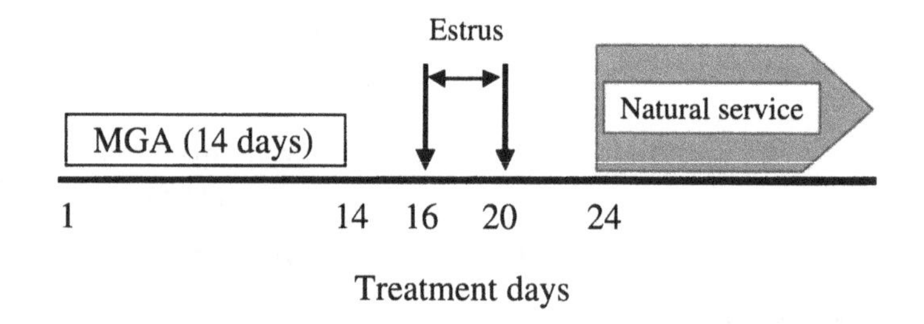

FIGURE 6.3 MGA and natural service. (Adapted from Patterson et al., 1991.)

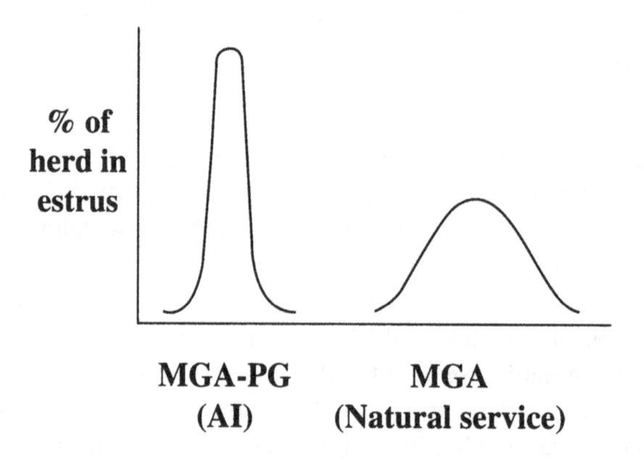

FIGURE 6.4 Distribution of estrus comparing the MGA–PG system to an MGA-only system. (Adapted from Patterson et al., 1991.)

Figure 6.4 (Patterson et al., 1991) illustrates the distribution of estrus comparing the MGA–PG system to an MGA-only system. The combined MGA–PG system is best suited for use with AI programs because of the high degree of synchrony that can be achieved during the synchronized period. This decreases the amount of time required for detection of estrus. Under natural mating conditions, there may be an advantage for bulls used in these programs to distribute estrus over several additional days. Table 6.4 provides a summary of field trials involving heifers where MGA was used in conjunction with natural service or MGA–PG was used prior to AI (Patterson, unpublished data). One of the major advantages in using MGA to control estrous cycles of cattle, as seen from the data presented in Table 6.4, is the flexibility in matching specific synchronization protocols with the particular management system involved.

POTENTIAL FOR INDUCED ESTROUS CYCLICITY WITH PROGESTINS

Progestins were used to induce estrus in peripubertal heifers (Gonzalez-Padilla et al., 1975) and are often combined with estrogens to mimic changes that occur in concentrations of blood hormones around the time of puberty. Increased progesterone is thought to be a prerequisite for the development of normal estrous cycles. Progesterone increases during the initiation of puberty in the heifer (Berardinelli et al., 1979) and before resumption of normal ovarian cyclicity in postpartum suckled beef cows (Prybil and Butler, 1978; Rawlings et al., 1980). Progestins stimulate an increase in follicular growth that results in increased production of estradiol-17β by ovarian follicles (Henricks et al., 1973; Wetteman and Hafs, 1973; Sheffel et al., 1982; Garcia-Winder et al., 1986). Melengestrol acetate initiates estrous cyclicity in peripubertal beef heifers (Patterson et al., 1990) and is associated with increased LH pulse frequency during the treatment period (Smith and Day, 1990; Imwalle et al., 1998). Recent studies suggest that the stimulatory effects of progestins on LH secretion are greatest after removal of the steroid (Anderson et al., 1996; Hall et al., 1997; Imwalle et al., 1998). Furthermore, improvements in observed pubertal induction response following treatment with a progestin occur with an increase in age (Hall et al., 1997). Progestins appear to hasten puberty only in those heifers in which estradiol-17β negative feedback has begun to decline (Anderson et al., 1996). Available data suggest that upon withdrawal of MGA, puberty onset is enhanced in response to an increase in pulsatile LH secretion that accelerates follicle growth to the preovulatory stage (Imwalle et al., 1998).

Burfening (1979) suggested that because puberty is a heritable trait, induced puberty in replacement heifers over several generations might result in situations in which attainment of puberty would be difficult without hormone treatment. This consideration cannot be overlooked. However, there is a need to explore treatments to induce puberty in breeds of cattle that are late-maturing but of sufficient age and weight at the time of treatment to permit successful application (Patterson et al., 1990). The decision to utilize this practice within a herd perhaps differs with various types of beef operations. For instance, the common goal of most managers of commercial cow–calf herds is to maximize weaning rate. In other words, the investment in time and resources in a heifer from weaning to breeding requires that management efforts be made to facilitate puberty onset and maximize the likelihood of early pregnancy. In this scenario, a method to induce puberty in heifers could serve as a valuable tool to improve reproductive performance of heifers retained for breeding purposes. On the other hand, seedstock managers should weigh the economic importance of puberty onset in their own herds, as well as their customers', and the associated potential and resulting implication of masking its true genetic expression.

USING REPRODUCTIVE TRACT SCORES IN HEIFERS TO TIME ESTROUS SYNCHRONIZATION PROGRAMS WITH MGA

Another practice (Anderson et al., 1991) can be used to assist beef producers with selection of potential herd replacements and support timing of estrous synchronization programs. A reproductive tract scoring (RTS) system was developed to estimate pubertal status (Table 6.5; Anderson et al., 1991). Scores are subjective estimates of sexual maturity, based on ovarian follicular development and palpable size of the uterus. An RTS of 1 is assigned to heifers with infantile tracts, as indicated by small, toneless uterine horns and small ovaries devoid of significant structures. Heifers scored with an RTS of 1 are likely the furthest from puberty at the time of examination. Heifers assigned an RTS of 2 are thought to be closer to puberty than those scoring 1, due primarily to larger uterine horns and ovaries. Those heifers assigned an RTS of 3 are thought to be on the verge of estrous cyclicity based on uterine tone and palpable follicles. Heifers assigned a score of 4 are considered to be cycling as indicated by uterine tone and size, coiling of the uterine horns, as well as presence of a preovulatory size follicle. Heifers assigned a score of 4 do not have an easily distinguished corpus luteum. Heifers with an RTS of 5 are similar to those scoring 4, except for the presence of a palpable corpus luteum (Table 6.5). Prebreeding examinations that include RTS furnish the opportunity to assess reproductive development, but further provide an appraisal of possible aberrant situations that may detract from a heifer's subsequent reproductive potential (infantile tracts, freemartins, pregnancy).

Collectively, prebreeding weight, reproductive tract score, pelvic height, pelvic width, and total pelvic area can be used to evaluate success of a heifer development

TABLE 6.5
Reproductive Tract Scores (RTS)

		Ovarian			
RTS	**Uterine Horns**	**Length (mm)**	**Height (mm)**	**Width (mm)**	**Structure**
1	Immature <20 mm diameter no tone	15	10	8	No palpable follicles
2	20–25 mm diameter no tone	18	12	10	8 mm follicles
3	20–25 mm diameter slight tone	22	15	10	8–10 mm follicles
4	30 mm diameter good tone	30	16	12	10 mm follicles CL possible
5	>30 mm diameter	>32	20	15	CL present

Adapted from Anderson et al. (1991).

TABLE 6.6
Prebreeding Weights, Measurements, and Subsequent Estrous Response after Synchronization of Estrous with MGA–PG

RTS	No.	Weight (lb)	Pelvic Height (cm)	Pelvic Width (cm)	Pelvic Area (cm^2)	Estrous Response (%)
1	61	594[a]	13.9[a]	10.9[a]	152[a]	54[a]
2	278	620[b]	14.1[a]	11.2[a]	158[a]	66[b]
3	1103	697[c]	14.5[b]	11.4[b]	166[b]	76[c]
4	494	733[d]	14.7[c]	11.7[c]	172[c]	83[d]
5	728	755[d]	14.7[c]	11.7[c]	172[c]	86[d]

Weights and measurements were taken within 2 weeks prior to the first day of MGA. Estrous response is the percentage of heifers that exhibited estrus and were inseminated within 144 hours after PG.
[a,b,c,d] Means within a column with different superscripts differ ($P < .05$).

Adapted from Patterson and Bullock (1995).

program. Timing these procedures is critical in determining whether heifers are ready to be placed on an estrous synchronization treatment, the type of treatment to be used, and the anticipated outcome of a particular treatment regarding estrous response and subsequent pregnancy. Table 6.6 summarizes prebreeding data that were collected on 2664 heifers (Patterson and Bullock, 1995). Measurements were obtained within 2 weeks prior to administration of a 14–17 day MGA–PG treatment. Reproductive tract score was correlated with prebreeding weight (r = 0.39), pelvic height (r = 0.30), pelvic width (r = 0.34), and total pelvic area (r = 0.39). Poor reproductive performance of heifers, with RTS of 1, points to the importance of identifying and culling these heifers before the breeding season begins (Table 6.5).

In situations where heifers are scheduled to begin an estrous synchronization treatment with MGA, we recommend that RTS be performed within 2 weeks prior to the initiation of treatment. We further recommend that heifers are ready to begin treatment with MGA if 50% of the heifers within a group are assigned an RTS of 4 or 5 (Patterson et al., 2000b). This indicates that these heifers have reached puberty and are estrous cycling. Based on the age and weight of prepubertal or peripubertal contemporaries, up to 70% of these heifers can be expected to exhibit estrous and ovulate after MGA withdrawal, so the potential estrous response during the synchronized period is up to 80% (Table 6.6). Estrous response among heifers that were assigned scores of 2 or 3 was lower than for those assigned scores of 4 or 5. However, as RTS increased, estrous response improved.

Inadequacies in nutritional development programs are often associated with situations in which the desired degree of estrous cyclicity has not been achieved. This necessitates reevaluation of the nutritional development program and in many cases a postponement of the breeding season. The results obtained from a prebreeding exam provide an objective assessment of the success or failure of a development program and are useful in determining the appropriate timing of estrous synchronization treatments (Anderson et al., 1991; Patterson and Bullock, 1995; Randle, 1999).

Estrous synchronization and artificial insemination contribute to a total heifer development program in several ways. Estrous synchronization improves time management for producers that use AI by concentrating the breeding and resulting calving periods. Managers are able to spend more time observing heifers as they calve because calving occurs over a shorter time period. Calf losses in many cases are reduced because of improved management during the calving period. Artificial insemination provides the opportunity to breed heifers to bulls selected for low birth weight—expected progeny differences (EPD)—with high accuracy. This practice minimizes the incidence and severity of calving difficulty and decreases calf loss that results from dystocia. In addition, heifers that conceive during a synchronized period typically wean calves that are older and heavier at weaning time (Schafer et al., 1990). Finally, heifer calves that result from AI can be an excellent source of future replacements, facilitating more rapid improvement in the genetic makeup of an entire herd (Patterson et al., 2000b).

EFFECT OF THE MGA–PG PROTOCOL ON ESTROUS SYNCHRONIZATION AND CONCEPTION RATE

Estrous synchronization programs designed for heifers and postpartum beef cows should be evaluated in relation to their effect on conception (Patterson et al., 1989b; Folman et al., 1990; Patterson et al., 1995). Until recently, there was little published evidence comparing methods of estrous cycle control that utilize PG alone to methods that utilize progesterone or progestins in conjunction with PG. Feeding MGA for 14 days, followed by PG injection 17 days after MGA feeding (MGA–PG protocol) is an effective method of estrous cycle control in heifers (Brown et al., 1988; Patterson and Corah, 1992). More recently, an increase in estrus response, synchronized conception and pregnancy rates, and fecundity in the postpartum cow was reported among cows treated with the MGA–PG protocol when compared with PG alone (Patterson et al., 1995; Fralix et al., 1996). An additional application of the MGA–PG protocol is for synchronization of donor cows prior to superovulation (Patterson et al., 1997).

The advantages of this treatment include: a) the potential for induced cyclicity in anestrous postpartum cows (Short et al., 1990); b) increased estrous response during the synchronized period; and 3) enhanced fertility (Patterson et al., 1995). We know from work with both dairy (Britt et al., 1972) and beef cows (Patterson et al., 1995) that the second synchronized estrus after MGA, whether spontaneous or induced with PG, may be inherently more fertile. Britt et al. (1974) reported improved reproductive performance in cows that received MGA for 2 weeks during the early postpartum period. Reported differences in conception rate for beef cows are shown in Figure 6.5B (Patterson et al., 1995). Treated cows in that study each received 0.5 mg of MGA or carrier without MGA for 14 days. All cows received PG 17 days after the last feeding day of MGA or carrier without MGA. Control and treated cows that failed to exhibit estrus within 6 days after the first injection of PG were reinjected 11 days later (Figure 6.5A). Many of the cows that failed to respond to PG on day 17 after withdrawal of MGA were cows that were anestrus prior to MGA treatment. In fact, up to 20% of anestrous cows experience short cycles prior to PG

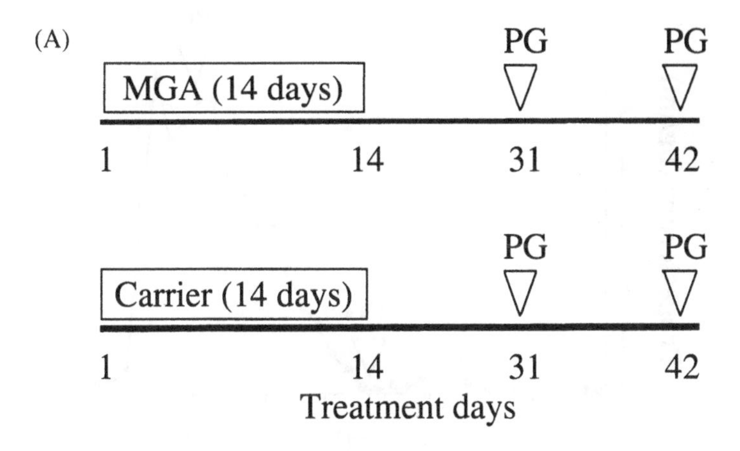

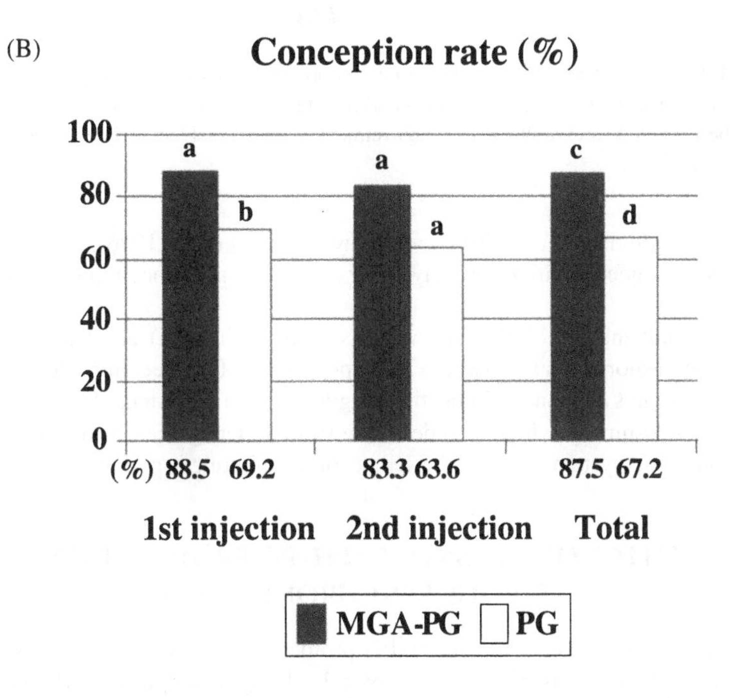

FIGURE 6.5 (A) MGA-treated cows received 0.5 mg MGA/cow/day for 14 days or carrier only, with PG administered 17 days after MGA or carrier withdrawal. Cows that failed to exhibit estrus within 6 days after PG were reinjected with PG a second time 11 days later. (Adapted from Patterson et al., 1991.) (B) Synchronized conception rates of cows exhibiting estrus after each of two PG injections. Cows pretreated with MGA prior to PG experienced a 20% improvement in synchronized conception and pregnancy rate, as well as a 15% twinning rate. (Adapted from Patterson et al., 1995.)

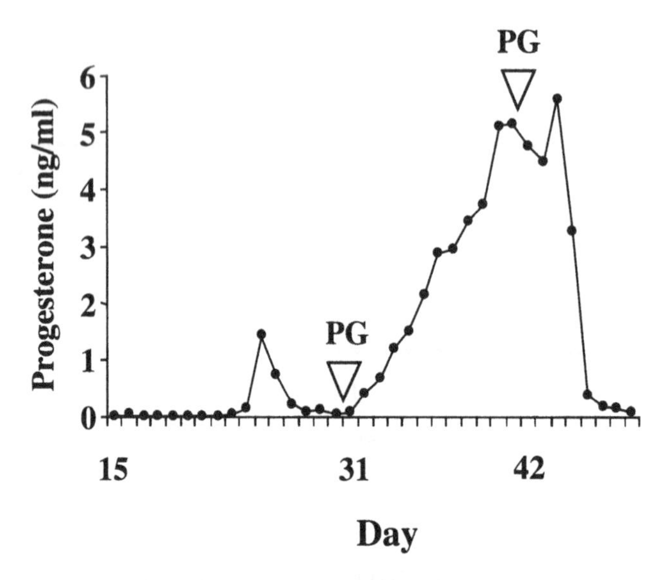

FIGURE 6.6 Progesterone profile depicting a short luteal phase subsequent to MGA withdrawal. This cow would not respond to PG administered 17 days after MGA withdrawal, but would be expected to exhibit estrus when reinjected with PG 11 days later. (Adapted from Fralix et al., 1996.)

administered on day 17 after MGA withdrawal (Fralix et al., 1996). Nonresponding cows may be injected with PG 11 days later (day 42 from the beginning of treatment; Figure 6.6).

The disadvantages of the MGA–PG system include: a) anestrous cows that experience a short luteal phase after the period of MGA feeding which in some cases necessitates a second PG injection (Figure 6.6); b) the potential for an increased incidence of twinning, which is undesirable from a management viewpoint in many beef production systems; and c) the length of the treatment period.

EFFECT OF THE MGA–GnRH–PG PROTOCOL ON ESTROUS SYNCHRONIZATION

As previously mentioned, the GnRH–PG protocol increased the estrus response in postpartum beef cows over PG-treated controls (Twagiramungu et al., 1995). Advantages of the GnRH–PG system include simplicity of administration and short treatment duration. The major disadvantage of the GnRH–PG protocol is the percentage of cows that exhibit estrus after GnRH and before PG, which in some cases may be as high as 15% of the total number of cows treated (Kojima et al., 2000). In order to inseminate all cows that respond to this treatment, estrous detection is required for a period of 10 days beginning 4 days before PG administration. Additionally, GnRH–PG protocols are not effective in synchronizing estrus in heifers (Pursley et al., 1997b; Stevenson et al., 1999).

Our general hypothesis is that progestin (MGA) treatment prior to the GnRH–PG protocol will successfully: 1) induce ovulation in anestrous cows and peripubertal beef heifers; 2) reduce the incidence of short luteal phases among anestrous cows induced to ovulate; 3) increase estrous response, synchronized conception, and pregnancy rate; and 4) increase the likelihood of successful fixed-time insemination.

To date, there have been few studies designed to evaluate progestin treatment prior to administration of the GnRH–PG protocol. We recently compared the addition of MGA to the GnRH–PG protocol to the standard MGA–PG protocol (Figure 6.7A). The design and preliminary results from that study are shown in Figures 6.7A and 6.7B (Patterson et al., 1999).

Synchrony of estrus was improved among MGA–GnRH–PG-treated cows, with over 80% of the cows in that treatment exhibiting estrus 48 to 96 hours after

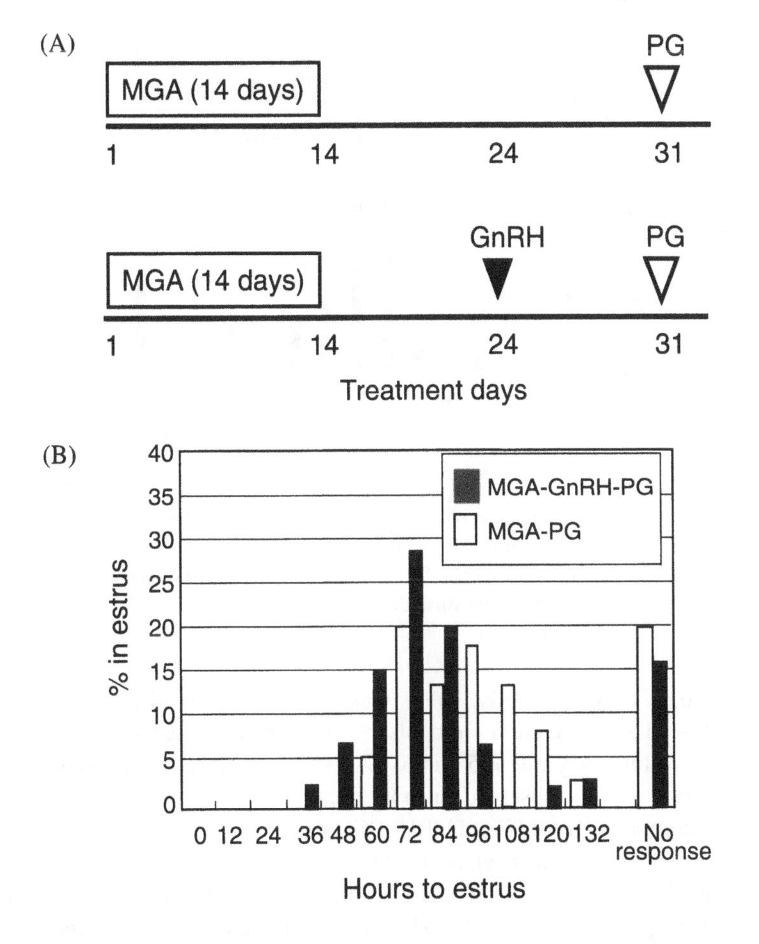

FIGURE 6.7 (A) Cows were fed MGA for 14 days. GnRH was administered to one half of the cows 10 days after MGA withdrawal, and all cows were injected with PG 7 days later. (B) Estrous response for MGA–PG- or MGA–GnRH–PG-treated cows. (From Patterson et al., 1999. With permission.)

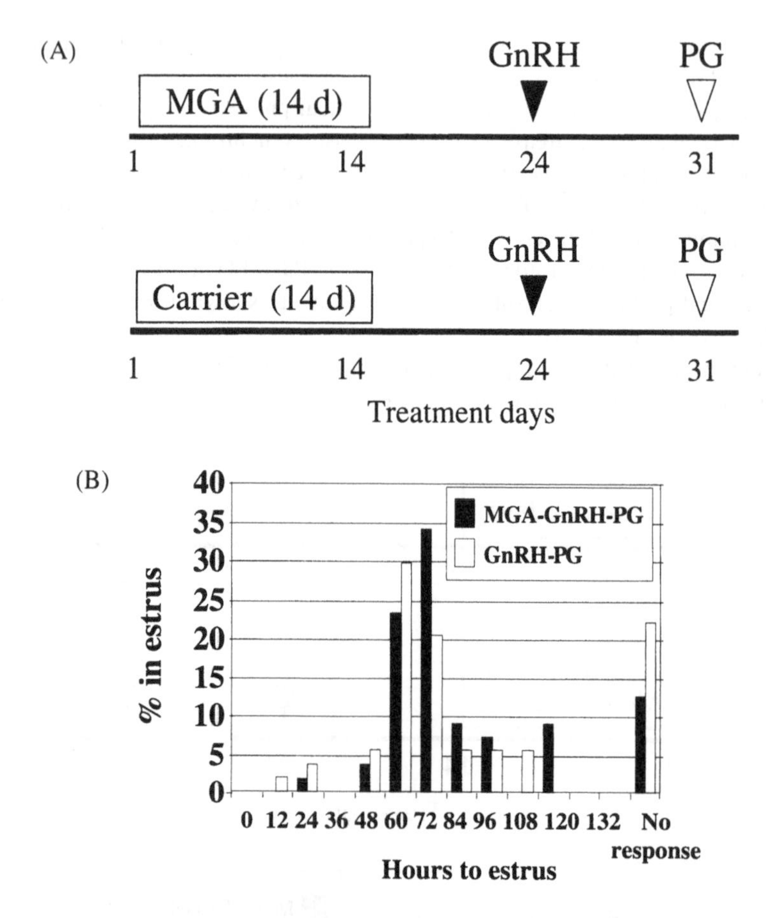

FIGURE 6.8 (A) Cows were fed MGA or carrier without MGA for 14 days. GnRH was administered to all cows 10 days after MGA or carrier withdrawal, and all cows were injected with PG 7 days later. (B) Estrous response for MGA–GnRH–PG- or GnRH–PG-treated cows. (From Patterson et al., 2000a. With permission.)

PG (Figure 6.7B). Additionally, there was no significant difference between MGA–GnRH–PG and MGA–PG protocols in synchronized conception (78 and 83%, respectively) or pregnancy rate (65 and 67%, respectively) during the synchronized period, with high fertility observed for both treatment groups.

We also compared the MGA–GnRH–PG protocol to the GnRH–PG protocol (Figure 6.8A) in postpartum beef cows. The design and preliminary results from that study are shown in Figures 6.8A and 6.8B (Patterson et al., 2000a). Synchrony of estrus was improved among MGA–GnRH–PG-treated cows compared to cows that did not receive MGA (GnRH–PG), while maintaining high pregnancy rates to AI (70 and 59%, respectively). Furthermore, the distribution of estrus among MGA–GnRH–PG-treated cows was almost identical to the distribution illustrated in Figure 6.7B, demonstrating the repeatability of response following this treatment.

A MODIFIED MGA PROTOCOL

Recent studies with heifers show that both synchrony of estrus and total estrous response are improved when PG is administered 19 days after MGA withdrawal compared with those of heifers injected on day 17 after MGA withdrawal (Nix et al., 1998; Deutscher et al., 2000). We evaluated a modified MGA–PG protocol for inducing and synchronizing a fertile estrus in beef heifers (Wood et al., 2001; Figure 6.9). The first modification changed the day of PG injection from day 31 to day 33 of treatment. The second modification was the addition of a GnRH injection on day 26 of treatment. We found that injection of GnRH on day 26 of the MGA–PG protocol induced luteal tissue formation and initiated a new follicular wave on approximately day 28 in cycling beef heifers (Figure 6.10B). The proportion of heifers with synchronized follicular waves on day 33 was increased significantly compared to heifers that did not receive GnRH (Wood et al., 2001; 6.10A and 6.10B).

More recently, Wood et al. (2000) reported differences in estrous response and synchrony of estrus during the synchronized period among heifers assigned to the treatments illustrated in Figure 6.6. This difference in estrous response and degree of synchrony was based on the percentage of heifers that were pubertal at the time treatment with MGA began. Figures 6.11A and 6.11B illustrate these differences (Wood et al., 2000). Figure 6.11A shows the distribution of estrus where only 30% of the heifers were pubertal at the time treatment with MGA began, whereas Figure 6.11B illustrates the distribution of estrus for heifers where 56% of the heifers were pubertal at the same time. The increased degree of cyclicity of heifers shown in Figure 6.11B was associated with a reduced variance in the interval to estrus among MGA–GnRH–PG-treated heifers. Synchronized conception and pregnancy rates remained high for both MGA–GnRH–PG- and MGA–PG-treated heifers and were not different (67 and 60%, respectively [Figure 6.11A], and 75 and 72%, respectively [Figure 6.11B]).

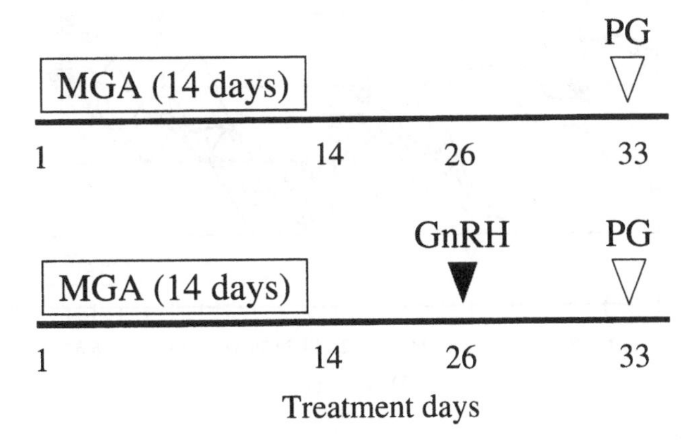

FIGURE 6.9 A modified long-term MGA protocol. Heifers were fed MGA for 14 days; 19 days after MGA withdrawal, PG was administered to all heifers. GnRH was administered to one half of the heifers 7 days prior to PG. (From Wood et al., 2001. With permission.)

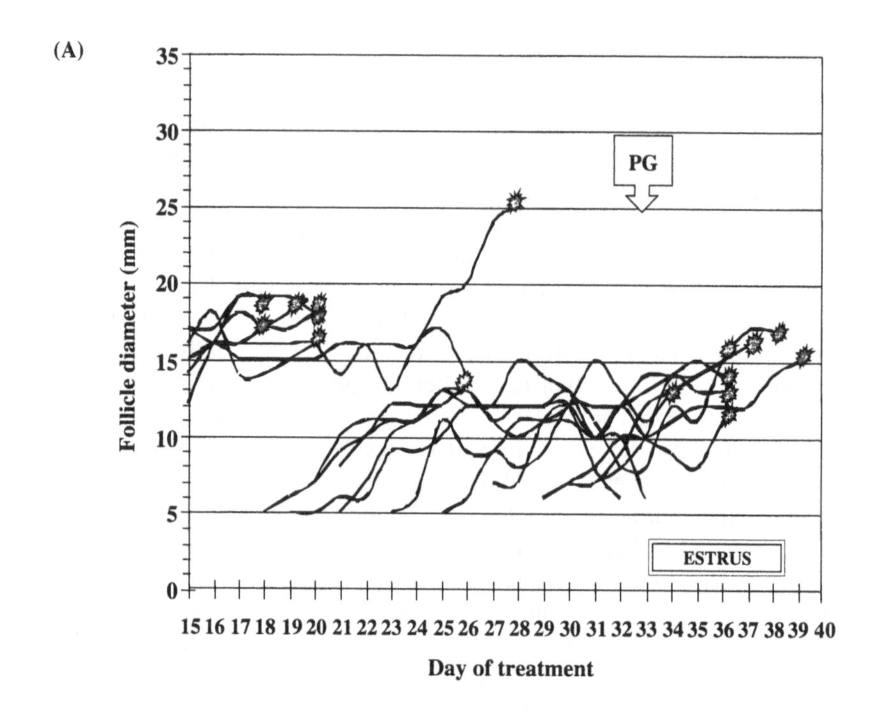

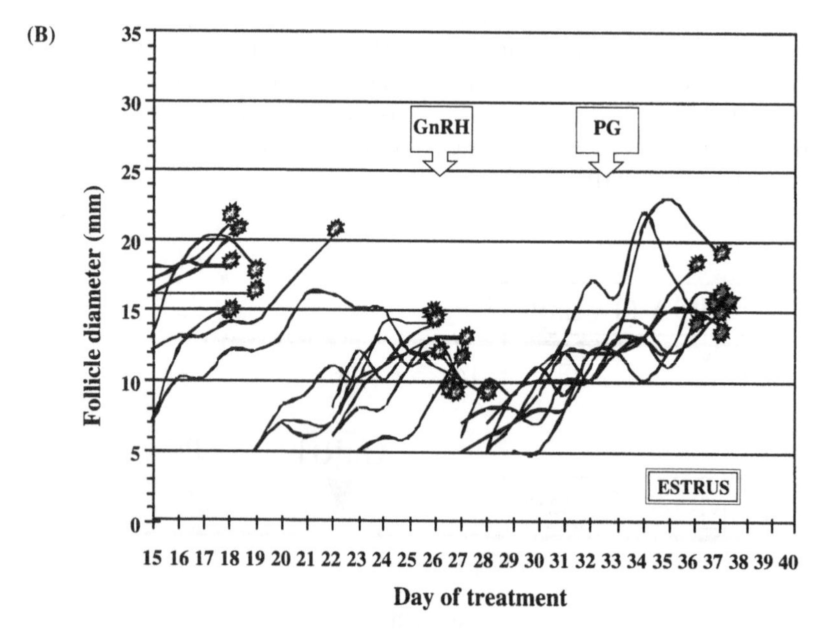

FIGURE 6.10 Patterns of dominant follicle development in control (MGA–PG; A) and GnRH-treated (MGA–GnRH–PG; B) heifers. Administration of GnRH (B) caused the synchronized development of a dominant follicle before PG injection. Follicular development in MGA–PG-treated heifers was poorly synchronized. (From Wood et al., 2001. With permission.)

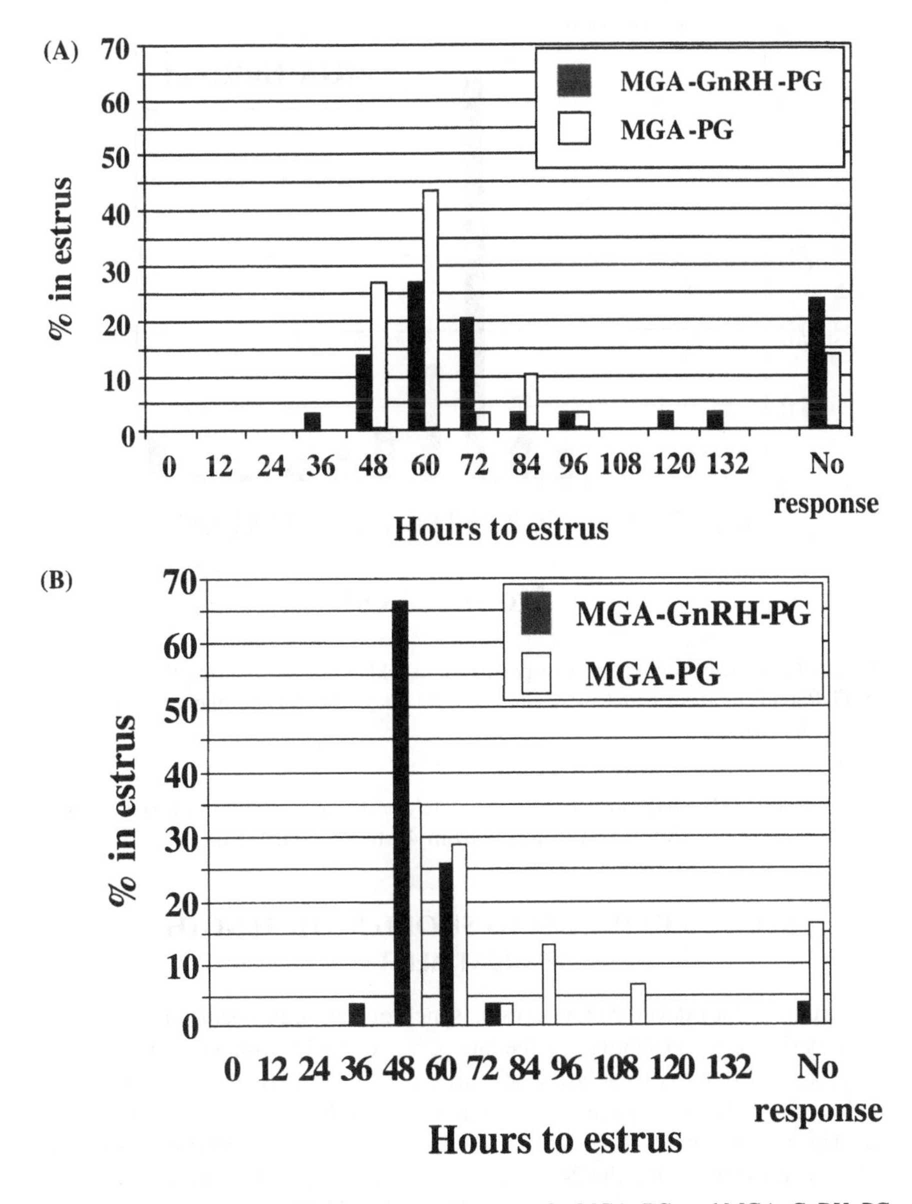

FIGURE 6.11 Percentage of heifers observed in estrus for MGA–PG- and MGA–GnRH–PG-treated heifers. Cyclicity rates were 30 and 56% for heifers at Location 1 (A) and 2 (B), respectively, at the time treatment with MGA began. (From Wood et al., 2000. With permission.)

We obtained preliminary data using this system in postpartum suckled beef cows during the 1999 fall breeding season (Patterson, unpublished data). Figure 6.12 shows the distribution of estrus in one herd of postpartum beef cows that were synchronized using the 14–19 day MGA–GnRH–PG protocol with GnRH administered on day 12 after MGA withdrawal. One hundred fifty of the 160 cows (94%)

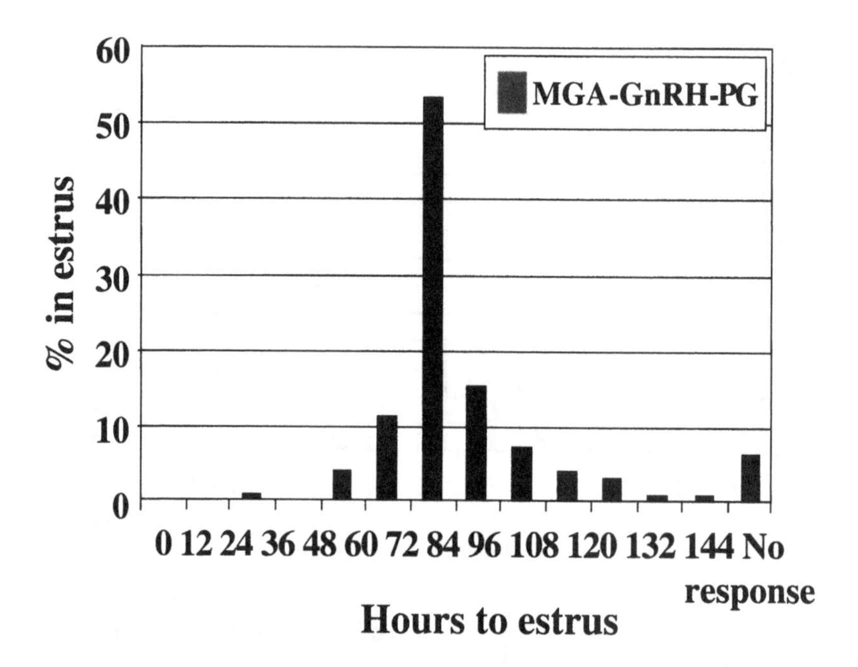

FIGURE 6.12 Distribution of estrus in postpartum suckled beef cows using the 14–19 day MGA–GnRH–PG protocol, with GnRH administered 12 days after MGA withdrawal (Patterson, unpublished data).

that were treated exhibited estrus during the synchronized period. Of the 150 cows that exhibited estrus, 134 (89%) were in heat from 48 to 96 hours after PG.

STUDIES DESIGNED TO SHORTEN THE LENGTH OF TREATMENT

Kojima et al. (2000) developed an estrous synchronization protocol for beef cattle that was designed to: 1) shorten the feeding period of MGA without compromising fertility; and 2) improve synchrony of estrus by synchronizing development and ovulation of follicles from the first wave of development (Figure 6.13A; Kojima et al., 2000). This new treatment, 7–11 Synch, was compared with the GnRH–PG protocol. Synchrony of estrus during the 24-hour peak response period (42- to 66-hour) was significantly higher among 7–11 Synch-treated cows. Furthermore, the distribution of estrus was reduced from 144 hours for GnRH–PG-treated cows to 60 hours for cows assigned to the 7–11 Synch treatment (Figure 6.13B; Kojima et al., 2000). The 7–11 Synch protocol resulted in a higher degree of estrous synchrony (91%) and greater AI pregnancy rate (68%) during a 24-hour peak response period compared to the GnRH–PG protocol (69 and 47%, respectively).

Based on our preliminary data, we believe that progesterone priming, either short- or long-term, prior to GnRH and PG provides the potential for improved synchrony with good fertility during the synchronized period.

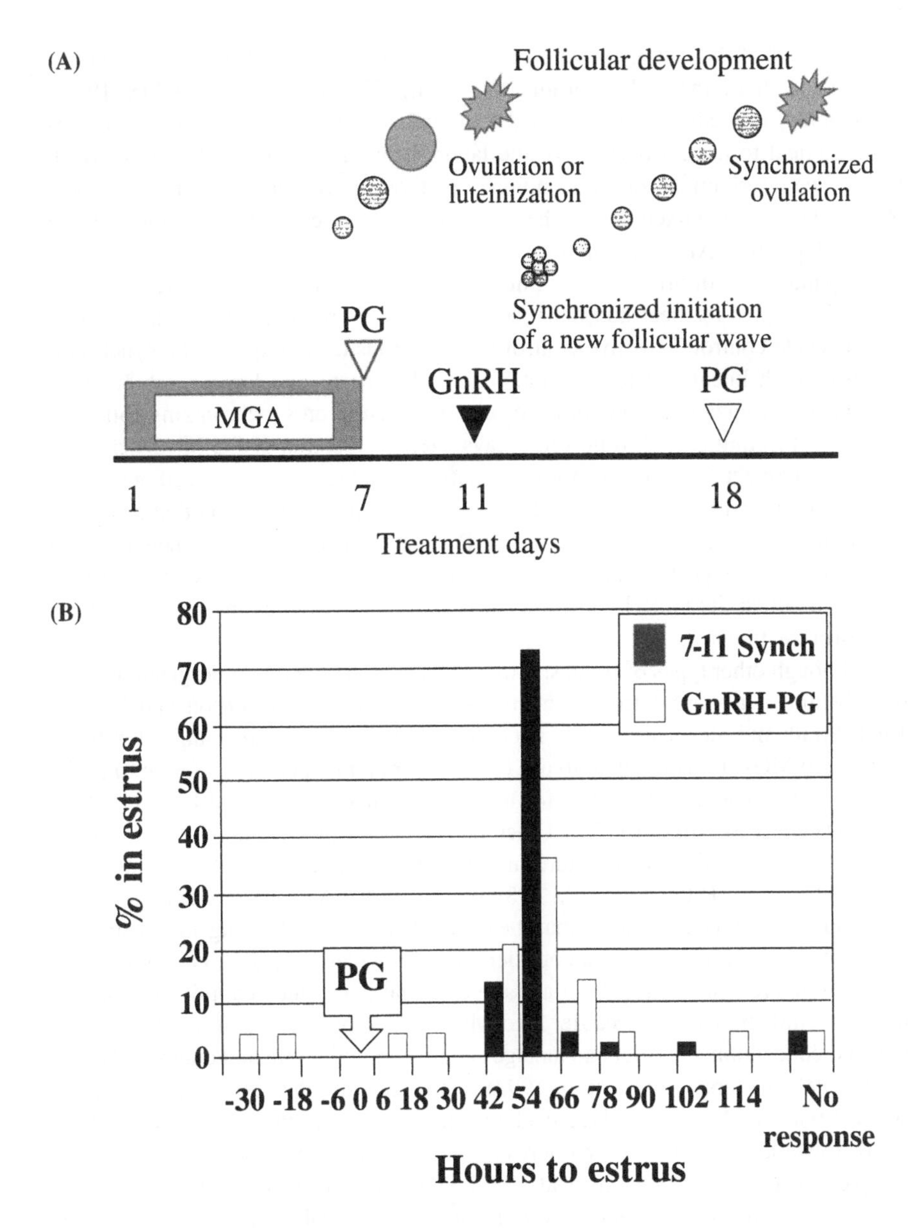

FIGURE 6.13 (A) Illustration of the treatment schedule and events associated with the 7–11 Synch protocol. (B) Estrous response of cows treated with the 7–11 Synch or GnRH–PG protocols. (Adapted from Kojima et al., 2000.)

SUMMARY

The percentage of beef cattle inseminated artificially is predicted to substantially increase with the advent of sexed semen (Seidel, 1998). Currently, however, surveys indicate that fewer than 5% of the beef cows in the United States are bred by AI and only half of the cattlemen who practice AI use any form of estrous synchronization to facilitate their AI programs (Corah and Kiracofe, 1989; NAHMS, 1994). The inability to predict time of estrus for individual females in a group often makes it impractical to use AI because of the labor required for estrous detection (Britt, 1987). The development of an economical method of artificially inseminating beef cows and heifers at a fixed time with high fertility would result in a dramatic increase in the adoption of AI in beef herds.

Expanded use of artificial insemination and/or adoption of emerging reproductive technologies for beef and dairy cows and heifers requires precise methods of estrous cycle control. Effective control of the estrous cycle requires the synchronization of both luteal and follicular functions. Efforts to develop a more effective estrous synchronization protocol have focused recently on synchronizing follicular waves by injecting GnRH, followed 7 days later by injection of PG (Ovsynch, CO-Synch, and Select Synch). A factor contributing to reduced synchronized pregnancy rates in dairy and beef cows treated with the preceding protocols is that 5 to 15% of cycling cows show estrus on or before PG injection. We propose a new protocol for inducing and synchronizing a fertile estrus in postpartum beef cows and beef heifers in which the GnRH–PG protocol is preceded by either short- or long-term progestin treatment.

Although other types of progestin treatments (CIDR, PRID, or norgestomet) can be substituted in these estrous synchronization protocols (Stevenson and Pursley, 1994; Smith and Stevenson, 1995; Xu and Burton, 1998), the advantages for MGA include: a) MGA is economical to use (2 cents per animal per day to feed); b) MGA was recently cleared by the FDA for use in reproductive classes of beef and dairy cattle (*Federal Register*, 1997); c) methodology and understanding of the use of MGA is documented in the literature since the 1960s (Zimbelman, 1963; Zimbelman and Smith, 1966; Patterson et al., 1989b); and d) MGA is easily administered in feed and does not require that animals be handled or restrained during administration. Collectively, these are important considerations for widespread use of any successful estrous synchronization treatment and are essential to expanded application of artificial insemination in beef cattle. The MGA–PG protocol avoids problems with reduced conception and offers advantages compared with untreated controls (Brown et al., 1988; Patterson and Corah, 1992). Until recently, there was no evidence to suggest that MGA would induce cyclicity in peripubertal heifers (Patterson et al., 1990; Imwalle et al., 1998), or improve conception and increase ovulation rate in postpartum beef cows (Patterson et al., 1995; Fralix et al., 1996; Patterson et al., 1997). Additionally, MGA is currently the only progestin available in the U.S. approved for use in reproductive classes of beef and dairy cattle.

Table 6.7 compares response and synchronized pregnancy rates for various estrous synchronization treatments reviewed in this chapter. These studies support a general hypothesis that treatment with a progestin (MGA) prior to the GnRH–PG estrous synchronization protocol provides the potential to: 1) induce ovulation in anestrous postpartum beef cows and peripubertal beef heifers; 2) reduce the incidence of a short luteal phase among anestrous cows induced to ovulate; 3) increase estrous response, synchronized conception, and pregnancy rate; and 4) increase the likelihood of successful fixed-time insemination. A major advantage in using MGA

TABLE 6.7
Estrous Response and Fertility in Postpartum Cows

Treatment	Estrous Response		Synchronized Pregnancy Rate	
	No.	%	No.	%
2 shot PG[a]	241/422	57	147/422	35
Select Synch[a]	353/528	67	237/528	45
MGA–PG[b]	305/408	75	220/408	54
MGA–2 shot PG[b]	327/348	93	243/348	70
MGA–GnRH-PG (14–17d)[c]	100/116	86	78/116	67
MGA–GnRH-PG (14–19d)[d]	150/160	94	N/A	
7–11 Synch[e]	40/44	91	30/44	68

[a] Adapted from Dejarnette and Wallace, unpublished data.
[b] Adapted from Patterson et al. (1993).
[c] Adapted from Patterson et al. (1999); Patterson et al. (2000b).
[d] Patterson, unpublished data.
[e] Adapted from Kojima et al. (2000).

to control estrous cycles of cattle and enhance efforts directed toward improvements in reproductive management is the flexibility in matching specific synchronization protocols with the particular management system involved.

REFERENCES

Adams, G. P. 1999. Comparative patterns of follicle development and selection in ruminants. *J. Reprod. Fertil. Suppl.* 54:17.

Anderson, K. J., D. G. Lefever, J. S. Brinks, and K. G. Odde. 1991. The use of reproductive tract scoring in beef heifers. *Agri-Practice* 12:123.

Anderson, L. H., C. M. McDowell, and M. L. Day. 1996. Progestin-induced puberty and secretion of luteinizing hormone in heifers. *Biol. Reprod.* 54:1025.

Beal, W. E., J. R. Chenault, M. L. Day, and L. R. Corah. 1988. Variation in conception rates following synchronization of estrus with melengestrol acetate and prostaglandin $F_{2\alpha}$. *J. Anim. Sci.* 66:599.

Berardinelli, J. G., R. A. Dailey, R. L. Butcher, and E. K. Inskeep. 1979. Source of progesterone prior to puberty in beef heifers. *J. Anim. Sci.* 49:1276.

Bo, G. A., G. P. Adams, R. A. Pierson, and R. J. Mapletoft. 1995. Exogenous control of follicular wave emergence in cattle. *Theriogenology* 43:31.

Brink, J. T. and G. H. Kiracofe. 1988. Effect of estrous cycle stage at Syncro-Mate-B treatment on conception and time to estrus in cattle. *Theriogenology* 29:513.

Britt, J. H. 1987. Induction and synchronization of ovulation. In: E.S.E. Hafez. (Ed.) *Reproduction in Farm Animals*. Lea and Febiger, Philadelphia, PA.

Britt, J. H., D. A. Morrow, R. J. Kittok, and B. E. Seguin. 1974. Uterine involution, ovarian activity, and fertility after melengestrol acetate and estradiol in early postpartum cows. *J. Dairy Sci.* 57:89.

Britt, J. H., E. Huertas Vega, and L. C. Ulberg. 1972. Managing reproduction in dairy cattle: progestogens for control of estrus in dairy cows. *J. Dairy Sci.* 55:598.

Brown, L. N., K. G. Odde, D. G. LeFever, M. E. King, and C. J. Neubauer. 1988. Comparison of MGA–$PGF_{2\alpha}$ to Syncro-Mate-B for estrous synchronization in beef heifers. *Theriogenology* 30:1.

Burfening, P. J. 1979. Induction of puberty and subsequent reproductive performance. *Theriogenology* 12:215.

Burke, J. M., R. L. d la Sota, C. A. Risco, C. R. Staples, E. J. P. Schmitt, and W. W. Thatcher. 1996. Evaluation of timed insemination using a gonadotropin-releasing agonist in lactating dairy cows. *J. Dairy Sci.* 79:1385.

Chenault, J. R., J. F. McAllister, and C. W. Kasson. 1987. Estrous synchronization with melengestrol acetate and prostaglandin $PGF_{2\alpha}$ in beef and dairy heifers. *J. Anim. Sci.* 65 (Suppl. 1):375.

Corah, L. R. and G. H. Kiracofe. 1989. Present status of heat synchronization in beef cattle. *Angus Journal.* June–July, pp. 628.

Deutscher, G., R. Davis, D. Colburn, and D. O'Hare. 2000. Refinement of the MGA–PGF synchronization program for heifers using a 19 day PGF injection. *Nebraska Beef Cattle Rep.* pp. 10.

Downing, E. R., D. G. Lefever, J. C. Whittier, J. E. Bruemmer, and T. W. Geary. 1998. Estrous and ovarian response to the Select Synch protocol. *J. Anim. Sci.* 81 (Suppl. 1): 373.

Dziuk, P. J. and R. A. Bellows. 1983. Management of reproduction in beef cattle, sheep and pigs. *J. Anim. Sci.* 57 (Suppl. 2):355.

Farin, P. W., P. J. Chenoweth, D. F. Tomky, L. Ball, and J. E. Pexton. 1989. Breeding soundness, libido and performance of beef bulls mated to estrus synchronized females. *Theriogenology* 32:717.

Federal Register. 1997. New animal drugs for use in animal feeds; melengestrol acetate. 62:(58)14304.

Folman, Y., M. Kaim, Z. Herz, and M. Rosenberg. 1990. Comparison of methods for the synchronization of estrous cycles in dairy cows. 2. Effects of progesterone and parity on conception. *J. Dairy Sci.* 73:2817.

Fralix, K. D., D. J. Patterson, K. K. Schillo, R. E. Stewart, and K. D. Bullock. 1996. Change in morphology of corpora lutea, central luteal cavities and steroid secretion patterns of postpartum suckled beef cows after melengestrol acetate with or without prostaglandin $F_{2\alpha}$. *Theriogenology* 45:1255.

Garcia-Winder, M., P. E. Lewis, D. R. Deaver, U. G. Smith, G. S. Lewis, and E. K. Inskeep. 1986. Endocrine profiles associated with the life span of induced corpora lutea in postpartum beef cows. *J. Anim. Sci.* 621:1353.

Geary, T. W., J. C. Whittier, E. R. Downing, D. G. LeFever, R. W. Silcox, M. D. Holland, T. M. Nett, and G. D. Niswender. 1998. Pregnancy rates of postpartum beef cows that were synchronized using Syncro-Mate-B or the Ovsynch protocol. *J. Anim. Sci.* 76:1523.

Geary, T. W., J. C. Whittier, D. M. Hallford, and M. D. MacNeil. 2001. Calf removal improves conception rates to the Ovsynch and CO-Synch protocols. *J. Anim. Sci.* 79:1.

Gonzalez-Padilla, E., R. Ruiz, D. Lefever, A. Denham, and J. N. Wiltbank. 1975. Puberty in beef heifers. III. Induction of fertile estrus. *J. Anim. Sci.* 40:1110.

Hall, J. B., R. B. Staigmiller, R. E. Short, R. A. Bellows, and S. E. Bartlett. 1997. Effect of age and pattern of gain on induction of puberty with a progestin in beef heifers. *J. Anim. Sci.* 75:1606.

Hansel, W. and E. M. Convey. 1983. Physiology of the estrous cycle. *J. Anim. Sci.* 57 (Suppl. 2):404.

Hansel, W., P. V. Malven, and D. L. Black. 1961. Estrous cycle regulation in the bovine. *J. Anim. Sci.* 20:61.

Heersche, G., Jr., G. H. Kiracofe, R. C. DeBenedetti, S. Wen, and R. M. McKee. 1979. Synchronization of estrus in beef heifers with a norgestomet implant and prostaglandin $F_{2\alpha}$. *Theriogenology* 11:197.

Henricks, D. M., J. R. Hill, and J. F. Dickey. 1973. Plasma ovarian hormone levels and fertility in beef heifers treated with melengestrol acetate (MGA). *J. Anim. Sci.* 37:1169.

Imwalle, D. B., D. J. Patterson, and K. K. Schillo. 1998. Effects of melengestrol acetate on onset of puberty, follicle growth, and patterns of luteinizing hormone secretion in beef heifers. *Biol. Reprod.* 58:1432.

Jainudeen, M. R. and E. S. E. Hafez. 1987. Reproductive cycles: cattle and water buffalo. In: E. S. E. Hafez (Ed.) *Reproduction in Farm Animals*. Lea and Febiger, Philadelphia, PA.

Kastelic, J. P., L. Knopf, and O. J. Ginther. 1990. Effect of day of prostaglandin $F_{2\alpha}$ treatment on selection and development of the ovulatory follicle in heifers. *Anim. Reprod. Sci.* 23:169.

King, M. E., G. H. Kiracofe, J. S. Stevenson, and R. R. Schalles. 1982. Effect of stage of the estrous cycle on interval to estrus after PGF in beef cattle. *Theriogenology* 8:191.

Kiracofe, G. H. 1988. Estrous synchronization in beef cattle. In: *Compendium on Continuing Education for the Practicing Veterinarian.* 10 (1) Article No. 5.

Kojima, F. N., B. E. Salfen, J. F. Bader, W. A. Ricke, M. C. Lucy, M. F. Smith, and D. J. Patterson. 2000. Development of an estrus synchronization protocol for beef cattle with short-term feeding of melengestrol acetate: 7–11 synch. *J. Anim. Sci.* 78:2186.

Lamond, D. R. 1964. Synchronization of ovarian cycles in sheep and cattle. *Anim. Breed. Abstr.* 32:269.

LeFever, D. G. and K. G. Odde. 1994. Pregnancy rate comparison for yearling heifers synchronized with MGA–$PGF_{2\alpha}$ versus MGA alone and bred by bulls under pasture conditions. *Proc. Western Sect. Am. Soc. Anim. Sci.* 45:35.

Macmillan, K. L. and H. V. Henderson. 1984. Analyses of the variation in the interval of prostaglandin $F_{2\alpha}$ to oestrus as a method of studying patterns of follicle development during diestrus in dairy cows. *Anim. Reprod. Sci.* 6:245.

McDowell, C. M., L. H. Anderson, J. E. Kinder, and M. L. Day. 1998. Duration of treatment with progesterone and regression of persistent ovarian follicles in cattle. *J. Anim. Sci.* 76:850.

McGuire, W. J., R. L. Larson, and G. H. Kiracofe. 1990. Syncro-Mate-B induces estrus in ovariectomized cows and heifers. *Theriogenology* 34:33.

Miksch, E. D., D. G. Lefever, G. Mukembo, J. C. Spitzer, and J. N. Wiltbank. 1978. Synchronization of estrus in beef cattle. II. Effect of an injection of norgestomet and an estrogen in conjunction with a norgestomet implant in heifers and cows. *Theriogenology* 10:201.

Moody, E. L, J. F. McAllister, and J. W. Lauderdale. 1978. Effect of $PGF_{2\alpha}$ and MGA on control of the estrous cycle in cattle. *J. Anim. Sci.* 47 (Suppl. 1):36.

NAHMS (National Animal Health Monitoring System). 1994. Sparse use of reproductive management technology for beef heifers and cows. pp. 1.

Nellor, J. E. and H. H. Cole. 1956. The hormonal control of estrus and ovulation in the beef heifer. *J. Anim. Sci.* 15:650.

Nix, D. W., G. C. Lamb, V. Traffas, and L. R. Corah. 1998. Increasing the interval to prostaglandin from 17 days to 19 days in an MGA–prostaglandin synchronization system for heifers. *Cattlemen's Day Proc. Rep. Prog.* 804. pp. 31.

Patterson, D. J., L. R. Corah, and J. R. Brethour. 1986. Effect of estrous synchronization with melengestrol acetate and prostaglandin on first service conception rates in yearling beef heifers. *J. Anim. Sci.* 63 (Suppl. 1):353.

Patterson, D. J., L. R. Corah, G. H. Kiracofe, J. S. Stevenson, and J. R. Brethour. 1989a. Conception rate in *Bos taurus* and *Bos indicus* crossbred heifers after postweaning energy manipulation and synchronization of estrus with melengestrol acetate and fenprostalene. *J. Anim. Sci.* 67:1138.

Patterson, D. J., G. H. Kiracofe, J. S. Stevenson, and L. R. Corah. 1989b. Control of the bovine estrous cycle with melengestrol acetate (MGA): a review. *J. Anim. Sci.* 67: 1895.

Patterson, D. J., L. R. Corah, and J. R. Brethour. 1990. Response of prepubertal *Bos taurus* and *Bos indicus* × *Bos taurus* heifers to melengestrol acetate with or without gonadotropin-releasing hormone. *Theriogenology* 33:661.

Patterson, D. J., J. T. Johns, N. Gay, and W. R. Burris. 1991. A 2-year summary of field studies utilizing MGA to synchronize estrus in AI or natural service breeding programs. *Univ. Kentucky Beef Cattle Res. Prog. Rep.* No. 337, pp. 48.

Patterson, D. J. and L. R. Corah. 1992. Evaluation of a melengestrol acetate and prostaglandin $F_{2\alpha}$ system for the synchronization of estrus in beef heifers. *Theriogenology* 38:441.

Patterson, D. J., B. L. Woods, N. W. Bradley, and R. B. Hightshoe. 1993. A 3-year summary of estrus synchronization results using the MGA double-injection prostaglandin system. *Univ. Kentucky Beef Cattle Res. Prog. Rep.* No. 353, pp. 39.

Patterson, D. J., J. B. Hall, N. W. Bradley, K. K. Schillo, B. L. Woods, and J. M. Kearnan. 1995. Improved synchrony, conception rate, and fecundity in postpartum suckled beef cows fed melengestrol acetate prior to prostaglandin $F_{2\alpha}$. *J. Anim. Sci.* 73:954.

Patterson, D. J. and K. D. Bullock. 1995. Using prebreeding weight, reproductive tract score, and pelvic area to evaluate prebreeding development of replacement beef heifers. *Proc. Beef Improv. Fed.*, Sheridan, WY. pp. 174.

Patterson, D. J., N. M. Nieman, L. D. Nelson, C. F. Nelson, K. K. Schillo, K. D. Bullock, D. T. Brophy, and B. L. Woods. 1997. Estrus synchronization with an oral progestogen prior to superovulation of postpartum beef cows. *Theriogenology* 48:1025.

Patterson, D. J., F. N. Kojima, and M. F. Smith. 1999. Addition of GnRH to a melengestrol acetate (MGA) prostaglandin $F_{2\alpha}$ (PG) estrous synchronization treatment improves synchrony of estrus and maintains high fertility in postpartum suckled beef cows. *J. Anim. Sci.* 77 (Suppl. 1):220.

Patterson, D. J., S. L. Wood, F. N. Kojima, and M. F. Smith. 2000a. Improved synchronization of estrus in postpartum suckled beef cows with a progestin–GnRH–prostaglandin $F_{2\alpha}$ (PG) protocol. *J. Anim. Sci.* 83 (Suppl. 1):218.

Patterson, D. J., S. L. Wood, and R. F. Randle. 2000b. Procedures that support reproductive management of replacement beef heifers. *Proc. Am. Soc. Anim. Sci.* 1999. Available at: http://www.asas.org/jas/symposia/proceedings/0902.pdf. Accessed August 3, 2000.

Pexton, J. E., P. W. Farin, R. A. Gerlach, J. L. Sullins, M. C. Shoop, and P. J. Chenoweth. 1989. Efficiency of single-sire mating programs with beef bulls mated to estrus synchronized females. *Theriogenology* 32:705.

Prybil, M. K. and W. R. Butler. 1978. The relationship between progesterone secretion and the initiation of ovulation in postpartum beef cows. *J. Anim. Sci.* 47 (Suppl. 1):383.

Pursley, J. R., M. O. Mee, and M. C. Wiltbank. 1995. Synchronization of ovulation in dairy cows using $PGF_{2\alpha}$ and GnRH. *Theriogenology* 44:915.

Pursley, J. R., M. W. Kosorok, and M. C. Wiltbank. 1997a. Reproductive management of lactating dairy cows using synchronization of ovulation. *J. Dairy Sci.* 80:301.

Pursley, J. R., M. C. Wiltbank, J. S. Stevenson, J. S. Ottobre, H. A. Garverick, and L. L. Anderson. 1997b. Pregnancy rates in cows and heifers inseminated at a synchronized ovulation or synchronized estrus. *J. Dairy Sci.* 80:295.

Pursley, J. R., R. W. Silcox, and M. C. Wiltbank. 1998. Effect of time of artificial insemination on pregnancy rates, calving rates, pregnancy loss, and gender ratio after synchronization of ovulation in lactating dairy cows. *J. Dairy Sci.* 81:2139.

Randle, R. F. 1999. The Missouri show-me-select replacement heifer program: production summary from the first two years. In: *Proc. Reprod. Tools and Techniques*. pp. 1. Univ. Missouri, Columbia, MO.

Rawlings, N. C., L. Weir, B. Todd, J. Manns, and J. Hyland. 1980. Some endocrine changes associated with the postpartum period of the suckling beef cow. *J. Reprod. Fertil.* 60:301.

Roche, J. F. 1974. Effect of short-term progesterone treatment on estrous response and fertility in heifers. *J. Reprod. Fertil.* 40:433.

Savio, J. D., M. P. Boland, N. Hynes, M. R. Mattiacci, and J. F. Roche. 1990. Will the first dominant follicle of the estrous cycle of heifers ovulate following luteolysis on day 7? *Theriogenology* 33:677.

Schafer, D. W., J. S. Brinks, and D. G. LeFever. 1990. Increased calf weaning weight and weight via estrus synchronization. *Beef Prog. Rep.* Colorado State Univ. pp. 115.

Schmitt, E. J. P., T. Diaz, M. Drost, and W. W. Thatcher. 1996. Use of a gonadotropin-releasing hormone agonist or human chorionic gonadotropin for timed insemination in cattle. *J. Anim. Sci.* 74:1084.

Seidel, G. E. Jr. 1998. Potential applications of sexed semen in cattle. *J. Anim. Sci.* 76: (Suppl. 1):219.

Sheffel, C. E., B. R. Pratt, W. L. Ferrell, and E. K. Inskeep. 1982. Induced corpora lutea in the postpartum beef cow. II. Effects of treatment with progestogen and gonadotropins. *J. Anim. Sci.* 54:830.

Short, R. E., R. A. Bellows, R. A. Staigmiller, J. G. Berardinelli, and E. E. Custer. 1990. Physiological mechanisms controlling anestrus and infertility in postpartum beef cattle. *J. Anim. Sci.* 68:799.

Sirois, J. and J. E. Fortune. 1988. Ovarian follicular dynamics during the estrous cycle in heifers monitored by real-time ultrasonography. *Biol. Reprod.* 39:308.

Smith, M. W. and J. S. Stevenson. 1995. Fate of the dominant follicle, embryonal survival, and pregnancy rates in dairy cattle treated with prostaglandin $F_{2\alpha}$ and progestins in the absence or presence of a functional corpus luteum. *J. Anim. Sci.* 73:3743.

Smith, R. K. and M. L. Day. 1990. Mechanism of induction of puberty in beef heifers with melengestrol acetate. *Ohio Beef Cattle Res. Industry Rep.* pp. 137.

Stevenson, J. S., M. K. Schmidt, and T. P. Call. 1984. Stage of estrous cycle, time of insemination and seasonal effects on estrus and fertility of Holstein heifers after prostaglandin. *J. Anim. Sci.* 67:1798.

Stevenson, J. S. and J. R. Pursley. 1994. Resumption of follicular activity and interval to postpartum ovulation after exogenous progestins. *J. Dairy Sci.* 77:725.

Stevenson, J. S., G. C. Lamb, J. A. Cartmill, B. A. Hensley, S. Z. El-Zarkouny, and T. J. Marple. 1999. Synchronizing estrus in replacement beef heifers using GnRH, melengestrol acetate, and $PGF_{2\alpha}$. *J. Anim. Sci.* 77 (Suppl. 1):225.

Thatcher, W. W., M. Drost, J. D. Savio, K. L. Macmillan, K. W. Entwistle, E. J. Schmitt, R. L. De La Sota, and G. R. Morris. 1993. New clinical uses of GnRH and its analogues in cattle. *Anim. Reprod. Sci.* 33:27.

Thimonier, J., D. Chupin, and J. Pelot. 1975. Synchronization of estrus in heifers and cyclic cows with progestogens and prostaglandin analogues alone or in combination. *Ann. Biol. Anim. Biochim. Biophys.* 15:437.

Twagiramungu, H., L. A. Guilbault, J. Proulx, and J. J. Dufour. 1992a. Synchronization of estrus and fertility in beef cattle with two injections of Buserelin and prostaglandin. *Theriogenology* 38:1131.

Twagiramungu, H., L. A. Guilbault, J. Proulx. P. Villeneuve, and J. J. Dufour. 1992b. Influence of an agonist of gonadotropin-releasing hormone (Buserelin) on estrus synchronization and fertility in beef cows. *J. Anim. Sci.* 70:1904.

Twagiramungu, H., L. A. Guilbault, and J. J. Dufour. 1995. Synchronization of ovarian follicular waves with a gonadotropin-releasing hormone agonist to increase the precision of estrus in cattle: a review. *J. Anim. Sci.* 73:3141.

Watts, T. L. and J. W. Fuquay. 1985. Response and fertility of dairy heifers following injection with prostaglandin $F_{2\alpha}$ during early, middle and late diestrus. *Theriogenology* 23:655.

Wettemann, R. P. and H. D. Hafs. 1973. Pituitary and gonadal hormones associated with fertile and nonfertile inseminations at synchronized and control estrus. *J. Anim. Sci.* 36:716.

Wood, S. L., M. C. Lucy, M. F. Smith, and D. J. Patterson. 2000. Estrus and fertility in yearling beef heifers after addition of GnRH to a melengestrol acetate (MGA)–Prostaglandin $F_{2\alpha}$ estrus synchronization protocol. *Theriogenology* 53:207.

Wood, S. L., M. C. Lucy, M F. Smith, and D. J. Patterson. 2001. Improved synchrony of estrus and ovulation with addition of GnRH to a melengestrol acetate—prostaglandin $F_{2\alpha}$ estrus synchronization treatment in beef heifers. *J. Anim. Sci.* 19:2210.

Xu, Z. Z. and L. J. Burton. 1998. Synchronization of estrus with $PGF_{2\alpha}$ administered 18 days after a progesterone treatment in lactating dairy cows. *Theriogenology* 50:905.

Zimbelman, R. G. 1963. Maintenance of pregnancy in heifers with oral progestogens. *J. Anim. Sci.* 22:868.

Zimbelman, R. G. and L. W. Smith. 1966. Control of ovulation in cattle with melengestrol acetate. I. Effect of dosage and route of administration. *J. Reprod. Fertil.* (Suppl. 1): 185.

7 A Vaginal Insert (CIDR) to Synchronize Estrus and Timed AI

Joel V. Yelich

CONTENTS

The cow–calf producer is faced with several challenges when trying to implement estrous synchronization and artificial insemination (AI) programs. The first challenge is dealing with the large percentage of anestrous or noncycling cows at the start of the breeding season. Adequate nutrition prior to calving and during the postpartum period can increase the percentage of cattle cycling at the start of the breeding season (Short et al., 1990). However, when nutrient requirements of the lactating cow are compromised during this period, the percentage of anestrous cattle at the start of the breeding season increases dramatically. Therefore, an ideal estrous synchronization program would initiate estrus cycles in some anestrous cattle. The second challenge deals with cattle management during the implementation of the synchronization program. Producers would like to handle the cattle as few times as possible during a synchronization and AI program and eliminate estrus detection altogether by inseminating cattle at a predetermined time, also known as timed AI (TAI). Using TAI would eliminate the costly and labor-intensive process of estrous detection. Therefore, an effective estrous synchronization system would require minimal cattle handling, induce estrous cycles in a majority of the anestrous cattle, and yield consistent pregnancy rates to a TAI.

Exogenous progestins such as melengestrol acetate (MGA) and norgestomet (Syncro-Mate-B®) have been used either alone or in combination with estrogen, gonadotropin- releasing hormone (GnRH), and/or prostaglandin $F_{2\alpha}$ ($PGF_{2\alpha}$) to synchronize estrus and induce estrous cycles in anestrous cattle with limited success (Patterson et al., 1989; Odde, 1990; Beal, 1995). Another progestin used for estrous synchronization and currently marketed in Australia, Canada, Europe, Mexico, and New Zealand but which has not been approved for use in the United States is the vaginal insert known as the CIDR® (Eazi-Breed™, controlled intravaginal progesterone-releasing device, InterAG, Hamilton, New Zealand). The CIDR is made of a silicone rubber impregnated

0-8493-1117-9/02/$0.00+$1.50

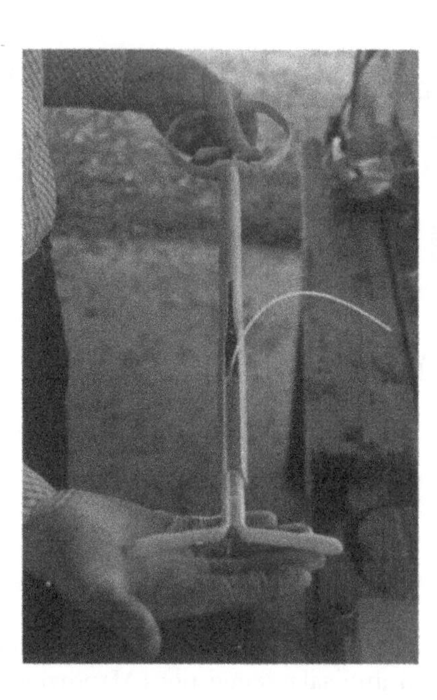

FIGURE 7.1 Controlled intravaginal progesterone-releasing device (CIDR) and the applicator used to insert the CIDR into the vagina.

with progesterone and molded over a nylon spine that is T-shaped. The wings of the CIDR fold upon themselves when placed in the applicator (Figure 7.1), which is inserted into the vagina. When the applicator is removed from the vagina, the wings of the CIDR fold out and apply pressure to the vaginal wall, which assists in retaining the CIDR in the vagina. A small nylon tail is attached to the end of the CIDR, which protrudes from the vulva allowing for easy removal of the CIDR from the vagina 7 or 8 days later. Progesterone is released from the CIDR at a controlled rate into the bloodstream of the animal. Two forms of the CIDR are currently marketed, containing either 1.3 or 1.9 grams of progesterone. The CIDR waiting for approval in the United States contains 1.3 grams of progesterone.

The major advantages of the CIDR include its ease of application and removal, and its high retention rates. Beef cattle can either be retained in a squeeze chute (it is not necessary to catch their heads) or in an AI breeding box to insert the CIDR. To remove the CIDR, cattle can be loaded into the alleyway leading into the squeeze chute and the CIDR is removed by pulling on the nylon tail exposed from the vulva. Concomitant with removal, cattle are given an injection of $PGF_{2\alpha}$ to initiate regression of any corpus luteum (CL) that may be present. Since there is no need to catch the cattle in the squeeze chute to remove the CIDR, stress associated with handling cattle is reduced considerably. Retentions rates are high, and it appears that less than 1% of the postpartum beef cows lose their CIDR during a 7-day duration (Lucy et al., 2001).

Like other progestins, the CIDR suppresses the expression of estrus and ovulation by elevating blood progesterone concentrations throughout its duration. A significant problem with using synthetic progestins in synchronization programs is manifested in the compromised fertility of the synchronized estrus after either short- (<8 days; Beal et al., 1988) or long-term (DeBois and Bierschwal, 1970; Zimbelman et al., 1970) treatments. The reduction in fertility is a result of one of two mechanisms. The first is the maintenance of a dominant follicle on the ovary under the influence of long-term progestins like MGA (Anderson and Day, 1994; Fike et al., 1999) and CIDR (Sirois and Fortune, 1990) in the absence of a functional CL resulting in the ovulation of subfertile oocytes (Revah and Butler, 1996). The second mechanism is the ovulation of aged oocytes (Mihm et al., 1994), which would be more prevalent during short-term progestin treatments. To prevent development of either a persistent follicle or ovulation of an aged oocyte after a progestin treatment, estrogens can be administered to regress follicles (Bo et al., 1995; Burke et al., 2000), or gonadotropin-releasing hormone (GnRH) can be administered to ovulate the dominant follicle (Macmillan et al., 1985a,b; Thatcher et al., 1989) at the initiation of a progestin treatment, resulting in emergence of a new cohort of follicles 2 to 5 days later (Bo et al., 1995; Thatcher et al., 1996; Burke et al., 2000). The occurrence of a persistent dominant follicle during a 7-day CIDR treatment, or ovulation of an aged oocyte following a 7-day CIDR is unclear and further investigation is required. Recent work by Lucy et al. (2001) suggests fertility of the estrus following a 7-day CIDR is not compromised.

Several basic CIDR synchronization protocols are shown in Figure 7.2. All programs consist of a CIDR treatment for 7 days, with $PGF_{2\alpha}$ administered at CIDR removal. Either estradiol benzoate (EB) or GnRH can be administered at CIDR insertion as a means of initiating follicle turnover. Following CIDR removal, multiple AI programs can be used. In option 1, cattle can be inseminated by observed estrus for approximately 5 days after CIDR removal. In options 2 and 3, EB can be injected 24 hours after CIDR removal and cattle can be inseminated either by observed estrus (option 2) or TAI approximately 24 to 36 hours after EB injection (Option 3). In option 4, cattle can be TAI and injected with GnRH 48 hours after CIDR removal. With options 3 and 4, 48-hour calf removal can also be implemented in the TAI programs.

The ability of the CIDR to synchronize estrus and shorten the interval to pregnancy was recently evaluated by Lucy et al. (2001) in an experiment conducted at seven locations across the United States with 851 lactating beef cows averaging 56 days postpartum, which were classified as either cycling or anestrous before treatment. The three treatments tested included untreated cattle (controls; $n = 285$), cattle treated with a single injection of $PGF_{2\alpha}$ ($PGF_{2\alpha}$; $n = 283$), and cattle treated with a 7-day CIDR, with $PGF_{2\alpha}$ on day 6 (CIDR + $PGF_{2\alpha}$; $n = 282$). Cows were AI by observed estrus for a 31-day breeding period. Estrous, conception, and pregnancy rates for the first 3 days of the synchronized breeding period are presented in Table 7.1. Treatment with the CIDR + $PGF_{2\alpha}$ significantly increased the percentage of cattle exhibiting estrus compared to controls and $PGF_{2\alpha}$-treated cattle. Although the cycling cows showed the greatest response to the CIDR + $PGF_{2\alpha}$ treatment, the CIDR + $PGF_{2\alpha}$ treatment also resulted in a greater number of anestrous cattle exhibiting estrus early in the breeding season compared to control and $PGF_{2\alpha}$-treated cattle. Nearly half of the cows determined to be anestrous at the start of the treatment exhibited estrus (45%) during the 3 days following CIDR treatment. These data plus a previous report by Fike et al. (1997) indicate the effectiveness of a 7-day CIDR

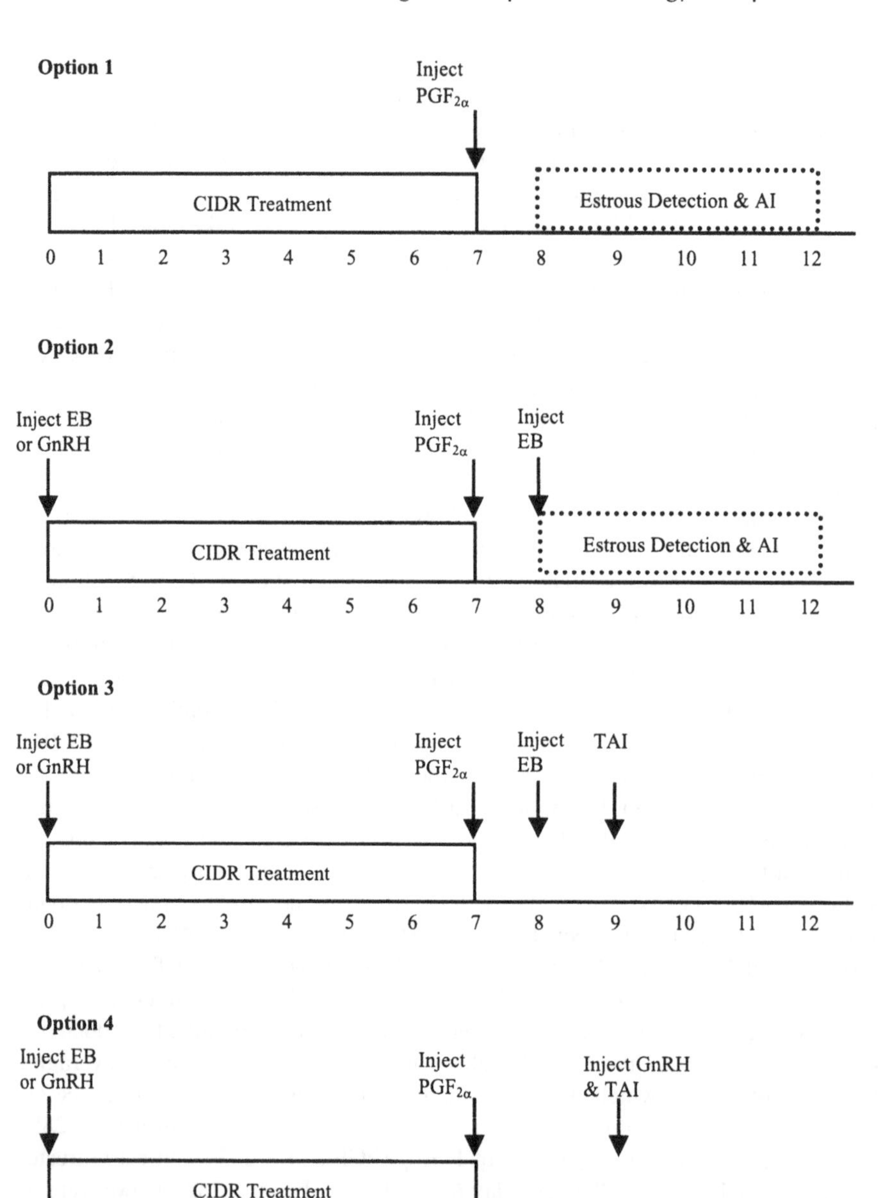

FIGURE 7.2 Different options used to synchronize estrous and/or ovulation in postpartum beef cows treated with a controlled intravaginal progesterone-releasing device (CIDR) and prostaglandin $F_{2\alpha}$ ($PGF_{2\alpha}$). Estradiol benzoate (EB) can be either administered on day 1 at a rate of 2 mg and/or on day 8 at a rate of 1 mg. Gonadotropin-releasing hormone (GnRH) is administered at a rate of 100 mg. Timed AI = TAI.

TABLE 7.1
Three-Day Estrous, Conception, and Pregnancy Rates for an Untreated Control, Synchronized with Prostaglandin $F_{2\alpha}$ ($PGF_{2\alpha}$) Alone or a Controlled Intravaginal Progesterone-Releasing Device (CIDR) + $PGF_{2\alpha}$ for a Seven-Location Study Using Anestrous (AN) and Cycling (CYC) Postpartum Beef Cows[a]

		Estrous Rate, %[b]		Conception Rate, %[c]		Pregnancy Rate, %[d]	
Treatment	Total No.	AN	CYC	AN	CYC	AN	CYC
Control	285	11 (16/151)	19 (26/134)	38 (6/16)	58 (15/26)	4 (6/151)	11 (15/134)
$PGF_{2\alpha}$	283	19 (30/154)	49 (63/129)	57 (17/30)	70 (44/63)	11 (17/154)	34 (44/129)
CIDR + $PGF_{2\alpha}$	283	45 (64/142)	72 (102/141)	57 (36/63)	63 (64/101)	26 (36/141)	46 (64/140)

[a] Cows were determined to be either cycling or anestrous before the treatment by blood progesterone concentrations. Treatments included control (no synchronization treatment), $PGF_{2\alpha}$ (single injection of 25 mg $PGF_{2\alpha}$), or CIDR + $PGF_{2\alpha}$ (CIDR for 7 days with a single injection of 25 mg $PGF_{2\alpha}$ on day 6).
[b] Number in estrus during the first 3 days of the breeding period/number treated.
[c] Number pregnant/number inseminated.
[d] Number pregnant during the first 3 days of the breeding period/number treated.

Adapted from Lucy et al. (2001)

treatment to induce estrous cycles in anestrous cattle, thus increasing the percentage of beef cows exhibiting estrus and eligible for insemination early in the breeding season. Across all locations and both cycling statuses, the percentage of cows exhibiting estrus during the first 3 days of the breeding period were 15, 33, and 59% for control, $PGF_{2\alpha}$, and CIDR + $PGF_{2\alpha}$ treatments, respectively.

First insemination conception rate (61%) for the 3-day synchronized period was similar across all treatments, cyclic statuses, and locations, suggesting that the CIDR treatment had no negative effects on conception rates in contrast to previous reports in cattle synchronized with short-term progestin treatments (Beal et al., 1988). Because of the increased estrous response and high conception rates, an increased percentage of CIDR + $PGF_{2\alpha}$ cows became pregnant early in the breeding season compared to controls and $PGF_{2\alpha}$-treated cows. Of the cycling cows treated with the CIDR + $PGF_{2\alpha}$, nearly half (46%) became pregnant in the first 3 days of the breeding period. Across all locations and both cycling statuses, the percentage of cows becoming pregnant during the first 3 days of the breeding period were 7, 22, and 36% for control, $PGF_{2\alpha}$, and CIDR + $PGF_{2\alpha}$ treatments, respectively. The percentage of cattle pregnant to a first insemination during the first 31 days of the breeding season tended ($P = 0.10$) to be affected by treatment, with pregnancy rates of 50, 55, and 58% for control, $PGF_{2\alpha}$, and CIDR + $PGF_{2\alpha}$ treatments, respectively. In summary, this study showed the effectiveness of the CIDR + $PGF_{2\alpha}$ estrous synchronization system for increasing estrous and pregnancy rates early in the breeding season of both cyclic and anestrous lactating beef cows across several different environments and genotypes of cattle. Furthermore, this study showed that the short-term progesterone treatment of the CIDR did not alter conception rates when cows were inseminated after an observed estrus. Additional research must be conducted to further optimize the effectiveness of the CIDR + $PGF_{2\alpha}$ estrous synchronization system for not only breeding by observed estrus but also for TAI programs.

To further evaluate the effectiveness of the CIDR + $PGF_{2\alpha}$ estrous synchronization system, we have conducted several experiments at the University of Florida during the past 4 years. To streamline the CIDR + $PGF_{2\alpha}$ synchronization system, the $PGF_{2\alpha}$ injection has been moved to the day of CIDR removal instead of day 6, hence eliminating an additional cattle handling. The major focus of this research has been to evaluate the effectiveness of the CIDR + $PGF_{2\alpha}$ synchronization system when used in combination with a TAI.

In a preliminary study, postpartum lactating Brangus cows were synchronized to evaluate breeding after an observed estrus or TAI. All cows were treated with a 7-day CIDR, with 25 mg $PGF_{2\alpha}$ administered at CIDR removal. The control group was inseminated 12 hours after an observed estrus. The other three groups received 1 mg estradiol benzoate (EB) 24 hours after CIDR removal and were either inseminated after an observed estrus or TAI 60 hours after CIDR removal. In addition, one of the three groups was injected with 2 mg EB at CIDR insertion to determine if follicle turnover at CIDR insertion would increase subsequent pregnancy rates. The results from this experiment are presented in Table 7.2 and provided the initial

TABLE 7.2
Estrous, Conception, and Pregnancy Rates of Lactating Postpartum Brangus Cows Synchronized with a Controlled Intravaginal Progesterone-Releasing Device (CIDR) with Prostaglandin $F_{2\alpha}$ ($PGF_{2\alpha}$) at CIDR Removal and Estradiol Benzoate (EB) and Either Inseminated after an Observed Estrus or Timed-Artificial Insemination (TAI)

Treatments[a]	No.	Estrous Response (%)[b]	Conception Rate (%)[c]	Pregnancy Rate (%)[d]
CIDR/$PGF_{2\alpha}$ (Estrous-AI)	48	75.0	47.2	35.4
CIDR/$PGF_{2\alpha}$ + EB (Estrous-AI)	52	90.4	40.4	36.5
CIDR/$PGF_{2\alpha}$ + EB (TAI)	56	—	32.1	32.1
EB/CIDR/$PGF_{2\alpha}$ + EB (TAI)	55	—	47.3	47.3

[a] Start of the experiment designated as day 0. All treatments received CIDR on day 0 with 25 mg $PGF_{2\alpha}$ i.m. on day 7. Treatment 4 received 2 mg EB i.m. on day 0, while treatments 2, 3, and 4 received 0.5 mg EB i.m. 24 hours after CIDR removal. Treatments 1 and 2 were AI 12 hours after observed estrus, and treatments 3 and 4 were TAI 60 hours after CIDR removal.
[b] Percentage of cows displaying estrous 5 days after CIDR of the total treated.
[c] Percentage of cows pregnant to AI during the 5 days after CIDR of the total AI.
[d] Percentage of cows pregnant to AI during the 5 days after CIDR of the total treated.

evidence that conception rates to a TAI protocol were similar to cattle inseminated by observed estrus, suggesting that TAI combined with the CIDR/$PGF_{2\alpha}$ has the potential to result in acceptable pregnancy rates. Although not significant, the administration of EB at CIDR insertion resulted in pregnancy rates that were approximately 12% greater compared with no-EB at CIDR insertion. Therefore, a larger field trial was conducted to determine if EB administered at CIDR insertion could significantly and consistently enhance pregnancy rates as part of a TAI protocol.

A second study (Barthle et al., 2000) used multiparous postpartum lactating cows of *Bos indicus* × *Bos taurus* breeding, which were stratified to treatment by days postpartum and body condition score at CIDR insertion (day 0 is start of the experiment). The CIDRs were removed on day 7 and all cows were administered 25 mg $PGF_{2\alpha}$. Estrus was detected for 5 days following CIDR removal in all treatments. At the insertion of the CIDR, half the cows were administered 2 mg EB, and the remaining cows were not. Cows were inseminated either after an observed estrus or TAI according to the following treatments: 1) 1 mg EB on day 8 with AI 12 hours after observed estrus (CIDR/$PGF_{2\alpha}$ + EB + Estrus-AI); 2) 1 mg EB on day 8 with TAI 60 hours after CIDR removal (CIDR/$PGF_{2\alpha}$ + EB + TAI); 3) TAI at 48 to 54 hours after CIDR removal with all cows receiving 100 µg GnRH at TAI (CIDR/$PGF_{2\alpha}$ + GnRH/TAI); 4) 2 mg EB on day 0 and 1 mg EB on day 8 with AI 12 hours after observed estrus (EB/CIDR/$PGF_{2\alpha}$ + EB + Estrus-AI); 5) 2 mg EB on day 0 with 1 mg EB on day 8 with TAI 60 hours after CIDR removal (EB/CIDR/$PGF_{2\alpha}$ + EB + TAI); 6) 2 mg EB on day 0 with TAI at 48 to 54 hours after CIDR removal with all cows receiving 100 µg GnRH at TAI (EB/CIDR/$PGF_{2\alpha}$ + GnRH/TAI). Calves remained with the cows up until the time of insemination.

Administration of EB at CIDR insertion significantly increased both conception and pregnancy rates compared to cows not receiving EB at CIDR insertion (Table 7.3), in agreement with an earlier report of Martinez et al. (1998). These data suggest that EB administered at the initiation of a 7-day CIDR probably turns over large follicles, preventing either the formation of a persistent follicle or the ovulation of aged oocytes after CIDR removal and improving pregnancy rates. Subsequent work in our lab has shown that 2 mg EB administered at CIDR insertion is nearly 100% effective in turning over follicles greater than 9 mm (Fullenwider et al., 1999). These data agree with other reports where turnover of large follicles, either at the initiation of or during a progestin treatment, results in increased pregnancy rates (Anderson and Day, 1994; Martinez et al., 1998; Fike et al., 1999). However, these data contradict a recent report by Lucy et al. (2001) where conception rates were not compromised following a 7-day CIDR treatment. The reason for the difference is unclear and will require further investigation. Within the no-EB and EB treatments of the present study, conception and pregnancy rates for cows that were inseminated by estrus or TAI were similar providing further data that support the potential for TAI to be used in conjunction with the CIDR/ $PGF_{2\alpha}$ synchronization system. An additional benefit of synchronizing cattle, which is often overlooked, is the increased percentage of cattle that get pregnant early in the breeding season. Greater than 70% of the cows became pregnant (Table 7.3) in the first 30 days of the breeding season regardless of whether or not they received EB at CIDR insertion. This ultimately results in shorter calving periods, leading to more uniform calves that are older and heavier at weaning.

TABLE 7.3
The Effect of Estradiol Benzoate (EB) at Delivery of a Controlled Intravaginal Progesterone-Releasing Device (CIDR) with Prostaglandin $F_{2\alpha}$ ($PGF_{2\alpha}$) at CIDR Removal on Estrous, Conception, and Pregnancy Rates of Lactating Postpartum Crossbred Cows of *Bos indicus* Breeding Inseminated after Either an Observed Estrus or Timed-AI (TAI)[a]

Variables	No.	Conception Rate (%)[c]	Pregnancy Rate (%)[d]	35-Day Pregnancy Rate (%)[e]
EB Effects[f]				
No EB at CIDR delivery	237	38.5*	37.6*	71.7
EB at CIDR delivery	241	53.8	53.5	76.3
Treatment Effects				
CIDR/$PGF_{2\alpha}$ + EB (Estrus)	80	42.7	38.8	71.6
CIDR/$PGF_{2\alpha}$ + EB (TAI)	77	35.9	35.9	67.9
CIDR/$PGF_{2\alpha}$ + GnRH (TAI)	80	38.3	38.3	76.5
EB/CIDR/$PGF_{2\alpha}$ + EB (Estrus)	69	59.1	58.0	79.7
EB/CIDR/$PGF_{2\alpha}$ + EB (TAI)	87	50.6	50.6	73.6
EB/CIDR/$PGF_{2\alpha}$ + GnRH (TAI)	85	52.9	52.9	76.5

[a] Start of the experiment designated as day 0. All cows received a CIDR with 25 mg $PGF_{2\alpha}$ i.m. on day 7. Treatments 2, 4, and 6 received 2 mg EB i.m. on day 0, while treatments 1, 3, and 5 did not receive EB. Treatments 1, 2, 3, and 4 received 1.0 mg EB i.m. 24 hours after CIDR removal. Treatments 1 and 2 were AI 12 hours after observed estrus, while treatments 3 and 4 were TAI 60 hours after observed estrus, and treatments 5 and 6 were TAI and received 100 μg GnRH i.m. 48 hours after CIDR removal.
[b] Percentage of cows displaying estrus 5 days after CIDR of the total treated.
[c] Percentage of cows pregnant to AI during the 5 days after CIDR of the total AI.
[d] Percentage of cows pregnant to AI during the 5 days after CIDR of the total treated.
[e] Percentage of cows pregnant during the 35 days after CIDR removal of the total treated.
[f] Significance level EB vs. No EB: (*P < .001)

The newly recruited follicle appears to need approximately 6 days to grow and develop to reach ovulatory size (Lucy et al., 1992). Administration of an estrogen results in a newly recruited follicle that appears approximately 3 days later (Bo et al., 1995; Burke et al., 2000), which would be day 3 or 4 of a CIDR treatment if it were administered at the initiation of the CIDR treatment. Therefore, the dominant follicle should be of ovulatory size approximately 9 days after CIDR insertion. Leaving the CIDR in for an additional day should allow the newly recruited follicle an extra day to mature endocrinologically and reach ovulatory size when TAI + GnRH are conducted 48 hours after CIDR removal or 9 days after CIDR insertion. Therefore, a third study was conducted to evaluate a 7- vs. 8-day CIDR treatment with and without EB at CIDR insertion. The experiment was conducted at five locations throughout Florida with several breed types represented, including Angus, Brangus, Brahman, Simbrah, Braford, Red Brangus, and crossbred (*Bos indicus* × *Bos taurus*) postpartum

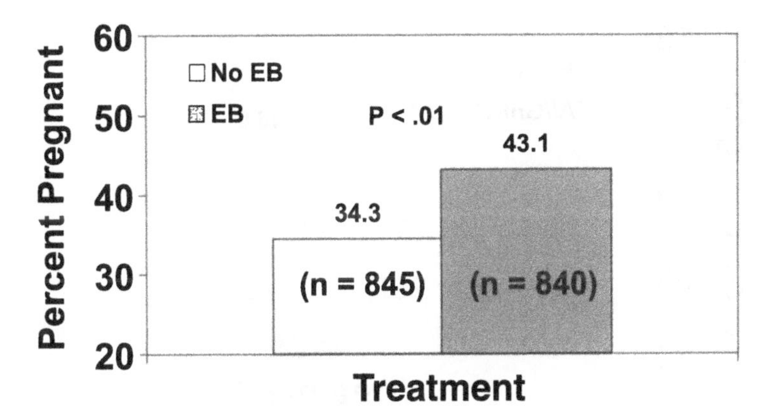

FIGURE 7.3 Pregnancy rates of lactating postpartum cows synchronized with a either a 7- or 8-day controlled intravaginal progesterone-releasing device (CIDR) with prostaglandin $F_{2\alpha}$ at CIDR removal and timed AI + GnRH 48 hours after CIDR removal. Within the 7- and 8-day CIDR treatments, half the cows were injected with 2 mg estradiol benzoate (EB) at CIDR insertion, and the remainder received nothing. (The main effect of 7-day CIDR vs. 8-day CIDR was not significant.)

lactating cows. When calving dates were known, cows were stratified to treatment by postpartum interval and body condition score. Cows received a CIDR for either 7 or 8 days with 25 mg $PGF_{2\alpha}$ at CIDR removal, with CIDR being removed on the same day for both the 7- and 8-day treatments. At CIDR insertion, half the cows received 2 mg EB and the remainder did not receive EB. All cows were injected with GnRH and TAI 48 to 54 hours after CIDR removal. Calves remained on the cows until the time of insemination.

As with the previous experiment (Table 7.3), EB at CIDR insertion significantly increased pregnancy rates to the TAI (Figure 7.3), and leaving the CIDR in for an additional day neither compromised nor enhanced pregnancy rates. The increased pregnancy rate in the second study (10%) was of a smaller magnitude compared to the first study (15%). The decrease in magnitude in the second study may have been due to an increased percentage of anestrous cattle at the start of the CIDR treatment in the second year compared to the first. Although, blood samples were not collected in either experiment, approximately 13% fewer cattle became pregnant in the first 30 days of the breeding season in experiment two than experiment one, suggesting that fewer cattle were cycling in the second experiment. However, both studies provide strong evidence that EB administered at CIDR insertion increases pregnancy rates and indirectly suggest that EB at CIDR insertion has a greater effect on increasing pregnancy rates in cycling than anestrous cows. Since the 8-day CIDR did not compromise pregnancy rates, it provides the producer management flexibility relative to time of CIDR removal and the subsequent TAI.

With the fairly strong evidence that EB at CIDR insertion enhances pregnancy rates, a fourth study was conducted to evaluate TAI pregnancy rates between two ovulation induction methods (Fullenwider et al., 2000). This study was conducted with nonlactating crossbred (*Bos indicus* × *Bos taurus*) cows randomly assigned to

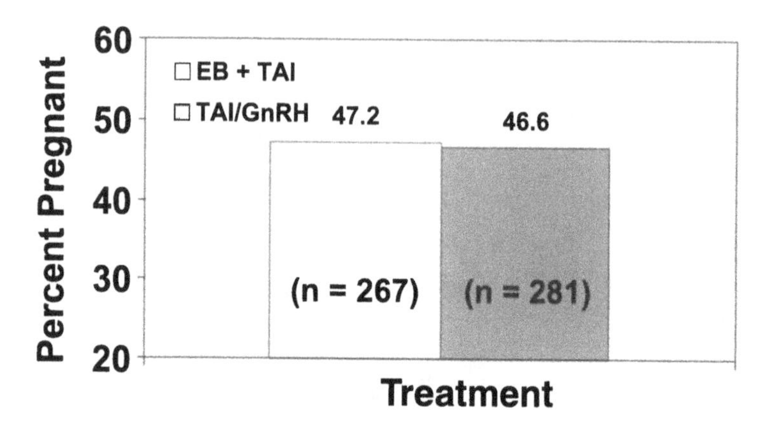

FIGURE 7.4 Pregnancy rates of lactating postpartum beef cows synchronized with a 7-day controlled intravaginal progesterone-releasing device (CIDR) + 2 mg estradiol benzoate (EB) at CIDR insertion with prostaglandin $F_{2\alpha}$ at CIDR removal. Half the cows received 1 mg EB 24 hours after CIDR removal and were timed AI (TAI) 24 to 30 hours later, and the remaining cows were TAI/GnRH 48 to 54 hours after CIDR removal.

two treatments at CIDR insertion. All cows received a 7-day CIDR with 2 mg EB at CIDR insertion and 25 mg $PGF_{2\alpha}$ at CIDR removal. Half the cows received 1 mg of EB 24 hours after CIDR removal followed by TAI 24 to 30 hours after EB injection, while the remaining cows were injected with GnRH and TAI 48 hours after CIDR removal. Ten sires were equally stratified across treatments to account for variation in sire fertility. Timed-AI pregnancy rates were similar between ovulation induction methods (Figure 7.4). However, there was significant sire effect ($P < .001$) on TAI pregnancy rate as TAI pregnancy rates ranged from a low of 25% to a high of 61% (Figure 7.5). Consequently, the rate-limiting factor associated with TAI appears to be fertility of the frozen/thawed semen and its ability to survive in the reproductive tract for an extended period of time and still allow fertilization to occur. Currently, there are no methods available to determine which bulls could be used in a TAI program to obtain maximum pregnancy rates.

Since EB at CIDR insertion consistently enhanced pregnancy rates in all our studies, a 2 × 2 factorial experiment was conducted to determine if administration of GnRH at CIDR insertion would also increase pregnancy rates to a TAI. The experiment was conducted with nonlactating crossbred (*Bos indicus* × *Bos taurus*) cows. On day 0 of the experiment, all cows received a 7-day CIDR, and half the cows were injected with GnRH and the remaining cows were not injected with GnRH. The CIDRs were removed on day 7, and all cows were injected with 25 mg $PGF_{2\alpha}$. Within both the GnRH and no-GnRH groups, half the cows were injected with GnRH and TAI 48 hours after CIDR removal, and the remaining cows were injected with GnRH 48 hours after CIDR removal and TAI 12 to 16 hours later. Administration of GnRH at CIDR insertion did not enhance pregnancy rates (Figure 7.6) in the present study, and there was no difference in pregnancy rates between the TAI protocols. These data are contradictory to all our previous data where TAI pregnancy rates can

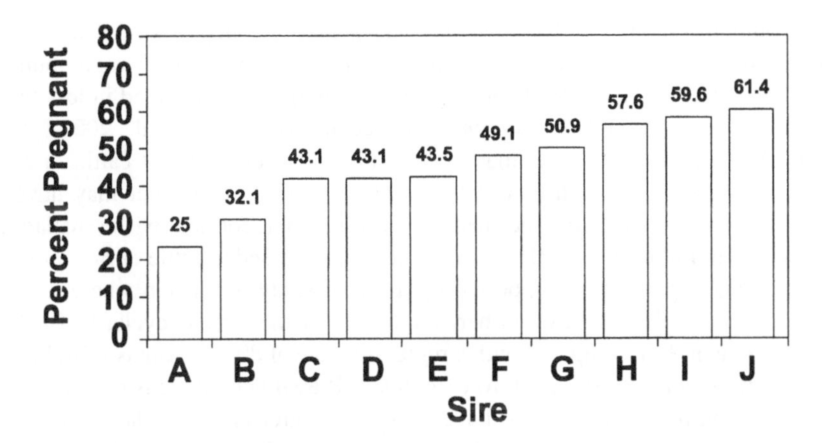

FIGURE 7.5 Timed-AI (TAI) pregnancy rates by sire for lactating postpartum beef cows synchronized with a 7-day controlled intravaginal progesterone-releasing device (CIDR) + 2 mg estradiol benzoate (EB) at CIDR insertion with prostaglandin $F_{2\alpha}$ at CIDR removal. Half the cows received 1 mg EB 24 hours after CIDR removal and were TAI 24 to 30 hours later, and the remaining cows were TAI + GnRH 48 to 54 hours after CIDR removal. Average number of cows inseminated for each sire was 55.

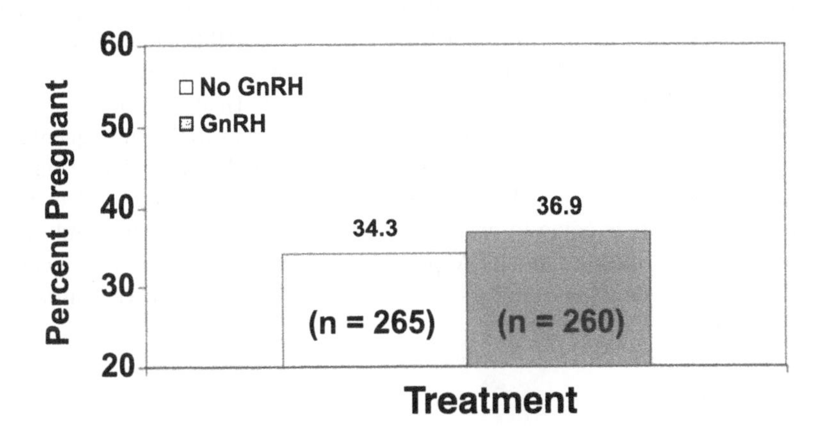

FIGURE 7.6 Pregnancy rates of lactating postpartum beef cows synchronized with a 7-day controlled intravaginal progesterone-releasing device (CIDR) with or without gonadotropin-releasing hormone (GnRH) at CIDR insertion with prostaglandin $F_{2\alpha}$ at CIDR removal. Half the cows with the GnRH and No-GnRH treatment were either timed AI with GnRH 48 hours after CIDR removal or received GnRH 48 hours after CIDR removal and were timed AI 12 to 16 hours later.

be enhanced by turning over the dominant follicle at CIDR insertion, and they are also contradictory to a recent report by Penny et al. (2000) who reported that pregnancy rates could be increased when GnRH was administered at CIDR insertion. The reason for a lack of response in our study is unclear and additional studies must be conducted to evaluate this.

The administration of GnRH followed 7 days later by $PGF_{2\alpha}$ has been shown to be an effective method of synchronizing cattle of *Bos taurus* breeding (Twagiramungu et al., 1995; Geary et al., 1998; Thompson et al., 1999) but has resulted in less than spectacular results in cattle of *Bos indicus* breeding (Lemaster et al., 2001). The reason for this is unclear but Lemaster et al. (2001) speculated that cattle of *Bos indicus* breeding treated with GnRH followed by $PGF_{2\alpha}$ 7 days later may have a reduced response to the $PGF_{2\alpha}$ because the accessory CL formed by the ovulation of the dominant follicle by GnRH is not effectively regressed in cattle of *Bos indicus* breeding. This hypothesis is supported by a recent study in our lab where Angus heifers injected with $PGF_{2\alpha}$, on either day 6 or 7 of the estrous cycle, had a CL regression rate of 96% compared with a regression rate of 80% in Angus × Brahman heifers (Portillo et al., 2001). Therefore, with an EB treatment, there is no accessory CL formed, and most cattle should have a CL that is greater than 7 days old at the time of $PGF_{2\alpha}$ administration. Other factors including differences in follicle growth and development and response to GnRH could also differ between cattle of *Bos indicus* and *Bos taurus* breeding.

In conclusion, treatment of postpartum beef cows with a 7-day CIDR results in an increased percentage of cattle showing estrus and getting pregnant early in the breeding season in both cycling and anestrous postpartum beef cows. The CIDR is most effective in cycling cows, but it does appear to induce estrus cycles in a significant percentage of anestrous cattle, allowing these cattle opportunity to become pregnant earlier in the breeding season than untreated controls. The CIDR has unlimited potential to be used in conjunction with TAI, although, its overall effectiveness may be limited by the fertility of the frozen/thawed semen used. Treatment of cows with estradiol benzoate at CIDR insertion enhances pregnancy rates to the subsequent AI, but whether a similar effect occurs with GnRH is unclear and requires further investigation. The addition of the CIDR to the arsenal of currently approved estrous synchronization agents is going to provide the producer with an effective and easy-to-use tool for synchronizing cattle for AI programs in lactating postpartum beef cows. Its ease of application and removal, plus its high retention rates also makes it attractive for producers to use.

REFERENCES

Anderson, L. H. and M. L. Day. 1994. Acute progesterone administration regresses persistent dominant follicles and improves fertility of cattle in which estrus was synchronized with melengestrol acetate. *J. Anim. Sci.* 72:2955.

Barthle, C. R., J. R. Kempfer, J. K. Fullenwider, J. W. Lemaster, C. L. Barnett, G. E. Portillo, and J. V. Yelich. 2000. Effect of estradiol benzoate (EB) administered at insertion of an intravaginal progesterone releasing insert (CIDR) on pregnancy rates in crossbred *Bos indicus* cows. *J. Anim. Sci.* 78:210.

Beal, W. E. 1995. Estrous synchronization programs: what works and what doesn't. In: *Twenty-Seventh Res. Symp. Ann. Mt.* pp. 13. Beef Improv. Fed., Colby, KS.

Beal, W. E., J. R. Chenault, M. L. Day, and L. R. Corah. 1988. Variation in conception rates following synchronization of estrus with melengestrol acetate and prostaglandin $F_{2\alpha}$. *J. Anim. Sci.* 66:599.

Bo, G. A., G. P. Adams, R. A. Pierson, and R. J. Mapletoft. 1995. Exogenous control of follicular wave emergence in cattle. *Theriogenology* 43:31.

Burke, C. R., M. L. Day, C. R. Bunt, and K. L. Macmillan. 2000. Use of a small dose of estradiol benzoate during diestrus to synchronize development of the ovulatory follicle in cattle. *J. Anim. Sci.* 78:145.

DeBois, C. H. W. and C. J. Bierschwal. 1970. Estrous cycle synchronization in dairy cattle given a 14-day treatment of melengestrol acetate. *Am. J. Vet. Res.* 31:1545.

Fike, K. E., M. E. Wehrman, B. R. Lindsey, E. G. Bergfield, E. J. Melvin, J. A. Quintal, E. L. Zanella, F. N. Kojima, and J. E. Kinder. 1999. Estrous synchronization of beef cattle with a combination of melengestrol acetate and an injection of progesterone and 17β-estradiol. *J. Anim. Sci.* 77:715.

Fike, K. E., M. L. Day, E. K. Inskeep, J. E. Kinder, P. E. Lewis, R. E. Short, and H. D. Hafs. 1997. Estrous and luteal function in suckled beef cows that were anestrous when treated with an intravaginal device containing progesterone with or without a subsequent injection of estradiol benzoate. *J. Anim. Sci.* 75:2009.

Fullenwider, J. K., J. R. Kempfer, C. L. Barnett, G. E. Portillo, C. R. Barthle, and J. V. Yelich. 2000. An estrous synchronization field study comparing estradiol benzoate (EB) and GnRH in combination with a intravaginal progesterone insert (CIDR) for timed-AI in crossbred *Bos indicus* cows. *J. Anim. Sci.* 78:217.

Fullenwider, J. K., J. R. Kempfer, C. L. Barnett, and J. V. Yelich. 1999. Effects of estradiol benzoate (EB) with an intravaginal progesterone insert (INSERT), PG and GnRH on follicle turnover and pregnancy/estrous rates in lactating crossbred *Bos indicus* cattle. *J. Anim. Sci.* 77 (Suppl. 1):224.

Geary, T. M., J. C. Whittier, E. R. Downing, D. G. LeFever, R. W. Silcox, M. D. Holland, T. M. Nett, and G. D. Niswender. 1998. Pregnancy rates of postpartum beef cows that were synchronized using Syncro-Mate-B or the Ovsynch protocol. *J. Anim. Sci.* 76:1523.

Lemaster, J. W., J. V. Yelich, J. R. Kempfer, J. K. Fullenwider, C. L. Barnett, M. D. Fanning, and J. F. Selph. 2001. Effectiveness of GnRH + Prostaglandin $F_{2\alpha}$ for synchronization of estrus in cattle of *Bos indicus* breeding. *J. Anim. Sci.* 79:309.

Lucy, M. C., H. J. Billings, W. R. Butler, L. R. Ehnis, M. J. Fields, D. J. Kesler, J. E. Kinder, R. C. Mattos, R. E. Short, W. W. Thatcher, R. P. Wetteman, J. V. Yelich, and H. D. Hafs. 2001. Efficacy of an intravaginal progesterone insert and an injection of $PGF_{2\alpha}$ for synchronizing estrous and shortening the interval to pregnancy in postpartum beef cows, peripubertal heifers, and dairy heifers. *J. Anim. Sci.* 79:982.

Lucy, M. C., J. D. Savio, L. Badinga, R. L. de la Sota, and W. W. Thatcher. 1992. Factors that affect ovarian follicular dynamics in cattle. *J. Anim. Sci.* 70:3615.

Macmillan, K. L., A. M. Day, V. K. Taufa, M. Gibb, and M. G. Pearce. 1985a. Effects of an agonist of gonadotropin-releasing hormone in cattle. I. Hormone concentrations and oestrous cycle length. *Anim. Reprod. Sci.* 8:203.

Macmillan, K. L., A. M. Day, V. K. Taufa, A. J. Peterson, and M. G. Pearce. 1985b. Effects of an agonist of gonadotropin-releasing hormone in cattle. II. Interactions with injected prostaglandin $F_{2\alpha}$ and unilateral ovariectomy. *Anim. Reprod. Sci.* 8:213.

Martinez, M. F., J. P. Kastelic, G. P. Adams, E. Janzen, W. Olson, and R. J. Mapletoft. 1998. Alternative methods of synchronizing estrus and ovulation for fixed-time insemination in cattle. *Theriogenology* 49:350.

Mihm, M., N. Curran, P. Hyttel, M. P. Boland, and J. F. Roche. 1994. Resumption of meiosis in cattle oocytes from preovulatory follicles with a short and a long duration of dominance. *J. Reprod. Fertil.* 13:14.

Odde, K. J. 1990. A review of synchronization of estrous in postpartum cattle. *J. Anim. Sci.* 68:817.

Patterson, D. J., G. H. Kiracofe, J. S. Stevenson, and L. R. Corah. 1989. Control of the bovine estrous cycle with melengestrol acetate (MGA): a review. *J. Anim. Sci.* 67:1895.

Penny, C. D., B. G. Lowman, N. A. Scott, and P. R. Scott. 2000. Repeated oestrous synchronization of beef cows with progesterone implants and the effects of gonadotropin-releasing hormone agonist at implant insertion. *Vet. Rec.* 146:395.

Portillo, G. E., E. A. Hiers, C. R. Barthle, M. K. V. Dahms, W. W. Thatcher, and J. V. Yelich. 2001. Luteolysis after $PGF_{2\alpha}$ on day 6 or 7 of the estrous cycle in Angus and Angus × Brahman heifers. *J. Anim. Sci.* (in press).

Revah, I. and W. R. Butler. 1996. Prolonged dominance of follicles and reduced viability of bovine oocytes. *J. Reprod. Fertil.* 106:39.

Short, R. E., R. A. Bellows, R. B. Staigmiller, J. G. Berardinelli, and E. E. Custer. 1990. Physiological mechanisms controlling anestrous and infertility in postpartum beef cattle. *J. Anim. Sci.* 68:799.

Sirois, J. and J. E. Fortune. 1990. Lengthening the bovine estrous cycle with low concentrations of exogenous progesterone: a model for studying ovarian follicular dominance. *Endocrinology* 127:916.

Thatcher, W. W., D. L. De la Sota, E. J. P. Schmitt, T. C. Diaz, L. Badinga, F. A. Simmen, C. R, Staples, and M. Drost. 1996. Control and management of ovarian follicles in cattle to optimize fertility. *Reprod. Fertil. Dev.* 8:203.

Thatcher, W. W., K. L. Macmillian, P. J. Hansen, and M. Drost. 1989. Concepts for regulation of corpus luteum function by the conceptus and ovarian follicles to improve fertility. *Theriogenology* 31:149.

Thompson, K. E., J. S. Stevenson, G. C. Lamb, D. M. Grieger, and C. A. Löest. 1999. Follicular, hormonal, and pregnancy responses of early postpartum suckled beef cows to GnRH, norgestomet, and $PGF_{2\alpha}$. *J. Anim. Sci.* 77:1823.

Twagiramungu, H., L. A. Guilbault, and J. J. Dufour. 1995. Synchronization of ovarian follicular waves with a gonadotropin-releasing hormone agonist to increase the precision of estrus in cattle: a review. *J. Anim. Sci.* 73:3141.

Zimbelman, R. G., J. W. Lauderdale, J. H. Sokolowski, and T. G. Schalk. 1970. Safety and pharmacologic evaluations of melengestrol acetate in cattle and other animals: a review. *J. Am. Vet. Med. Assoc.* 157:1528.

8 Management of Follicular Growth with Progesterone and Estradiol within Progestin-Based Estrous Synchrony Systems

M. L. Day and C. R. Burke

CONTENTS

HISTORICAL PERSPECTIVE ON THE USE OF PROGESTINS FOR ESTROUS SYNCHRONIZATION

Original attempts to synchronize estrus in cattle were based on the knowledge that progesterone prevents the occurrence of estrus and ovulation (Christian and Casida, 1948). Various forms and methods of administration of progestins were subsequently tested and generally shown to be effective at synchronizing estrus (Trimberger and Hansel, 1955; Hansel et al., 1961; Roche, 1974). Relatively long durations of progestin

0-8493-1117-9/02/$0.00+$1.50

treatment (greater than 14 days) were necessary to allow for the spontaneous occurrence of luteolysis before treatment withdrawal (Roche, 1979). These treatment regimes gained the reputation as producing a highly synchronized, but "infertile" estrus (Trimberger and Hansel, 1955; Wishart, 1977).

Attempts were made to shorten the duration of progestin treatment by using pharmacological doses of estradiol to induce luteal regression rather than relying on the spontaneous occurrence of luteolysis (Wiltbank and Kasson, 1968; Mauleon, 1974). Synchrony and fertility tended to be variable because responses to estradiol were different between the various stages of the estrous cycle (Lemon, 1975; Miksch et al., 1978; Pratt et al., 1991). The identification of prostaglandin $F_{2\alpha}$ ($PGF_{2\alpha}$) as the natural luteolysin (McCracken et al., 1972) led to the development of synthetic forms of $PGF_{2\alpha}$ that were substantially more effective than estradiol at inducing luteolysis during diestrus (Lauderdale, 1975). However, estrous synchrony and fertility responses continued to be variable with progestins and $PGF_{2\alpha}$ (Wishart, 1974; Thimonier et al., 1975). The precision of synchrony of estrus remained vulnerable to the stage of the estrous cycle at which luteal regression was induced (King et al., 1982; Stevenson et al., 1984). Furthermore, it was noted that conception rates to the synchronized estrus were reduced when short-term progestin treatments, combined with $PGF_{2\alpha}$, were initiated in the last third of the estrous cycle (Beal et al., 1988; Brink and Kiracofe, 1988).

At this point, the capacity to prevent estrus until a desired time with progestins, and to predictably induce regression of the corpus luteum with $PGF_{2\alpha}$ had been achieved. However, the efficacy of synchrony programs using this approach was limited. Although it had been known for some time that in cattle ovarian follicular development was not a random event (Rajakoski, 1960), its relevancy to synchrony of estrus was not realized, and the means to control follicular development was not a deliberate component of synchrony systems. It was not until the technique of ovarian ultrasonography was applied to monitor daily changes throughout the estrous cycle that the wave-like pattern of follicle growth was characterized in detail (Savio et al., 1988; Sirois and Fortune, 1988; Ginther et al., 1989).

The problems associated with the use of progestins to synchronize estrus became apparent shortly after the characterization of the normal pattern of follicular growth that occurs during the estrous cycle. The wave-like pattern of follicular development was interrupted by administration of progestins in the absence of a corpus luteum, resulting in prolonged life span of the dominant follicle present at the time of luteal regression (Sirois and Fortune, 1990). It has since been demonstrated that low concentrations of progesterone or administration of progestins at doses typically used to synchronize estrus in cattle results in a greater frequency of luteinizing hormone (LH) pulses than when a corpus luteum is present (for review, see Kinder et al., 1996). The greater LH secretion is associated with elevated concentrations of estradiol (Roberson et al., 1989), extended life span of the dominant ovarian follicle, and development of abnormally large, "persistent" follicles (Savio et al., 1993). Termination of progestin treatment permits occurrence of the LH surge (Kesner et al., 1982) and ovulation of persistent follicles (Stock and Fortune, 1993). While the timing of estrus is typically very precise following removal of a progestin in cattle with persistent follicles, fertility at the resultant estrus is reduced (Sanchez et al., 1993).

It has been demonstrated that, if the life span of a dominant follicle is prolonged 4 days beyond normal, fertility is reduced (Mihm et al., 1994).

Fertilization rate of oocytes released from persistent follicles does not appear to be affected, but the resulting embryo dies before it reaches the 16-cell stage (Ahmad et al., 1995). As a result of this inherent characteristic associated with the use of progestins, the "compromising" situation existed that acceptable conception rates (i.e., >60%) could not be achieved even though the use of progestins could produce a highly precise estrus (i.e., >85% in 24 hours).

This primary disadvantage to the use of progestins for estrous synchrony in cattle is balanced against the many benefits that progestins provide for estrous control. Advantages to the use of progestins include: induction of onset of estrous cycles in prepubertal heifers and anestrous, postpartum cows; reduction of the incidence of short-lived corpora lutea associated with the first ovulation in anestrous cattle; increased number of cattle responding with luteal regression in response to a single injection of $PGF_{2\alpha}$; and suppression of estrus in cattle administered exogenous estradiol or subjected to superovulation for the purposes of embryo transfer (see Day, 1998). As a result of the numerous positive aspects of the use of progestins, considerable research has been performed to develop technologies that garner the advantages provided by progestins, while avoiding the negative aspects of altered follicular growth associated with their use. The underlying objective of most of these investigations has been to develop methods to regulate follicle growth and synchronize estrus to avoid ovulation of persistent follicles.

In our program, we have addressed this problem from two fundamentally different perspectives. One approach we have taken is to accept that persistent follicles will form when progestins are used, and to manipulate this anomaly to assist in the development of processes to regulate follicular growth and control estrus. This concept is based on the use of progesterone to induce atresia of persistent dominant follicles. In the second approach, we have investigated methods to altogether prevent the development of persistent follicles as the fundamental process to regulate follicular growth and control estrus. Work in this area has centered on the use of estradiol to induce atresia of dominant follicles at the initiation of a progestin treatment. The remainder of this chapter will address our work and that of others as it relates to these approaches to regulation of follicular development within estrous control systems.

SYNCHRONIZATION OF ESTRUS THROUGH SYNCHRONOUS ATRESIA OF PERSISTENT FOLLICLES WITH EXOGENOUS PROGESTERONE

Following the discovery that persistent follicles form when progestins are administered (in the absence of a corpus luteum in the ovary) at doses typically used to synchronize estrus, it was reported that atresia of this persistent follicle could be induced by administration of a second progestin-releasing device (intravaginal progesterone insert, Stock and Fortune, 1993; subcutaneous norgestomet implant, Savio et al., 1993). These findings indicated that persistent follicles were acutely responsive to increased concentrations of a progestin, leading to atresia of the persistent follicle and emergence of a new wave of follicular development in the next few days

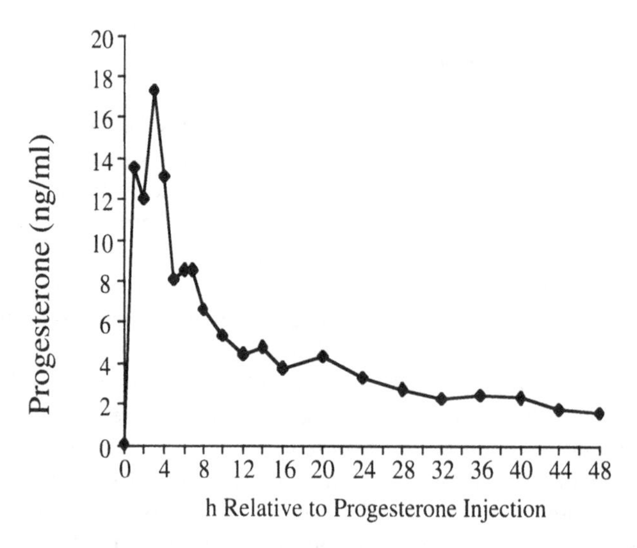

FIGURE 8.1 Average concentrations of progesterone in heifers (n = 4) receiving an intramuscular injection containing 200 mg progesterone (20 mg/ml in sesame oil) at 0 hours. (McDowell and Day, unpublished.)

(follicle turnover). Based upon these findings, we performed a series of experiments to investigate the process of programmed persistent follicle turnover with progesterone, the requirements to predictably achieve this endpoint, and the application of these findings to the development of estrous control systems.

An initial objective was to determine if atresia of persistent follicles could be induced with a short-term increase in progesterone resulting from an injection. We had previously determined that an intramuscular injection containing 200 mg progesterone would increase peripheral progesterone to concentrations greater than 3 ng/ml (mid-luteal phase concentrations) for more than 24 hours (Figure 8.1; McDowell and Day, unpublished). This dosage was used in a series of subsequent experiments.

In the first study, heifers that were provided a low level of a synthetic progestin in the absence of a corpus luteum and had developed persistent ovarian follicles were injected with 200 mg progesterone (Anderson and Day, 1994). Injection of progesterone induced atresia of the persistent dominant follicle in all heifers (n = 6) and a relatively synchronous emergence of a new wave of follicular development 3.5 ± .3 days later. Findings from this experiment indicated that coordinated growth of follicular waves among animals could be achieved through synchronously inducing atresia of persistent follicles with progesterone. Implications of this initial study, in regard to estrous synchronization, were that this approach could provide a method to manipulate follicular development to achieve a precise synchrony and avoid the depression in fertility associated with the ovulation of persistent follicles.

A subsequent experiment was performed to further define the requirements of progesterone treatment to induce turnover of persistent follicles and the endocrine basis for this action (McDowell et al., 1998a). In this experiment, heifers with persistent follicles were administered two intravaginal progesterone-releasing inserts for either 6 or 24 hours. The objective of this experiment was to determine the

TABLE 8.1
Frequency of LH Pulses[a]

	Period[b]					
	0	1	2	3	4	5
Frequency						
Control[c]	5.0[h]	6.2[h]	6.4[h]	6.4[h]	6.0[h]	5.6[h]
6-hours[d]	4.8[h]	3.0[i]	4.2[i]	5.8[h]	5.8	
24-hours[e]	5.0[h]	3.8[j]	2.6[j]	3.0[j]	3.2[j]	3.8[j]

[a] Number of LH pulses per 6 hours (frequency; treatment × hours; P < .05; SEM = 4).
[b] Periods of 6 hours relative to initiation of progesterone treatments in heifers treated with PRID. (Periods 0 to 5 = hours –6 to 0, 0 to 6, 6 to 12, 12 to 18, 18 to 24, and 24 to 30, respectively).
[cde] Heifers received either two progesterone-releasing intravaginal devices (PRID) for 6-hours (6-hours), two PRID for 24 hours (24-hours) or no PRID (control).
[hij] Means lacking a common superscript within a 6-hour period differ (P < .05).

Data from McDowell et al., 1998a.

duration of treatment with luteal-phase concentrations of progesterone necessary to induce turnover of persistent follicles and the impact of these treatments on LH and estradiol secretion. In this experiment, treatment with progesterone for 6 hours resulted in follicle wave turnover in 1 of 5 heifers, whereas treatment with progesterone for 24 hours induced follicle wave turnover in 4 of 5 heifers. An immediate decrease in frequency of LH pulses was detected in both the 6-hour and 24-hour treatments during Period 1 (Table 8.1). This suppression was transient for the 6-hour treatment, whereas suppression of LH secretion continued throughout the 30-hour period following initiation of treatments in the 24-hour group. Similar responses were noted for amplitude of LH pulses, mean LH concentration, and concentrations of estradiol. This experiment suggested that a period of at least 24 hours of elevated progesterone was essential to predictably induce atresia of persistent follicles and that the action of progesterone was elicited through suppression of secretion of LH. Further, this experiment confirmed that the dosage of progesterone injected in previous experiments (Anderson and Day, 1994) was providing an appropriate level of progesterone to induce turnover of persistent follicles.

The injection of 200 mg progesterone to synchronously induce turnover of persistent ovarian follicles was incorporated into a series of experiments to determine the utility of this approach for synchronization of estrus. In the first experiment, postpartum cows and heifers were fed the oral progestin, MGA, for 14 days and injected with 200 mg progesterone on the 12th day of MGA feeding (Anderson and Day, 1994). In females that did not have a corpus luteum on the 12th day of MGA feeding (i.e., those anticipated to have persistent follicles), conception rate at the subsequent, synchronized estrus was increased in response to injection of progesterone in both heifers and postpartum cows (Table 8.2). In this preliminary experiment, progesterone presumably induced turnover of persistent follicles in these females, resulting in enhanced fertility at the synchronized estrus. In females included in this study

TABLE 8.2
Conception Rate of Nonluteal Heifers and Postpartum Cows Fed MGA for 14 Days

		Reproductive Variable	
Treatment	**No.**	**Conception Rate[a] (%)**	**Interval to Breeding[b] (days)**
Heifers			
Control	12	16.7[c]	5.2±4
Progesterone	16	50.0[d]	6.3±.4
Postpartum cows			
Control	12	25.0[c]	3.7±.1
Progesterone	12	66.7[d]	4.6±.3

[a] Conception rates to a synchronized estrus of heifers or postpartum cows fed MGA for 14 days and administered either no further treatment (control) or 200 mg progesterone on the 12th day of MGA feeding (progesterone). All females had concentrations of progesterone <1 ng/ml on the 12th day of MGA feeding.
[b] Number of days from the last day of MGA feeding to insemination for females detected in estrus.
[c,d] Within a column, means within a classification lacking a common superscript differ ($P < .05$).

Data from Anderson and Day, 1994.

that had luteal levels of progesterone on the 12th day of MGA feeding, persistent follicles would not be expected to be present. Accordingly, injection of progesterone had no influence on fertility at the synchronized estrus.

Another study was performed to address the question of whether conception rates following programmed, persistent follicle turnover with progesterone were equivalent to those at a spontaneous estrus in postpartum and nonlactating cows (Anderson and Day, 1998). In two experiments, cyclic, nonlactating ($n = 30$) and a combination of cyclic and anestrous postpartum ($n = 113$) cows were either fed MGA for 20 days, injected with 200 mg progesterone on the 18th day of MGA and inseminated at the synchronized estrus following cessation of MGA feeding (progesterone) or inseminated at a spontaneous estrus (control). Conception rates in both nonlactating (85 and 87%) and postpartum cows (69 and 68%) did not differ between the control and progesterone treatments, respectively. In anestrous postpartum cows, estrus was synchronized in 43% and ovarian activity initiated in 70% with the progesterone treatment. It was concluded from these experiments that conception rates following synchronization of estrus with this approach is normal, and that these systems have the potential to synchronize estrus in some anestrous females.

A next step was to compare an estrous synchronization program utilizing two injections of $PGF_{2\alpha}$ given at 14-day intervals ($2 \times PGF_{2\alpha}$) to a system of synchronization utilizing the concept of coordinated turnover of persistent follicles with progesterone in four groups of heifers or postpartum cows (McDowell et al., 1998b).

TABLE 8.3
Synchronization, Conception, and Pregnancy Rates during the Synchronization Period in Experiments 1 and 2

Treatment[a]	No.	Synchronization Rate[b]	Conception Rate[c] (no.)	Pregnancy Rate[d]
		Experiment 1		
Cows				
MGA/P_4	67	70.1*	63.8(47)	44.8*
$2XPGF_{2\alpha}$	66	42.4	60.7(28)	25.8
Heifers				
MGA/P_4	29	96.6	57.1(28)	55.2
$2XPGF_{2\alpha}$	28	92.9	53.8(26)	50.0
		Experiment 2		
Cows				
MGA/P_4	59	79.7	57.4(47)	45.8
$2XPGF_{2\alpha}$	60	88.3	67.9(53)	60.0
Heifers				
MGA/P_4	42	69.0	55.2(29)	38.1
$2XPGF_{2\alpha}$	42	69.0	62.1(28)	42.9

[a] Cows were either fed melengestrol acetate (MGA) from day 13 to 0, injected with 200 mg progesterone (P_4) on day 6 and injected with $PGF_{2\alpha}$ on day 13 and day 0 (MGA/P_4), or injected with $PGF_{2\alpha}$ on day 13 and day 0 ($2XPGF_{2\alpha}$).
[b] Percentage of cows or heifers observed in estrus during the synchronization period (day 1 to 5).
[c] Percentage of females that conceived to an insemination during the synchronization period.
[d] Percentage of females in each treatment that conceived during the synchronization period.
* Effects of treatment, $P < .05$.

Data from McDowell et al., 1998b.

Females in the MGA/progesterone treatment were fed MGA for 14 days and given an injection of $PGF_{2\alpha}$ at both the initiation and termination of MGA feeding. Progesterone (200 mg) was injected, i.m., on the 8th day of MGA feeding. In groups of females that were cyclic at the time of synchronization of estrus, similar synchronization, conception, and pregnancy rates were achieved with each of these systems (Table 8.3). However, in the group of cows in which a majority (77.5%) were anestrous at the initiation of treatments (Experiment 1, postpartum cows), the synchronization rate, precision of estrus, and pregnancy rate during the synchrony period were enhanced with the MGA/progesterone treatment. Data from this study confirms that fertility of female cattle in which estrus is synchronized with this approach is normal. More importantly, in postpartum cows in which a majority were anestrous, this approach substantially increased the reproductive performance at the synchronized estrus. The improvement in reproductive performance in this group of females appeared to result from a combination of influences to increase the induction and synchronization of estrus in anestrous females, suppress the occurrence of spontaneous estrus during the interval between the two injections of $PGF_{2\alpha}$, prevent ovulation of persistent follicles and coordinate follicular turnover in a portion of the females treated.

In summary, with this approach to estrous control, the primary historical weakness of the use of progestins for estrous control is used as an advantage to provide a predictable means to control follicular development and the timing of estrus. The addition of an injection of estradiol or GnRH following cessation of progestin treatment to synchronize/induce ovulation, as is common to other systems of synchrony, has not been applied with this concept of estrous control. It is anticipated that this would further enhance the utility of this approach through increased opportunity for conception in anestrous beef cows. Challenges of this approach to estrous control are the length of time required to permit development of persistent follicles and the necessity to provide appropriate treatments to ensure that most females develop persistent follicles. A practical limitation that exists at present is that a source of injectable progesterone is not currently available to beef cattle producers in the United States.

SYNCHRONIZATION OF ESTRUS THROUGH SYNCHRONOUS ATRESIA AND OVULATION OF DOMINANT FOLLICLES WITH ESTRADIOL

The alternative approach we have taken to overcome the problem of development of persistent follicles when progestins are used to synchronize estrus is to manage the "normal" population of follicles that exist in the ovary. This approach is designed to wholly prevent development of persistent follicles and coordinate synchronous follicle wave turnover. An excellent tool for this purpose is the use of estradiol to reset follicular development through the induction of atresia of ovarian follicles (Bo et al., 1995) before development of persistent follicles can occur with progestin-based treatments. However, the application of estradiol to estrous control systems extends beyond the capacity to regulate the process of follicular turnover. Estradiol can also be used to synchronize the events of estrus and ovulation after progestin withdrawal. For either indication, the effects of estradiol are mediated through the gonadotropic axis, with the presence or absence of progesterone determining the type of response elicited by estradiol. We discuss each of these applications in more detail in the following sections.

USE OF ESTRADIOL TO SYNCHRONIZE OVARIAN FOLLICULAR DEVELOPMENT

One of the earliest reports suggesting that estradiol may be inducing atresia of large ovarian follicles came from a study using injections of 10 mg estradiol valerate while investigating ovarian follicular cysts in dairy cows (Engelhardt et al., 1989). The subsequent development of transrectal ultrasonography in cattle allowed researchers to conveniently monitor *in vivo* ovarian responses to estradiol. Initial studies focused on the early phase of the estrous cycle, during the development of the first dominant follicle (Bo et al., 1995). It was demonstrated that injected estradiol-17β induced atresia of the dominant follicle and emergence of a new follicle wave 4.3 ± 0.2 days later. The interval to new wave emergence was consistent regardless of the stage of growth of the first dominant follicle at the time of estradiol treatment.

The development and commercialization of progesterone-based treatments (i.e., PRID and CIDR) routinely included a gelatin capsule containing 10 mg estradiol benzoate affixed to the insert to promote luteolysis. It was later found that with the inclusion of estradiol, developers were unwittingly perturbing follicle wave development on the ovaries. This was clearly demonstrated in a field trial involving more than 1000 dairy cows (Macmillan and Burke, 1996). Intravaginal progesterone inserts (CIDR) were inserted for 7 or 12 days, with or without a gelatin capsule containing 10 mg estradiol benzoate, so that inserts were removed from 20 to 24 days after the preceding estrus. Cows were subsequently inseminated on detection of estrus. Inclusion of estradiol benzoate decreased the insemination rate at 48 hours in those treated with a CIDR for 7 days, but increased the insemination rate at 48 hours in those that were treated with a CIDR for 12 days. Fertility tended to decline as the interval between insert removal and the preceding estrus increased (i.e., from 20 to 24 days), while the inclusion of estradiol benzoate increased fertility irrespective of this interval.

This field study prompted an investigation of the underlying ovarian responses to intravaginal progesterone/estradiol treatments. The animal model involved initiating treatments 13 days after synchronized estrus in nonlactating dairy cows: the "day 13 follicle wave turnover" model. The large field trial described previously suggested that the effect of including estradiol with initiation of a progesterone treatment was most pronounced when treatment was initiated on day 13 of the estrous cycle. To this point, ovarian responses to treatments initiated in the mid to late stages of the estrous cycle had not been previously reported despite this stage being perhaps the most difficult and critical time to achieve control of follicle wave dynamics. The "day 13 follicle turnover model" became the basis of a series of experiments with the overall aim of characterizing the effects of exogenous progesterone and estradiol benzoate in the latter phases of the estrous cycle with some consideration of the interactive effects of luteolysis.

The initial study (Burke et al., 1999) included four treatments in a 2 × 2 factorial design. Treatments were initiated on day 13 of the estrous cycle and involved insertion of a CIDR insert for 5 days with or without a capsule containing 10 mg estradiol benzoate, the capsule alone, and an untreated control group. Follicle wave turnover occurred in all animals treated with estradiol benzoate and was characterized by the emergence of a new wave of follicles 4.0 ± 0.3 days after treatment initiation. Thus all cows receiving an estradiol capsule had estrous cycles containing three waves of follicular development. In contrast, the ratio of two and three wave cycles was 1:1 when estradiol was not included in the treatment regime. This study showed that 10 mg estradiol benzoate was effective in promoting follicle wave turnover of a healthy growing dominant follicle during a time of high concentrations of plasma progesterone in nonlactating cows.

However, in a field study with multiple herds of cyclic, lactating dairy cows (Xu et al., 1996), cows receiving an intravaginal, 10 mg estradiol benzoate capsule in conjunction with a CIDR insert had lower conception rates than cows inseminated at a spontaneous estrus. These results suggested that the estradiol capsule did not always prevent development of persistent follicles in lactating cows. Further, it appeared that improved fertility could be gained by administering estradiol as an injection rather than via the intravaginal route (Ryan et al., 1996). A subsequent report (O'Rourke et al., 1997) confirmed that maximal blood concentrations of estradiol following treatment with a capsule containing 10 mg estradiol benzoate were equivalent to those following intramuscular injection of 0.5 mg estradiol benzoate. Cyclic, lactating dairy cows, injected with 2 mg estradiol benzoate at the initiation of a 7-day CIDR treatment, had conception rates that were equivalent to those for cows inseminated at a spontaneous estrus (Day et al., 2000). Furthermore, conception rates for cows receiving an injection containing 1 mg estradiol benzoate tended to be lower than for cows receiving 2 mg estradiol benzoate in the same study. Collectively, these results suggest that the 10 mg capsule provided an inadequate signal for follicle turnover in lactating dairy cows but that the dosage received from this route of estradiol administration may be sufficient in nonlactating animals.

It had previously been suggested from field studies that injection of estradiol benzoate would effectively induce follicular turnover in lactating dairy cows. Injection of 1 mg estradiol benzoate to lactating dairy cows at 12, 13, or 14 days after a synchronized estrus accompanied by artificial insemination partially synchronized return-to-service 9 to 10 days later (Macmillan et al., 1997) as compared to cows not receiving estradiol benzoate. These findings suggested that most cows had three waves of follicular development as a result of estradiol-induced follicle turnover. The follicular response to this treatment was subsequently tested using the "day 13 follicle turnover model" (Burke et al., 2000). Injection of 1 mg estradiol benzoate was found to be totally effective in promoting atresia of the second dominant follicle of the estrous cycle and emergence of a third wave 4 to 5 days later in nonlactating dairy cows (Burke et al., 2000). In a more recent experiment, this model was used to compare the efficacy of estradiol benzoate doses of 0.5, 1, or 2 mg/1100 lb body weight to induce follicle wave turnover in nonlactating beef cows (Bogacz et al., 1999). In this study, both 1 and 2 mg of estradiol benzoate were highly effective in inducing follicle wave turnover using the "day 13" model. Timing of emergence of the new wave of follicles was later with 2 vs. 1 mg estradiol benzoate. Follicle wave turnover also occurred following injection of 1 mg estradiol benzoate in conjunction with insertion of a CIDR in heifers in the early, mid, and late stages of the estrous cycle (Bogacz et al., 2000).

The ability of estradiol to induce follicle wave turnover depends on elevated concentrations of progesterone in circulation. This was clearly demonstrated in a study using the "day 13 follicle wave turnover" model in which cows (n = 24) received an injection of 1 mg estradiol benzoate on day 13 (Burke et al., 1997) and then received no further treatment, or a luteolytic dose of $PGF_{2\alpha}$ either 0, 24, or 48 hours after the injection of estradiol benzoate. The $PGF_{2\alpha}$ was effective in inducing a precipitous decline in concentrations of progesterone within 24 hours of administration in all animals treated. Follicle wave turnover as indicated by incidence of three-wave cycles was induced in all animals that did not receive $PGF_{2\alpha}$ and in all but one of those receiving $PGF_{2\alpha}$ at 48 hours after estradiol benzoate. In contrast, those receiving $PGF_{2\alpha}$ at either 0 or 24 hours after estradiol benzoate had 2-, 3-, and 4-wave cycles and some prevalence of the development cystic follicles which failed to ovulate. We concluded from this study that concentrations of circulating progesterone needed to remain elevated for at least 48 hours following estradiol treatment in order for follicle wave turnover to be induced. This study raised the possibility that the chance occurrence of spontaneous luteolysis around the time that progesterone/estradiol synchrony treatments are initiated could prevent the ability of estradiol to induce follicle wave turnover.

Accordingly, a subsequent study investigated whether the addition of an exogenous progesterone source with an injection of 1 mg estradiol benzoate during induced luteolysis in the "day 13 follicle wave model" would be sufficient to facilitate estradiol-induced follicle wave turnover (Burke et al., 1998). Each of 24 lactating cows received a luteolytic dose of $PGF_{2\alpha}$ on day 13 of the estrous cycle. Animals were then allocated to receive either 1 mg estradiol benzoate, a 6-day CIDR insert, or a 6-day CIDR insert and an injection of 1 or 2 mg estradiol benzoate. All cows not receiving the combination of estradiol benzoate and a CIDR insert had estrous cycles with two waves of follicular development. In contrast, all cows receiving the combination of either 1 or 2 mg estradiol benzoate by injection and a CIDR insert had estrous cycles with three waves of follicular development. We concluded that progesterone, provided exogenously via a CIDR insert, was sufficient to facilitate an estradiol-induced turnover of follicle waves even with simultaneous induction of luteolysis. A follow-up study found that an injection of 200 mg progesterone, rather than insertion of the CIDR in the design described above, was not sufficient to facilitate estradiol-induced follicle turnover (C.R. Burke, unpublished).

All evidence to date suggests that the mechanisms by which estradiol induces follicle wave turnover is mediated through systemic mechanisms involving the gonadotropic axis. In ovariectomized cows, treatment with progesterone provided an effective, but transient inhibition of LH secretion (i.e., 36 hours) and had no measurable effect on follicle-stimulating hormone (FSH) secretion (Burke et al., 1996). The progesterone-induced suppression of LH was extended several days by addition of estradiol benzoate at the initiation of the CIDR treatment, and FSH secretion was also depressed. Follicular development is initially supported by FSH, while a greater dependency on LH is acquired as dominant follicles mature (Findlay et al., 1996; Gong et al., 1996). Because progesterone suppresses LH secretion, but not FSH secretion directly, newly developed FSH-dependent follicles may be less affected by progesterone in the absence of estradiol. In contrast, effective suppression of both FSH and LH secretion by administration of progesterone and estradiol in combination probably accounts for induction of atresia regardless of age or diameter of the dominant follicle. A recent study has shown that estrogenic function of the dominant follicle is profoundly reduced within 36 h following administration of 1 mg estradiol benzoate/500 kg body weight (Burke et al., 2001a), although it remains unclear whether such follicles have undergone apoptosis and are truly atretic at this time interval. It seems unlikely that estradiol would have a direct atretogenic effect on ovarian follicles since follicular concentrations of estradiol in non-atretic follicles are typically many times greater than in peripheral circulation (Ginther et al., 1997). Further, direct placement of estradiol into the ovarian stroma adjacent to the dominant follicle failed to induce follicle wave turnover (Bo et al., 1996).

In summary, a series of progressive experiments have shown that estradiol is profoundly capable of controlling ovarian follicular development. However, this atretogenic action is critically dependent on a synergistic relationship with circulating progesterone and is probably mediated through removal of gonadotropic support for the follicle. The timing and variability in timing of new follicle wave emergence may influence the degree to which onset of estrus is synchronized in a group of female cattle. The length of the progestin treatment must account for the timing of follicle wave turnover, and the subsequent period required for the new dominant follicle to attain the capacity for ovulation.

USE OF ESTRADIOL TO SYNCHRONIZE ESTRUS AND OVULATION

A detectable display of estrous behavior in cattle is requisite in breeding programs that use artificial insemination on the basis of detected estrus. Estrous detection efficiency varies widely between production systems and is a major limitation to reproductive competence in many classes of female cattle (Senger, 1994; Stevenson et al., 1995; Anderson and Day, 1998; Xu et al., 1998). Several factors may contribute to these variations, including the reproductive and metabolic status of the animal, the competency of those responsible and the system for detecting estrus, the environment in which the animal is managed, and the responsiveness or opportunity of other animals to interact with an animal in estrus (Helmer and Britt, 1985).

The use of exogenous estradiol after termination of progesterone treatment was developed as a treatment regime for anestrous dairy cows to facilitate the expression of estrus (McDougall et al., 1992). Estrous response following several days of progesterone exposure by insertion of an intravaginal progesterone insert was increased from 70% to over 90% by the addition of an injection of 1 mg estradiol benzoate 24 hours after progesterone withdrawal. Much of the improvement was attributed to the elimination of "silent ovulations" (i.e., ovulation unaccompanied by a detected estrus), since conception rates were not altered even though more animals were inseminated. Thus, the overall improvement due to the addition of estradiol was considered largely due to an increase in animals being detected in estrus and submitted for artificial insemination (Macmillan et al., 1995; Day et al., 2000). Similar improvements in estrous behavior have resulted in the application of estradiol to anestrous beef cows (Fike et al., 1997) and beef heifers (Rasby et al., 1998) previously treated with a CIDR insert.

The rationale for the use of estradiol benzoate after progesterone withdrawal described above for treatment of anestrous cows has been extended to produce a more precisely synchronized estrus in cycling cows (Day et al., 2000). In especially high-producing dairy cows, the use of estradiol after progesterone treatment is beneficial in alleviating the problem of poor estrous expression that is characteristic of animals under high metabolic demand (K.L. Macmillan, personal communication).

An issue associated with the use of estradiol for facilitating an expressive estrus after progesterone treatment has been the possibility that this estrus is not accompanied by a timely ovulation. A recent study involving anestrous dairy cows in New Zealand reported the incidence of "false heats" to be 21.4%, and highly variable between herds (5.4 to 46.2%; F.M. Rhodes, unpublished). One possible reason for a false heat may be that follicular wave development is inadequately controlled during the progesterone treatment because of the presence of an immature and unresponsive dominant follicle at the time estradiol is administered. A recent series of experiments evaluated the effects of estradiol benzoate after progesterone withdrawal with respect to the age of maturity of the "preovulatory" dominant follicle in prepubertal heifers, cycling heifers, and anestrous postpartum beef cows (Burke et al., 2001b). Behavioral estrus was readily induced by estradiol benzoate irrespective of animal type or maturity of the potential ovulatory dominant follicle. In heifers, the induced estrus was accompanied with timely ovulation in most heifers, even when the injection of estradiol benzoate was administered when the dominant follicle had emerged just 1 to 2 days previously. In contrast, recently emerged dominant follicles in anestrous cows were less likely to respond to the ovulatory cue provided by estradiol benzoate, while luteal size and function were compromised in those that did ovulate. The results imply that hormonal regimes for treating anestrous cattle should also include a component that regulates follicular development to ensure the ultimate presence of a mature and responsive ovulatory follicle.

SUMMARY

It is apparent that regulation of the wave-like pattern of follicular development is, or will be, a requirement for most approaches to estrous synchronization in cattle. We have described the investigation and application of two fundamentally different methods to achieve this endpoint. With each method, ovulation of persistent dominant follicles can be prevented, and the timing of emergence of a new wave of follicular development can be coordinated. The achievement of each of these endpoints is critical to the success of estrous control systems in cattle. Prevention of the ovulation of persistent follicles is critical for the attainment of normal conception rates at a synchronized estrus. To apply treatments that induce and/or synchronize ovulation and estrus for the purposes of timed-artificial insemination, it is essential that the wave-like pattern of follicular development be coordinated across all animals.

REFERENCES

Ahmad, N., F.N. Schrick, R.L. Butcher, and E.K. Inskeep. 1995. Effect of persistent follicles on early embryonic losses in beef cows. *Biol. Reprod.* 52:1129.

Anderson, L.H. and M.L. Day. 1994. Acute progesterone administration regresses persistent dominant follicles and improves fertility of cattle in which estrus was synchronized with melengestrol acetate. *J. Anim. Sci.* 72:2955.

Anderson, L.H. and M.L. Day. 1998. Development of a progestin-based estrus synchronization program: I. Reproductive response of cows fed melengestrol acetate for 20 days with an injection of progesterone. *J. Anim. Sci.* 76:1267.

Beal, W.E., J.R. Chenault, M.L. Day, and L.R. Corah. 1988. Variation in conception rates following synchronization of estrus with melengestrol acetate and prostaglandin $F_{2\alpha}$. *J. Anim. Sci.* 66:599.

Bo, G.A., G.P. Adams, M. Caccia, M. Martinez, R.A. Pierson, and R.J. Mapletoft. 1995. Ovarian follicular wave emergence after treatment with progestogen and estradiol in cattle. *Anim. Reprod. Sci.* 39:193.

Bo, G.A., D.R. Bergfelt, G.M. Brogliatti, R.A. Pierson, and R.J. Mapletoft. 1996. Systemic versus local effects of exogenous estradiol on follicular development in heifers. *Theriogenology* 45:333.

Bogacz, V.L., J.E. Huston, D.E. Grum, and M.L. Day. 1999. Identification of the optimal dose of estradiol benzoate in combination with a progestin to program follicular turnover in cyclic cattle. *J. Anim. Sci.* 77 (Suppl. 1):124.

Bogacz, V.L., J.E. Huston, D.E. Grum, and M.L. Day. 2000. Effect of estradiol benzoate in combination with progesterone to induce follicular turnover at various stages of the estrous cycle. *J. Anim. Sci.* 78 (Suppl. 1.):211.

Brink, J.T. and G.H. Kiracofe. 1988. Effect of estrous cycle stage at Syncro-Mate-B treatment on conception and time to estrus in cattle. *Theriogenology* 29:513.

Burke, C.R., K.L. Macmillan, and M.P. Boland. 1996. Oestradiol potentiates a prolonged progesterone-induced suppression of LH release in ovariectomised cows. *Anim. Reprod. Sci.* 45:13.

Burke, C.R., M.L. Day, B.A. Clark, C.R. Bunt, M.J. Rathbone, and K.L. Macmillan. 1997. Effect of luteolysis on follicle wave control using oestradiol benzoate in cattle. *Proc. N. Z. Soc. Endocrinol.* 40:134.

Burke, C.R., S. Morgan, B.A. Clark, and F.M. Rhodes. 1998. Effect of luteolysis on control of ovarian follicles using oestradiol benzoate and progesterone in cattle. *Proc. N. Z. Soc. Anim. Prod.* 58:89.

Burke, C.R., M.P. Boland, and K.L. Macmillan. 1999. Ovarian responses to progesterone and oestradiol benzoate administered intravaginally during dioestrus in cattle. *Anim. Reprod. Sci.* 55:23.

Burke, C.R., M.L. Day, C.R. Bunt, and K.L. Macmillan. 2000. Use of a low dose of estradiol benzoate during diestrus to synchronize development of the ovulatory follicle in cattle. *J. Anim. Sci.* 78:145.

Burke, C.R., S. Morgan, M.L. Mussard, D.E. Grum, and M.L. Day. 2001a. Estradiol benzoate (EB) inhibits secretion of LH and induces atresia of dominant follicles within 36 hours in cyclic heifers. *J. Anim. Sci.* 79:137.

Burke, C.R., M.L. Mussard, D.E. Grum, and M.L. Day. 2001b. Effects of maturity of the potential ovulatory follicle on induction of oestrus and ovulation in cattle with oestradiol benzoate. *Anim. Reprod. Sci.* 66:161.

Christian, R.E. and L.E. Casida. 1948. The effect of progesterone in altering the estrual cycle of the cow. *J. Anim. Sci.* 7:540 (Abstr.).

Day, M.L. 1998. Estrous control and management of follicular growth with progesterone-based synchrony systems. *Proc. Seventeenth Ann. Conv. Am. Embryo Trans. Assoc.* pp. 10. October 16, 1998, San Antonio TX.

Day, M.L., C.R. Burke, V.K. Taufa, A.M. Day, and K.L. Macmillan. 2000. The strategic use of estradiol to enhance fertility and submission rates of progestin-based estrus synchronization programs in seasonal dairy herds. *J. Anim. Sci.* 78:523.

Englehardt, H., J.S. Walton, R.B. Miller, and G.J. King. 1989. Estradiol-induced blockade of ovulation in the cow: effects of luteinizing hormone release and follicular fluid steroids. *Biol. Reprod.* 40:12877.

Fike, K.E., M.L. Day, E.K. Inskeep, J.E. Kinder, P.E. Lewis, R.E. Short, and H.D. Hafs. 1997. Estrus and luteal function in suckled beef cows that were anestrous when treated with an intravaginal device containing progesterone with or without a subsequent injection of estradiol benzoate. *J. Anim. Sci.* 75:2009.

Findlay, J.K., A.E. Drummond, and R.C. Fry. 1996. Intragonadal regulation of follicular development and ovulation. *Anim. Reprod. Sci.* 42:321.

Ginther, O.J., L. Knopf, and J.P. Kastelic. 1989. Temporal associations among ovarian events in cattle during oestrous cycles with two and three follicular waves. *J. Reprod. Fertil.* 87:223.

Ginther, O.J., K. Kot, L.J. Kulick, and M.C. Wiltbank. 1997. Sampling follicular fluid without altering follicular status in cattle: oestradiol concentrations early in the follicular wave. *J. Reprod. Fertil.* 109:181.

Gong, J.G., B.K. Campbell, T.A. Bramley, C.G. Gutierrez, A.R. Peters, and R. Webb. 1996. Suppression in the secretion of follicle-stimulating hormone and luteinizing hormone, and ovarian follicular development in heifers continuously infused with a gonadotropin-releasing hormone agonist. *Biol. Reprod.* 55:68.

Hansel, W., P.V. Malven, and D.L. Black. 1961. Estrous cycle regulation in the bovine. *J. Anim. Sci.* 20:621.

Helmer, S.D. and J.H. Britt. 1985. Mounting behavior as affected by stage of estrous cycle in Holstein heifers. *J. Dairy Sci.* 68:1290.

Kesner, J.S., V. Padmnabhan, and E.M. Convey. 1982. Estradiol induces and progesterone inhibits the preovulatory surges of luteinizing hormone and follicle-stimulating hormone in heifers. *Biol. Reprod.* 26:271.

Kinder, J.E., F.N. Kojima, E.G.M. Bergfeld, M.E. Wehrman, and K.E. Fike. 1996. Progestin and estrogen regulation of pulsatile LH release and development of persistent ovarian follicles in cattle. *J. Anim. Sci.* 74:1424.

King, M.E., G.H. Kiracofe, J.S. Stevenson, and R.R. Schalles. 1982. Effect of stage of the estrous cycle on interval to estrus after $PGF_{2\alpha}$? in beef cattle. *Theriogenology* 18:191.

Lauderdale, J.W. 1975. The use of prostaglandin in cattle. *Ann. Biol. Anim. Bioch. Biophys.* 15:419.

Lemon, M. 1975. The effect of oestrogens alone or in association with progestagens on the formation and regression of the corpus luteum of the cyclic cow. *Ann. Biol. Anim. Bioch. Biophys.* 15:243.

Macmillan, K.L. and C.R. Burke. 1996. Effects of oestrous cycle control on reproductive efficiency. *Anim. Reprod. Sci.* 42:307.

Macmillan, K.L., V.K. Taufa, and A.M. Day. 1997. Manipulating ovaries' follicle wave patterns can partially synchronize returns to service and increases the pregnancy rate to second insemination. *Proc. N. Z. Soc. Anim. Prod.* 57:237.

Macmillan, K.L., V.K. Taufa, A.M. Day, and S. McDougall. 1995. Some effects of using progesterone and oestradiol benzoate to stimulate oestrus and ovulation in dairy cows with anovulatory anoestrus. *Proc. Aust. Soc. Reprod. Biol.* 26:74.

Mauleon, P. 1974. New trends in the control of reproduction in the bovine. *Livestock Prod. Sci.* 1:117.

McCracken, J.A., J.C. Carlson, M.E. Glew, J.R. Goding, D.T. Baird, K. Green, and B. Samuelsson. 1972. Prostaglandin $F_{2\alpha}$ identified as a luteolytic hormone in sheep. *Nature New Biol.* 238:129.

McDougall, S., C.R. Burke, K.L. Macmillan, and N.B. Williamson. 1992. The effect of pretreatment with progesterone on the oestrous response to oestradiol-17β benzoate in the post-partum dairy cow. *Proc. N. Z. Soc. Anim. Prod.* 52:157.

McDowell, C.M., L.H. Anderson, J.E. Kinder, and M.L. Day. 1998a. Duration of treatment with progesterone and regression of persistent ovarian follicles in cattle. *J. Anim. Sci.* 76:850.

McDowell, C.M., L.H. Anderson, R.P. Lemenager, D.A. Mangione, and M.L. Day. 1998b. Development of a progestin-based estrus synchronization program: II. Reproductive response of cows fed melengestrol acetate for 14 days with injections of progesterone and prostaglandin $F_{2\alpha}$. *J. Anim. Sci.* 76:1273.

Mihm, M., A. Baguisi, M.P. Boland, and J.R. Roche. 1994. Association between the duration of dominance of the ovulatory follicle and pregnancy rate in beef heifers. *J. Reprod. Fertil.* 102:123.

Miksch, E.D., D.G. LeFever, G. Mukembo, J.C. Spitzer, and J.N. Wiltbank. 1978. Synchronization of estrus in beef cattle. II. Effect of an injection of norgestomet and an estrogen in conjunction with a norgestomet implant in heifers and cows. *Theriogenology* 10:201.

O'Rourke, M., M.G. Diskin, J.M. Shreenan, and J.F. Roche. 1997. The effect of dose and method of oestradiol administration on plasma concentrations of E_2 and FSH in long-term ovariectomized beef heifers. *J. Reprod. Fertil. Abstract Series* 19:56.

Pratt, S.L., J.C. Spitzer, G.L. Burns, and B.B. Plyler. 1991. Luteal function, estrous response, and pregnancy rate after treatment with norgestomet and various dosages of estradiol valerate in suckled cows. *J. Anim. Sci.* 69:2721.

Rajakoski, E. 1960. The ovarian follicular system in sexually mature heifers with special reference to seasonal, cyclical, and left-right variations. *Acta Endocrinologica,* (Suppl. 52):4.

Rasby, R.J., M.L. Day, S.K. Johnson, J.E. Kinder, J.M. Lynch, R.E. Short, R.P. Wettemann, and H.D. Hafs. 1998. Luteal function and estrus in peripubertal beef heifers given an intravaginal progesterone releasing insert with or without a subsequent injection of estradiol. *Theriogenology* 50:553.

Roberson, M.S., M.W. Wolfe, T.T. Stumpf, R.J. Kittok, and J.E. Kinder. 1989. Luteinizing hormone secretion and corpus luteum function in cows receiving two levels of progesterone. *Biol. Reprod.* 41:997.

Roche, J.F. 1974. Synchronization of oestrus in heifers with implants of progesterone. *J. Reprod. Fertil.* 41:337.

Roche, J.F. 1979. Control of oestrus in cattle. *World Rev. Anim. Prod.* 15:49.

Ryan, D.P., J.A. Galvin, and K.J. O'Farrell. 1996. Effect of various oestrous synchronization regimens applied to lactating dairy cows on oestrus synchrony and pregnancy rates. In: *Proc. Thirteenth Int. Congr. Anim. Reprod.,* Sydney, Australia. 3:1910.

Sanchez, T., M.E. Wehrman, E.G. Bergfeld, K.E. Peters, F.N. Kojima, A.S. Cupp, V. Mariscal, R.J. Kittok, R.J. Rasby, and J.E. Kinder. 1993. Pregnancy rate is greater when the corpus luteum is present during the period of progestin treatment to synchronize time of estrus in cows and heifers. *Biol. Reprod.* 49:1102.

Savio, J.D., L. Keenan, and M.P. Boland. 1988. Pattern of growth of dominant follicles during the oestrous cycle of heifers. *J. Reprod. Fertil.* 83:663.

Savio, J.D., W.W. Thatcher, G.R. Morris, K. Entwistle, M. Drost, and M.R. Mattiacci. 1993. Effects of induction of low plasma progesterone concentrations with a progesterone-releasing device on follicular turnover and fertility in cattle. *J. Reprod. Fertil.* 98:77.

Senger, P.L. 1994. The estrus detection problem: new concepts, technologies, and possibilities. *J. Dairy Sci.* 77:2745.

Sirois, J. and J.E. Fortune. 1988. Ovarian follicular dynamics during the estrous cycle in heifers monitored by real-time ultrasonography. *Biol. Reprod.* 39:308.

Sirois, J. and J.E. Fortune. 1990. Lengthening the bovine estrous cycle with low levels of exogenous progesterone: a model for studying ovarian follicular dominance. *Endocrinology* 127:916.

Stevenson, J.S., M.K. Schmidt, and E.P. Call. 1984. Stage of estrous cycle, time of insemination, and seasonal effects on estrus and fertility of Holstein heifers after prostaglandin $F_{2\alpha}$. *J. Dairy Sci.* 67:1798.

Stevenson, J.S., M.W. Smith, J.R. Jaeger, L.R. Corah, and D.G. LeFever. 1995. Detection of estrus by visual observation and radiotelemetry in peripubertal, estrus-synchronized beef heifers. *J. Anim. Sci.* 74:729.

Stock, A.E. and J.E. Fortune. 1993. Ovarian follicular dominance in cattle: relationship between prolonged growth of the ovulatory follicle and endocrine parameters. *Endocrinology* 132:1108.

Thimonier, J., D. Chupin, and J. Pelot. 1975. Synchronization of oestrus in heifers and cyclic cows with progestagens and prostaglandin analogues alone or in combination. *Ann. Bio. Anim. Biochem. Biophys.* 15:437.

Trimberger, G.W. and W. Hansel. 1955. Conception rate and ovarian function following estrus control by progesterone injections in dairy cattle. *J. Anim. Sci.* 14:224.

Wiltbank, M.C. and C.W. Kasson. 1968. Synchronization of estrus in cattle with an oral progestational agent and an injection of an oestrogen. *J. Anim. Sci.* 27:113.

Wishart, D.F. 1974. Synchronization of oestrus in cattle using a progestogen (SC21009) and $PGF_{2\alpha}$. *Theriogenology* 1:87.

Wishart, D.F. 1977. Synchronization of oestrus in heifers using steroid (SC 5914, SC 9880 and SC 21009) treatment for 21 days: the effect of treatment on the ovum collection and fertilization rate and the development of the early embryo. *Theriogenology* 8:249.

Xu, Z.Z., L.J. Burton, and K.L. Macmillan. 1996. Reproductive performance of lactating dairy cows following oestrus synchronization with progesterone, oestradiol and prostaglandin. *N. Z. Vet. J.* 44:99.

Xu, Z.Z., D.J. McKnight, R. Vishwanath, C.J. Pitt, and L.J. Burton. 1998. Estrus detection using radiotelemetry or visual observation and tail painting for dairy cows on pasture. *J. Dairy Sci.* 81:2890.

9 The Freezing, Thawing, and Transfer of Cattle Embryos

John F. Hasler

CONTENTS

The development of successful techniques for the freezing and thawing of cattle embryos had a profound impact on the field of embryo transfer (ET), which, in turn, has had a significant influence on the cattle industry. In a 1998 worldwide census of the ET industry, it was shown that 50% of cattle embryos were frozen for transfer at a later date. In the North American survey, 48% of the approximately 200,000 embryos were frozen after recovery from donors. At Em Tran, Inc., a dramatic and increasing demand for freezing embryos during the 1990s culminated with over 90% of the embryos having been frozen in 1999. Although not all cattle breed associations keep records of the animals produced by ET, the records of some associations indicate that ET is being widely used by breeders of genetically superior cattle. This is evidenced by the fact that 86% of the top 100 Holstein bulls (as evaluated by Holstein Type-Production Index in 2000) were produced by ET methodology.

Prior to the development of reliable freezing techniques, movement of bovine embryos from one country to another was a difficult proposition. Lawson et al. (1972) showed that cattle embryos could be cultured for several days in the oviducts of follicular-phase rabbits and then successfully transferred back into recipient cattle.

0-8493-1117-9/02/$0.00+$1.50

This technique was used on a small scale for moving cattle embryos between countries during the mid to late 1970s. In the late 1970s, I transported cattle embryos several times between the United States and Hungary in my pocket. These embryos were maintained in liquid media for 30 to 36 hours prior to transfer, and the resulting pregnancy rates were low.

The first published report of success in freezing and thawing mammalian embryos for the production of liveborn young involved the use of dimethyl sulfoxide (DMSO) or glycerol as cryoprotectants for laboratory mouse embryos (Whittingham et al., 1972). Shortly thereafter, cattle (Wilmut and Rowson, 1973) and sheep (Willadsen et al., 1974) embryos were successfully frozen and thawed with production of offspring. Following these pioneering efforts, however, a number of years elapsed before embryo freezing protocols were developed to a high level of efficiency and the technology adopted by the ET industry. During the early days of the commercial ET industry, prior to the availability of freezing technology, practitioners depended on having enough recipients available to utilize all of the embryos recovered on a given day. Obviously, this was not always possible and at Em Tran from 1978 to 1983, a yearly average of 200 freshly recovered embryos were discarded for which recipients were not available.

EMBRYO FREEZING

Preparing Embryos for Freezing

A critical component in doing a good job of freezing embryos is choosing which embryos should be frozen. Whenever possible, the embryos should be recovered between 6.5 and 7.5 days after the onset of estrus in the donor. Embryos that are either younger or older than this range of age may have a lower survival rate when subjected to the freezing and thawing processes. Embryos should be at a stage of development that is commensurate with their age. Recoveries performed between days 6.5 to 7.5 will normally yield embryos ranging from late morulae to mid-blastocysts. However, variability in either the time of ovulation among donors or in the accuracy of estrous detection may result in the collection of embryos that are either more or less developed than what is described. The grade of embryos, as described in Table 9.1, is a highly significant factor in survival following freezing/thawing. These embryo grades were established by the International Embryo Transfer Society (IETS) and have been universally adopted for use on the permanent records of all embryos frozen for international export. Many ET practitioners, however, continue to use in their own records a slightly more refined grading system in which embryos meeting the IETS grade 1 status are divided into grade 1 (excellent) and grade 2 (good) classes, with IETS grade 2 (fair) embryos being downgraded to 3. The data from several commercial ET programs show that although the pregnancy rate of grade 1 vs. 2 embryos is very similar, transfer of grade 1 embryos usually results in a significantly higher pregnancy rate: 57.8 vs. 54.3% (Em Tran, Inc., unpublished data, embryos exported to the Netherlands); 67 vs. 61% (Hasler, 1986); 68 vs. 61% (Hasler, 1989); 57.1 vs. 52.9% (Arreseigor et al., 1998); 55.9 vs. 48.4% (Otter et al., 1998). Pregnancy rates of IETS grade 1 embryos, which is the grade that

TABLE 9.1
The IETS Codes for Embryo Quality Grades

Grade Code	Description
1: Excellent or Good	Symmetrical and spherical embryo mass with individual blastomeres (cells) that are uniform in size, color, and density. This embryo is consistent with its expected stage of development. Irregularities should be relatively minor, and at least 85% of the cellular material should be an intact, viable embryonic mass. This judgment should be based on the percentage of embryonic cells represented by the extruded material in the perivitelline space. The zona pellucida should be smooth and have no concave or flat surfaces that might cause the embryo to adhere to a Petri dish or a straw.
2: Fair	Moderate irregularities in overall shape of the embryonic mass or in size, color, and density of individual cells. At least 50% of the cellular material should be an intact, viable embryonic mass.
3: Poor	Major irregularities in shape of the embryonic mass or in size, color, and density of individual cells. At least 25% of the cellular material should be an intact, viable embryonic mass.
4: Dead or degenerating	Degenerating embryos, oocytes or 1-cell embryos: nonviable.

Adapted from Stringfellow and Seidel, 1998.

should be used on all packaging and in sales and export documents, would reflect the average of the previous figures, taking into account the sample sizes of each grade.

There is a general reluctance within the ET industry to market embryos for export sale that are IETS grade 2 (fair) quality. In the author's opinion, this is a prudent policy if the ET industry is going to experience a continued growth and success of frozen embryo sales. Fair quality embryos should certainly be frozen if the straws are clearly labeled as IETS grade 2 quality, with a clear understanding between buyer and seller. In an international export program, a large number of IETS grade 2 embryos frozen at Em Tran, Inc., have resulted in a 38% pregnancy rate (unpublished data).

Upon recovery from donor females, embryos should be handled in as clean a manner as possible. Sterile, plastic Petri dishes should be used for embryo searching and storage. These dishes should be disposed of after a single use. Washing and resterilizing disposable Petri dishes is usually not practical because the plastics will not withstand the heat of autoclaving and ethylene oxide residues have been shown to be persistent and toxic for considerable lengths of time. Phosphate buffered saline or a variety of commercial embryo holding media containing either 10% fetal calf serum or 0.4% bovine serum albumen are suitable for rinsing and holding embryos prior to freezing. The medium should also contain antibiotics. Suitable antibiotics such as penicillin/streptomycin or gentamicin can be purchased at the appropriate concentrations from commercial ET suppliers. If embryos are intended for domestic use, they should be rinsed in embryo-holding media 10 times, with at least a 100-fold dilution of each previous wash.

TABLE 9.2
The IETS Protocol for the Trypsin Treatment of Embryos

No. Rinses	Embryo Treatment
5	PBS + broad-spectrum antibiotics and 0.4% BSA
2	0.25% trypsin in Hank's balanced salt solution without Ca^{++} and Mg^{++} for 60 to 90 seconds of total exposure
5	PBS + antibiotics and 2% serum

PBS = phosphate buffered saline.
BSA = bovine serum albumin.

Adapted from Stringfellow and Seidel, 1998.

If the embryos are destined for export to another country, it is essential that the ET practitioner know the import requirements for the country in question. Variations among the requirements of different countries may include health testing of donor females at a specific interval before and/or after flushing and specific requirements for handling and rinsing the embryos prior to freezing. Virtually all import protocols require that the embryos be carefully examined at a minimum magnification of 50 × to ensure that the zona pellucida is intact and free of defects and that no material, such as cumulus cells, is adhering to the zona. Most countries currently require trypsin treatment of embryos prior to freezing according to the protocol shown in Table 9.2. It has been proven that this treatment is effective in removing or inactivating certain viruses from the surface of the zona pellucida. In the past, each of the countries in the European Union maintained separate and different embryo import requirements. Fortunately, all countries in the European Union now share a common protocol. Each country's protocol for embryo importation can be obtained from the Animal and Plant Health Inspection Service (APHIS) division of the U.S. Department of Agriculture and is available on the Internet.

If an ET practitioner intends to freeze embryos for sale or conveyance to another party, the embryos and cane containing them should be labeled with the unique ET practitioner freeze code number that is assigned upon request by the IETS (1111 North Dunlap Ave, Savoy, IL, 61874).

TRADITIONAL METHOD OF FREEZING EMBRYOS

Protocols for freezing embryos have changed dramatically over the last 20 years. The initial successes with freezing cattle embryos involved adding the cryoprotectant DMSO in six steps of increasing concentration, and a two-step program of temperature decreasing at a rate of 0.3° and then 0.1°C per minute. These types of freezing programs required more than 2 hours from start to finish. Improved protocols utilized 10% (1.4 *M*) glycerol as the cryoprotectant. When embryos are placed in this solution, there is an initial shrinking of the cells because the difference in osmotic pressure inside vs. outside the cells results in water exiting the cells faster than glycerol goes into the cells. However, it was discovered that there was no detectable damage to the embryos if they were equilibrated directly in the glycerol at the final concentration. Very satisfactory results have been achieved when the embryos are seeded at –6°, then cooled in a one-step freezing curve at a rate between 0.3° and 0.6°C per minute. A sample freezing protocol that utilizes glycerol follows:

1. Equilibrate embryos in 10% glycerol for 10 to 20 minutes.
2. Load embryos in 0.25 cc straws; seal and label straws. (Embryos can be loaded while they are equilibrating.)
3. Place straws in freezer that is maintaining a holding temperature of –6°C.
4. Seed straws and maintain at –6°C for 15 minutes.
5. Decrease temperature at 0.5°C per minute to –32°C.
6. Plunge straws into liquid nitrogen.

Although most semen in the United States is packaged in 0.5 cc straws, embryos are usually frozen in 0.25 cc straws, the vast majority of which are manufactured by one company. These straws are available from a majority of ET supply firms and can be purchased in bulk or in presterilized packages. To load cryoprotectant and embryos into straws, it is necessary to use a pipetter or a 1-ml syringe with an adapter that will fit with a tight seal either inside or outside the end of the straw. The cryoprotectant medium can either be drawn into or expelled from the straw by operating the pipetter or syringe. Embryos that are to be frozen in glycerol or those in ethylene glycol that are intended for direct transfer can be loaded as shown in Figure 9.1. In this diagram, the first column of cryoprotectant actually seals the plug by penetrating the cotton portion and saturating the polyvinyl alcohol powder that forms the inner section of the plug. Once saturated, the polyvinyl forms a solid, elastic plug that seals the end of the straw against the loss of medium or the entrance of liquid nitrogen. The second short column of medium serves to separate the column of medium containing the embryo from contacting the medium next to the plug. In addition, the two air bubbles function as pistons to help ensure that the embryo is expelled when the plug is pushed forward after thawing. Another air bubble at the other end of the straw

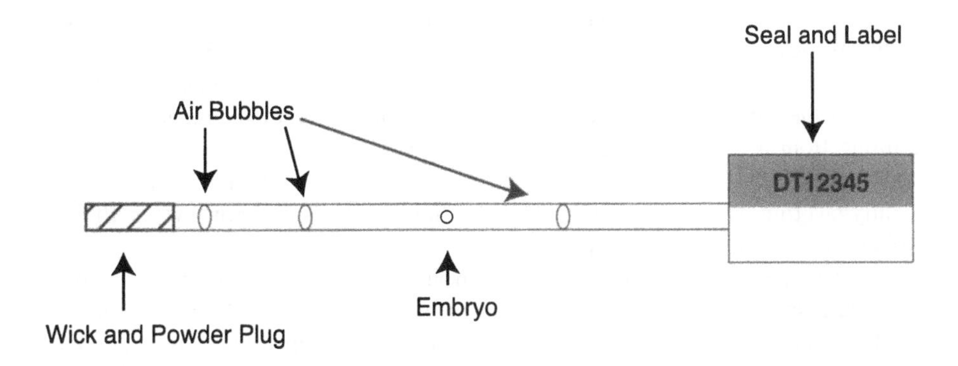

FIGURE 9.1 Diagram of straw loaded with an embryo and columns of either glycerol or ethylene glycol cryoprotectant solutions separated by air bubbles.

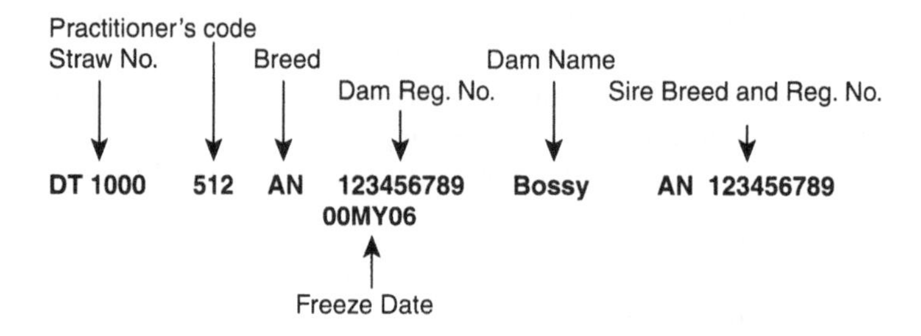

FIGURE 9.2 Identification as specified by the IETS on a straw containing a bovine embryo frozen for direct transfer. (From Stringfellow and Seidel, 1998. With permission.)

serves to keep the embryo in the center column of cryoprotectant medium. The straw can be sealed by heat, polyvinyl powder, or by using one of several types of plastic sealing devices that are commercially available. Some of the plastic sealing devices provide a surface on which the label can be either applied with indelible ink or a printed paper label. An example of how each straw should be labeled is shown in Figure 9.2. This labeling format conforms to the style mandated by the IETS and recognized by various cattle breed associations.

DIRECT TRANSFER METHOD OF FREEZING EMBRYOS

In the late 1980s and early 90s, a number of investigators demonstrated that ethylene glycol was an effective cryoprotectant for the deep freezing of the embryos of a number of mammalian species. Voelkel and Hu (1992) reported acceptable pregnancy rates following the transfer of bovine embryos frozen in ethylene glycol and transferred after thawing directly from the straws in which the embryos had been frozen. This technique has come to be known in the industry as the "direct transfer" procedure and is now widely used on an international basis. A patent for this methodology was awarded by the U.S. Patent Office to W.R. Grace & Co. (Voelkel, 1992). However, there has been no enforcement of this patent in the years following. The growing popularity of the direct transfer method was apparent in a survey of the North American ET industry (Leibo and Mapletoft, 1998). In 1997, 55.4% of frozen/thawed embryos transferred in the United States and 87.6% of those in Canada were frozen in the direct transfer method. A comparison of reported pregnancy rates resulting from transfer of embryos frozen in glycerol vs. those frozen in ethylene glycol is shown in Table 9.3. Although there were differences within some studies slightly favoring one cryoprotectant, there was no significant difference when all the studies were analyzed as a group.

Ethylene glycol, used to prevent freezing in automobile radiators, is a more suitable cryoprotectant than glycerol for direct transfer because it is a smaller molecule and penetrates the cell membranes of embryo cells faster. This property is more important for the thawing stage than for the freezing stage of the process. This is further explained in the following "Thawing" section. Embryos can be equilibrated in ethylene glycol in the same manner as in glycerol, i.e., they can be placed directly in the final ethylene glycol concentration of 1.5 *M* at room temperature and loaded in

TABLE 9.3
Comparison of Pregnancy Rates Achieved after Transfer of Bovine Embryos that Were Frozen in Glycerol or Ethylene Glycol

Glycerol		Ethylene Glycol		
No.	**% Preg.**	**No.**	**% Preg.**	**References**
92	48.9	189	58.2	Lange (1995)
185	56.8	780	59.6	Hinshaw et al. (1996)
56	69.6	56	50.0	Looney et al. (1996)
423	45.9	97	47.4	Arreseigor et al. (1998)
838	55.3	228	51.8	Holland Genetics (unpublished data)
1609	58.0	11,376	59.2	Leibo and Mapletoft (1998)
225	53.8	218	51.8	Em Tran, Inc., (unpublished data)[a]
Total				
3428	55.4	12,944	58.8	$x^2 = P > 0.05$

[a] Embryos from each donor were split evenly between the two cryoprotectants.

straws during the equilibration period. Equilibration involves attaining equal concentrations of ethylene glycol outside and inside the cells and is probably completed within 5 minutes. In my opinion, however, equilibration can be continued for as long as 20 minutes to allow for the loading and sealing of a number of straws. I hold this opinion, in spite of the fairly widespread view, perhaps best expressed as a "clinical impression," that ethylene glycol is toxic to bovine embryos if exposure prior to freezing exceeds 10 minutes. There are no published data to support this view; in fact several studies contradict it (Takagi et al., 1993; Hasler et al., 1997; Otter et al., 1998). Nevertheless, practitioners should be cautious about procedures involving embryo freezing and thawing at high ambient temperatures. Large numbers of embryos have been equilibrated in ethylene glycol in the laboratory at Em Tran for 10 or more minutes at 25±3°C with very acceptable pregnancy rates. It is very possible, however, that some of the reported problems with ethylene glycol may result from equilibration in on-farm ET programs with much higher summer environmental temperatures. Ethylene glycerol may well be more toxic to embryos during extended exposure at higher temperatures.

Some practitioners prefer to combine sucrose (0.1 to 0.25 *M*) with ethylene glycol for freezing direct transfer embryos. Experimental comparisons (Hasler et al., 1997) and surveys of practitioners (Leibo and Mapletoft, 1998) have not shown any benefit from the inclusion of sucrose. Embryos to be frozen for direct transfer can be cooled with the same temperature decrease curve described above for glycerol. As specified by the IETS, embryos for direct transfer should be frozen in yellow straws and stored in yellow goblets with a yellow tab on top of the cane.

EMBRYO THAWING

Whether embryos are frozen in the traditional method with glycerol or in ethylene glycol for direct transfer, rapid thawing is necessary for optimal survival. A typical protocol involves thawing straws for 5 to 10 seconds in the air, followed by submerging them in 25 to 30°C water until the ice is completely melted. There is evidence that completely thawing in air results in reduced embryo viability, whereas thawing directly in water results in a higher incidence of cracked zona pellucidas.

CRYOPROTECTANT REMOVAL (DILUTION)

Following thawing, it is necessary to remove the cryoprotectant from inside the embryo. Actually, the cryoprotectant is diluted rather than completely removed. For embryos frozen in glycerol, removal is accomplished prior to transfer of the embryo, whereas for direct transfer embryos, the ethylene glycol is diluted out of the embryo within the uterus of the recipient. During the early years of freezing cattle embryos, glycerol was usually removed in a six-step protocol involving small, progressively decreasing concentrations of glycerol. This was considered necessary to avoid the risk of cell membranes rupturing due to osmotic swelling caused by water rushing into the cells faster than the glycerol could exit. The use of sucrose during the thawing procedure eliminated the need for multiple concentrations of glycerol. Sucrose is a molecule that does not normally permeate embryo cell membranes in response to a lack of osmotic balance. Consequently, the sucrose prevents osmotic swelling while the glycerol flows out of the cells to be replaced by water.

GLYCEROL REMOVAL

Following thawing, embryos frozen in glycerol should be removed from the straw by cutting the sealed end and gently pushing the wick and powder plug forward with a stilette that fills the bore of the straw. The embryo can be placed directly from the straw into an empty Petri dish or into the first dilution solution. A widely used protocol for diluting embryos frozen in 10% glycerol is as follows:

1. 6.6% glycerol + 0.3 *M* sucrose for 5 minutes.
2. 3.3% glycerol + 0.3 *M* sucrose for 5 minutes.
3. 0.3 *M* sucrose for 5 minutes.
4. Three rinses in embryo-holding media, load embryo in straw and transfer to recipient.

Some practitioners have reported very satisfactory results by diluting embryos directly in a 1.0 *M* sucrose solution followed by rinsing in embryo-holding media and transfer. Embryos should be transferred as soon as possible after they are diluted and loaded into a transfer instrument.

ETHYLENE GLYCOL REMOVAL (DIRECT TRANSFER SYSTEM)

After thawing, the sealed end of the straw is cut, and the straw is loaded into a transfer gun and transferred into a recipient as quickly as possible. As was discussed previously regarding temperatures at freezing, embryos exposed to high ambient temperatures at thawing for an extended period may result in reduced pregnancy rates.

TRANSFER OF EMBRYOS

It is important to match the age of the embryo to the stage of the estrous cycle of the recipient. Satisfactory results are achieved by using recipients that were in estrus within ±1 day of the age of the embryo. In other words, a 7-day embryo can be transferred into a recipient that was in estrus between 6 and 8 days earlier. At this stage of the estrous cycle, cattle are more susceptible to infection than during the period of estrus in which they are normally inseminated. Consequently, it is important to maintain a high level of sanitation during the transfer procedure. There is another distinct difference between artificial insemination and the transfer of an embryo. Semen is normally deposited into the body of the uterus just ahead of the cervix, whereas, for optimal results, an embryo should be placed half way up the uterine horn on the same side as the corpus luteum of ovulation. To accomplish this, the practitioner must be able to rectally palpate the corpus luteum prior to transferring the embryo. For pregnancy to be established, it is essential that the recipient have a corpus luteum. However, it has been reported in numerous studies that the size of the corpus luteum is not related to the probability of establishing a pregnancy. There is convincing evidence that in most programs, higher pregnancy rates are achieved in heifers than in cows.

SUMMARY

Although most of us often refer only to the best way(s) to freeze embryos, it is no less important to consider the best ways to store and handle frozen embryos, thaw, rehydrate, and, finally, transfer them into suitable recipients. Each of these steps is equally important in contributing to the ultimate goal of making pregnancies.

Anyone serious about performing ET and freezing embryos should consider joining the American Embryo Transfer Association (2727 W. 2nd St., Hastings, NE 68902). With a membership of over 300, the AETA holds an annual convention at which the members can share business and technical information. The AETA also administers a certification program, started in 1985, that is recognized by many countries. To gain certification, ET businesses or individuals must pass a comprehensive exam and complete yearly continuing education requirements. In addition, the U.S. Department of Agriculture conducts physical inspections of the facilities of certified businesses. Practitioners should also seriously consider joining the more than 1000 members of the IETS. Benefits include an annual conference attended by as many as 700 members, a periodic newsletter, and an annual volume of the journal *Theriogenology* devoted exclusively to ET. It is also very important for all ET practitioners to use the current third edition of the *Manual of the IETS.*

A common problem encountered in the ET industry involves inaccurate or inadequate labeling of embryo straws or goblets containing straws. When the IETS standards are followed correctly, an ET practitioner anywhere in the world should be able to read all essential information on the straw and packaging of any embryo. It is also essential that ET practitioners handle frozen embryos appropriately and maintain them at −196°C until they are actually thawed. Straws should not be removed from goblets and sorted between nitrogen tanks without the use of a Dewar thermos full of nitrogen. Irreversible damage to frozen embryos may result if the temperature is allowed to rise above –80°C, which is about the temperature of dry ice.

There has been dramatic improvement in the success rates achieved with frozen bovine embryos over the last 20 years. When grade 1 embryos are frozen in glycerol, more than 99% appear to be of transferable quality following freezing/thawing. In many ET programs, the pregnancy rate of frozen/thawed embryos is within 10% of that with comparable quality fresh embryos. Success rates with direct transfer embryos are similar. Although there is room for improvement, future increases will be in very small increments. All of us would welcome any technical change resulting in a 5% increase in pregnancy rate. Experiments that statistically demonstrate small improvements in pregnancy rate are expensive and difficult to conduct. In what may have been the largest well-controlled field trial ever conducted involving the comparison of two freezing protocols on subsequent pregnancy rates, van Wagtendonk-de Leeuw et al. (1997) reported no difference between the two methods. This very expensive field trial involved the transfer of 728 embryos. In a field trial of this sort, it would have been necessary to transfer more than 1700 embryos to have a 90% certainty of detecting a 5% true difference in the pregnancy rate of the two freezing methods. Consequently, continued progress in this field will depend on carefully recording all data possible from commercial programs and making prudent evaluations of all quantifiable variables.

REFERENCES

Arreseigor, C. J., A. Sisul, A. E. Arreseigor, and R. C. Stahringer. 1998. Effect of cryoprotectant, thawing method, embryo grade and breed on pregnancy rates of cryopreserved bovine embryos. *Theriogenology* 49:160 (Abstr.).

Hasler, J. F. 1986. Results of bovine embryo freezing programs in two commercial embryo transfer companies during a one year period. *Proc. Fifth Ann. Convention AETA,* Fort Worth, TX. pp. 90.

Hasler, J. F. 1989. Results of the first frozen bovine embryos exported from the USA to Argentina. *Theriogenology* 31:230 (Abstr.).

Hasler, J. F. 1997. Survival of IVF-derived bovine embryos frozen in glycerol or ethylene glycol. *Theriogenology* 48:563.

Hinshaw, R. H., W. E. Beal, S. S. Whitman, and S. Roberts III. 1996. Evaluating embryo freezing methods and electronic estrus detection in a contract recipient bovine embryo transfer program. *Proc. Fifteenth Ann. Convention AETA,* Portland, OR. pp. 11.

Lange, H. 1995. Cryopreservation of bovine embryos and demi-embryos using ethylene glycol for direct transfer after thawing. *Theriogenology* 43:258 (Abstr.).

Lawson, R. A. S., L. E. A. Rowson, and C. E. Adams. 1972. The development of cow eggs in the rabbit oviduct and their viability after re-transfer to heifers. *J. Reprod. Fertil.* 28:313.

Leibo, S. P. and R. J. Mapletoft. 1998. Direct transfer of cryopreserved cattle embryos in North America. *Proc. Seventeenth Ann. Convention AETA,* San Antonio, TX. pp. 91.

Looney, C. R., D. M. Broek, C. S. Gue, D. J. Funk, and D. C. Faber. 1996. Field experiences with bovine embryos frozen-thawed in ethylene glycol. *Theriogenology* 45:170 (Abstr.).

Otter, T., H. W. Flapper, L. M. T. E. Kaal, and A. M. Van Wagtendonk-de Leeuw. 1998. Pregnancy rates of bovine embryos frozen in 1.5 *M* ethylene glycol: a comparison between direct transfer versus transfer after a one-step dilution. *Proc. Fourteenth Sci. Mt. AETE,* Venice, Italy. pp. 222 (Abstr.).

Stringfellow, D. A. and S. M. Seidel (Eds.). 1998. *Manual of the International Embryo Transfer Society.* IETS, Savoy, IL.

Takagi, M., A. Boediono, S. Saha, and T. Suzuki. 1993. Survival of frozen-thawed bovine IVF embryos in relation to exposure time using various cryoprotectants. *Cryobiology* 30:306.

van Wagtendonk-de Leeuw, A. M., J. H. G. den Daas, and W. F. Rall. 1997. Field trial to compare pregnancy rates of bovine embryo cryopreservation methods: vitrification and one-step dilution versus slow freezing and three-step dilution. *Theriogenology* 48:1071.

Voelkel, S. A. 1992. Cryopreservation process for direct transfer of embryos. W. R. Grace and Co., assignee. U.S. Pat. No. 5,160,312.

Voelkel, S. A. and Y. X. Hu. 1992. Use of ethylene glycol as a cryoprotectant for bovine embryos allowing direct transfer of frozen-thawed embryos to recipient females. *Theriogenology* 37:687.

Whittingham, D. G., S. P. Leibo, and P. Mazur. 1972. Survival of mouse embryos frozen to −196° and −269°C. *Science* 178:411.

Willadsen, S. M., C. Polge, L. E. A. Rowson, and R. M. Moor. 1974. Preservation of sheep embryos in liquid nitrogen. *Cryobiology* 11:60.

Wilmut, I. and L. E. A. Rowson. 1973. Experiments on the low-temperature preservation of cow embryos. *Vet. Rec.* 92:686.

10 Application of Embryo Transfer to the Beef Cattle Industry

W. E. Beal

CONTENTS

The discovery of methods to collect and transfer mouse embryos in 1891 (Heape, 1891) had little impact on the cattle industry until Willett et al. (1951) successfully collected and transferred bovine embryos 60 years later (Table 10.1). By comparison, the development of nonsurgical techniques for embryo collection (Sugie et al., 1972) and transfer (Sugie, 1965) followed quickly. Nonsurgical techniques changed bovine embryo transfer (ET) from a clinical to an on-farm technique—a move that markedly reduced the cost.

Successful freezing of cattle embryos in 1973 facilitated the expanded use of embryo transfer by eliminating the need to maintain recipient cows in synchrony with donors at the time of embryo collection (Wilmut and Rowson, 1973). Improvements in freezing and thawing procedures brought the pregnancy rates following transfer of frozen embryos within 10% of those achieved with fresh embryos (Hasler, 1992). The most recent development that allows embryos frozen in ethylene glycol to be transferred directly to synchronous recipients, without the need for repeated washing of embryos prior to transfer, has simplified the ET process (Voelkel and Hu, 1992). Just as the process of artificial insemination of cows progressed from being performed only by veterinarians, then by professional artificial insemination technicians, and now is performed largely by producers, the transfer of embryos may become a common on-farm task performed by producers. In fact, one Internet website

0-8493-1117-9/02/$0.00+$1.50

TABLE 10.1
Historical Development of Embryo Transfer and Related Techniques

Event	Scientific Reference
Successful transfer of rabbit embryo	Heape, 1891
Successful transfer of cattle embryo	Willett et al., 1951
Nonsurgical transfer of cattle embryo	Sugie, 1965
Nonsurgical recovery of cattle embryos	Sugie et al., 1972
Frozen/thawed cattle embryos	Wilmut and Rowson, 1973
Direct transfer frozen/thawed cattle embryos	Voelkel and Hu, 1992

(www.selectsires.com/embryos/skills.html) provides instructions to producers for "skills needed for successful direct transfer of embryos."

The changes in ET that have occurred during the past 50 years have positioned this technique to be used widely in the beef industry. However, if the history of the use of artificial insemination is a preview of how widespread the use of embryo transfer will become, the future is not bright. Artificial insemination is recognized as the most powerful tool available for enhancing genetic selection; however, most estimates indicate that fewer than 5% of the beef cattle in the United States are sired by artificial insemination bulls. Therefore, the question remains: "If ET is to become a commonly used tool in the beef industry, what will it be used for and what obstacles must be overcome?" The purpose of this chapter is to speculate on the potential uses of ET in the beef industry and to propose methods of overcoming some of the practical obstacles that stand in the way of its use.

POTENTIAL USES OF EMBRYO TRANSFER

Purebred Beef Herds

The use of ET has largely been confined to the purebred segment of the beef industry. There has been an increase in the use of ET. For example, the registration of Angus cattle derived from ET has increased fivefold over the past 15 years (Table 10.2). However, the percentage of registered cattle derived from ET is still only a small proportion (7.1%) of the 270,000 Angus cattle registered each year. In that respect, the impact of ET is minimal.

The increased use of ET in purebred operations has been for two purposes: to improve genetic selection by increasing the number of progeny from superior dams and to multiply the number of cattle in a program in order to expand the herd or meet marketing demands.

Superovulation, collection and transfer of embryos from superior cows, should increase the rate of genetic improvement by increasing selection intensity. Likewise, increasing the number of offspring from a cow should provide more progeny information and increase the accuracy of selection. There are problems with both of these theoretical concepts (Bourdon, 1997). First, superior cows are harder to identify

TABLE 10.2
Registrations of Angus cattle 1984–1999: Total Registrations and Registrations of Cattle Born by Embryo Transfer (ET)

Year	Total Registrations	ET Registrations[a]	ET Registrations, %[b]
1984	174,539	3298	1.9
1985	156,150	4206	2.7
1986	133,475	4307	3.2
1987	141,239	5105	3.6
1988	143,520	5339	3.7
1989	156,697	5850	3.7
1990	159,036	5359	3.4
1991	166,769	6073	3.6
1992	174,414	6453	3.7
1993	193,401	6965	3.6
1994	214,261	7582	3.5
1995	224,710	9242	4.1
1996	220,586	10,963	5.0
1997	239,476	13,564	5.7
1998	252,969	15,078	6.0
1999	260,907	18,456	7.1

[a] Registrations of cattle born by embryo transfer.
[b] Calculated as ET registrations/total registrations × 100.

From American Angus Association, St. Joseph, MO. With permission.

accurately than superior bulls. Cows, even cows with ET calves, have fewer progeny than bulls used in artificial insemination programs. Therefore, breeding values of cows are less accurate and more prone to change. Second, the performance of ET calves is "masked" by the maternal ability (good or bad) of the recipient female that raises the calf. Hence, performance data on ET calves are not credited to the performance evaluation of the donor cow. Therefore, even though a donor cow has more progeny, those data are not useful in calculating her estimated breeding values. The "bottom line" according to animal geneticists is that identifying genetically superior cows is "risky."

Contrary to the academic wisdom of geneticists, most breeders of purebred cattle argue that they *know* the best females in their herd and can successfully choose superior donor candidates. In theory, they are wrong—but in a practical sense, they might be right. The reason is that most breeders choose donors based on a variety of qualities (EPDs, performance, pedigree, visual appraisal, marketability, etc.). In most cases, breeders, unlike geneticists, are not solely concerned with the rate of genetic improvement.

Bulls that are the product of ET have had a significant impact on the purebred industry. Again using Angus as an example, three of the top 10 sires in the Angus breed based on number of registrations in 1999 were ET calves themselves (Table 10.3).

TABLE 10.3
Type of Birth for Angus Sires with the Most Registered Progeny in 1999

Angus Sire	Type of Birth	Progeny Registered
NBar Emulation Ext	Natural	7798
TC Stockman 365	Natural	4543
RR Scotchcap 9440	Natural	3912
Krugerrand of Donamere 490	Natural	3524
SAF Fame	Natural	3112
Vdar Lucys Boy	ET	2986
Sitz Traveler 8180	Natural	2787
Rito 9FB3 of 5H11 Fullback	ET	2202
RR Scotchcap 1483	Natural	1909
Leachman Right Time	ET	1806

From American Angus Association, St. Joseph, MO. With permission.

TABLE 10.4
Percentage of Angus Sires in AI Studs Born by Embryo Transfer (ET)

AI Company[a]	No. Angus Sires[b]	No. Sires by ET	% ET Sires
American Breeders Service	69	16	23%
Alta Genetics	36	7	19%
Integrated Genetics	43	16	37%
Select Sires	27	10	37%

[a] Mention of commercial establishments is not intended to serve as an endorsement, nor is failure to mention a specific company an indication of lack of endorsement.
[b] Angus sires presented in 2000 semen sales catalogs from each company.

Because the rate of ET among the top 10 bulls (30%) is higher than the rate of ET used in the entire Angus breed (7%), it is obvious that ET has had a significant impact in producing at least some of the "most popular" bulls, if not the bulls with the greatest genetic merit.

A survey of the proportion of Angus bulls in artificial insemination bull studs that are ET progeny supports the concept that use of ET on selected donor cows is more likely to produce Angus bulls that are used widely (Table 10.4). When the birth status of the lineup of Angus sires advertised by four major artificial insemination studs was investigated, it revealed that 28% of the artificial insemination bulls were the product of embryo transfer. Angus bulls selected by artificial insemination companies are still more likely to have been "natural" calves than ET calves; however, the use of ET is certainly more highly represented among artificial insemination sires than among the general population of Angus cattle.

Both the ET representation among the 10 most popular Angus bulls and the proportion of Angus sires chosen by AI studs that are ET progeny point to the fact that breeders may be able to use ET to produce more desirable bulls. Hence, despite the pitfalls of identifying genetically superior cows, ET appears to work for producing progeny that are viewed as industry leaders by artificial insemination studs and purebred Angus breeders.

The second and probably the more common use of ET in purebred beef herds is to multiply the number of calves born from the breeding program. If an unsaturated market exists for bulls and heifers from a purebred program, then collection and transfer of embryos from a population of donors believed to be representative of the program is justified. Likewise, if the purpose is to simply increase the size of an existing herd, ET is a viable option for rapid expansion. In either case, identification of one or two "mystical donors" is less important than replicating the gene pool. The multiplier effect is accomplished if a large group of above average cows in the herd are used as embryo donors. Often the embryos are transferred to recipient cows in cooperator herds, thereby allowing a purebred program to produce progeny in excess of the resource limitations of its own operation. Two excellent examples of using ET for marketing or expansion purposes are Gardiner Angus Ranch in Ashland, KS, and Wehrmann Angus in New Market, VA.

The ET program at Gardiner Angus Ranch (GAR) is an excellent example of multiplying the genetics of a herd to meet marketing demands (Table 10.5). GAR calved approximately 300 cows naturally per year in 1997 through 1999. During the same period they raised or contracted 840 to 1250 ET calves. Those ET calves allowed them to replace the cow herd and market 1150 to 1550 bulls and females per year without increasing their registered cow herd to more than 300 cows.

Wehrmann Angus has had an intensive ET program since 1995, but their purpose recently has been to increase herd size rather than to increase the cattle available for marketing. In 1997, they began to use ET to multiply the number of animals in the cow herd (Table 10.6). The number of ET calves peaked in 1998 at 425 and most of the females have been retained in the herd to increase herd size. Their plan

TABLE 10.5
Number of Calves at Gardiner Angus Ranch Born Naturally or by Embryo Transfer (ET)

Year	Registered Cows[a]	Calves Born by ET
1997	310	1150
1998	292	1272
1999	294	1254
2001[b]	≤350	1150

[a] Natural calving.
[b] Projection.

TABLE 10.6
Number of Calves at Wehrmann Angus Born Naturally or by Embryo Transfer

Year	Registered Cows[a]	Calves Born by ET
1997	200	125
1998	220	425
1999	250	260
2001[b]	≤500	125

[a] Natural calving.
[b] Projection.

is to build the cow herd to 600 head by 2001 and to reduce the number of ET calves to 125 per year by that time.

Using ET in order to multiply the herd for expansion or marketing purposes is common. The opportunities for increasing genetic improvement in herds that use ET to multiply progeny numbers depend on the rearing and marketing strategy for the ET calves. In herds that market all ET bulls and heifers, the opportunity for improving the nucleus herd is lost. Conversely, if heifers raised in cooperator herds are returned to the nucleus herd to calve, their performance can be evaluated and the superior females can be incorporated into the nucleus herd.

Commercial Beef Herds

The most common use of ET in commercial beef herds has been for commercial cows to serve as recipients for purebred embryos. In this case, the embryo owner provides the recipient herd owner with purebred embryos, and the embryo owner buys back the ET calves at weaning at a premium price. The contract ET recipient herd business has been a lucrative one. Fee schedules for ET calves raised by two large contract recipient herds in Virginia during 1999 are shown in Table 10.7.

The use of ET to produce commercial calves has been nonexistent. For ET to be justifiable in a commercial beef herd, the ET progeny must impart some advantage that is not attainable with artificial insemination or natural service of the commercial cows. For example, if the commercial cow herd consisted of Holstein crossbred cows, but the market discriminated against lightly muscled feeder calves with dairy character, ET could be used to transfer embryos of a beef breed(s) into the Holstein crossbred cows. If return on investment was justified, the same tactic could be used to transfer embryos of larger-framed, more heavily muscled breeds into moderate-framed cows that were more efficient and less expensive to maintain. This strategy would be limited by birth weight constraints needed to avoid dystocia among the moderate-framed cows.

Embryo transfer could also be used to increase consistency of calves born in a commercial herd, especially if the herd were a heterogeneous mixture of crossbred cow types. This strategy could be important in herds where retained ownership and

TABLE 10.7
Examples of Fees for Transfer and Pre-Weaning Development of Embryo Transfer Calves

Herd #1		Herd #2	
Embryos[a]	Fee, $/calf[b]	Embryos[a]	Fee, $/calf[b]
10–25	950	10–50	950
26–50	900	26–100	900
51–100	850	>100	850
101–200	800		
>200	750		

[a] Number of embryos transferred.
[b] Transfer and development fee.

grid pricing of slaughter cattle rewarded quality and consistency. However, the added cost of securing and transferring embryos would probably outweigh the premiums paid for slaughter cattle. Furthermore, if cattle were sold at weaning or slaughter cattle were sold on pen averages, the advantages of a consistent, high quality calf crop would be negligible.

REDUCING OBSTACLES TO THE USE OF EMBRYO TRANSFER

The two most common objections to the use of ET are the cost and the intensive labor required. The labor associated with embryo transfer can be reduced with the use of new techniques to synchronize estrus/ovulation in recipient cows and by methods of resynchronizing nonpregnant recipients for a repeated opportunity at embryo transfer.

OVULATION CONTROL TO ELIMINATE HEAT DETECTION OF RECIPIENTS

The synchrony of estrus and timing of ovulation following traditional estrous synchronization treatments has been too variable to allow the transfer of fresh or frozen/thawed embryos without heat detection. Synchronization treatments developed more recently, however, combine traditional methods of controlling estrous cycle length with the manipulation of follicular development in order to "program" or "select" the ovulatory follicle. The new methods control ovulation with enough repeatability and precision to use for timing the transfer of embryos to recipient cows without the need for visual detection of estrus.

In a cooperative experiment with Ashby Embryos and Mossy Creek Farm, we evaluated methods for controlling the estrous cycles of recipient cows and eliminating the need for visual heat detection. Four hundred ninety-nine primiparous (n = 90) or multiparous (n = 409) beef cows nursing calves (28 to 92 days post partum)

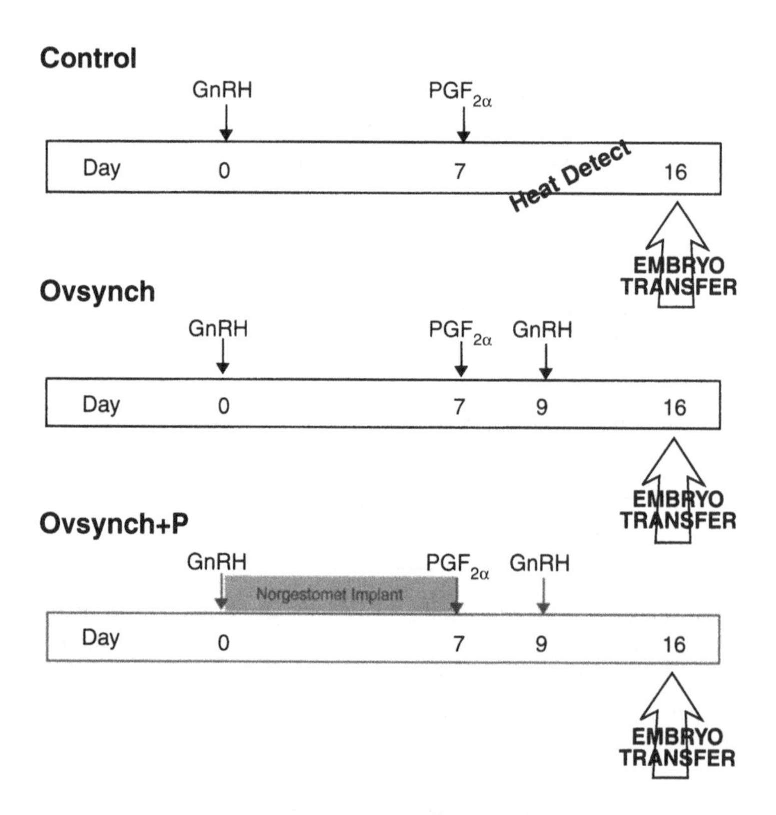

FIGURE 10.1 Diagram of estrous synchronization and ovulation control treatments.

were assigned to one of three treatments to synchronize estrus or to synchronize estrus and control ovulation (Figure 10.1). Approximately one third of the animals received 100 μg of GnRH (Cystorelin®; Merial, Iselin, NJ) via i.m. injection, followed 7 days later by 25 mg of $PGF_{2\alpha}$ (i.m. Lutalyse®; Upjohn Co., Kalamazoo, MI). This group (control, n = 169) was monitored continuously for signs of behavioral estrus using an electronic heat detection system (HeatWatch™; DDx Inc., Boulder, CO). Recipients in the control group were designated for ET 6 to 8 days following detection of estrus, and embryos were assigned based on the stage of embryo development (stage 3 to 6).

The second one third of the animals (Ovsynch, n = 165) were treated as described above, but received a second injection of GnRH 48 hours after the administration of $PGF_{2\alpha}$ to control the timing of ovulation (Figure 10.1). The final one third of the animals (Ovsynch+P, n = 165) were injected with 100 μg of GnRH and fitted with an s.c. hydron implant containing 6 mg norgestomet (Syncro-Mate-B®; Merial, Iselin, NJ) at the beginning of treatment. Seven days later the subcutaneous implant was removed and each cow received 25 mg of $PGF_{2\alpha}$. A second injection of GnRH was administered 48 hours after implant removal to control the timing of ovulation. All cows in the Ovsynch and Ovsynch+P groups were presented for ET 9 days after the injection of $PGF_{2\alpha}$. Following palpation to verify the presence of a corpus luteum (CL), embryos were thawed and transferred to suitable recipients in this group without

TABLE 10.8
Pregnancy Rate in Recipients Treated to Synchronize Estrus and Control the Timing of Ovulation

Treatment[a]	No.	Pregnancy Rate/Transfer	Pregnancy Rate/Recipient
Control	169	67/108 62%[b]	67/169 40%
Ovsynch	165	72/150 48%[c]	72/165 44%
Ovsynch+P	165	82/152 54%[c]	82/165 50%

[a] See Figure 10.1 and text for explanation of treatments.
[b,c] Means in same column with different superscripts are significantly different ($P < .07$).

regard to the occurrence of estrus. Embryos frozen at different stages of development were assigned for transfer to recipients in the ovulation control groups (Ovsynch and Ovsynch+P) based on an "assigned" time of estrus that was coincident with the second injection of GnRH.

The overall pregnancy rates (pregnancies/recipients; Table 10.8) were not significantly different among the three treatment groups. The higher transfer rate among cows in the ovulation control groups compensated for the lower pregnancy rate per transfer recorded in those groups. Therefore, the use of an ovulation control treatment in addition to a method for controlling corpus luteum function and synchronizing estrus was successful in enabling the transfer of embryos without the need for heat detection in this experiment. In a second field trial at the same location, pregnancy rate/transfer was 65% (105/161) when recipients were detected in estrus and 66% (183/276) when embryos were transferred to recipients after ovulation control treatment (Ovsynch+P) without heat detection.

Resynchronizing Nonpregnant Recipients

In many herds, commercial cows are synchronized once at the beginning of the breeding season to be used as recipients, then a bull is turned in for natural service. In this case, recipient cows have only one opportunity to become pregnant with an ET calf. For the past 3 years, we have synchronized the recipients a second time if they do not become pregnant after transferring an embryo in order to increase the number of embryos we can transfer.

The most successful program for resynchronizing recipients that have received an embryo but failed to become pregnant involves inserting a 6-mg norgestomet implant (Syncro-Mate-B) into all the recipients 10 days after the first embryo has been transferred (equal to day 17; Figure 10.2). The implant is removed 7 days later (equivalent of day 24). Recipients that are not pregnant from the first ET come in heat (usually 2 days after the implant is removed), whereas recipients that became pregnant to the first ET maintain pregnancy and do not show signs of heat after the implant is removed. Ultrasonography can be used at the time of implant removal to identify and segregate pregnant recipients if desired.

TABLE 10.9
Resynchronization Rate of Embryo Transfer and Pregnancy Following to Control the Timing of Estrus in Nonpregnant Recipients

	No.	Transfer Rate	Pregnancy Rate/Transfer	Pregnancy Rate/Recipient
Resynchronized	116	80/116 69%[a]	53/80 66%	53/116 46%

[a] See Figure 10.2 and text for explanation of treatments.

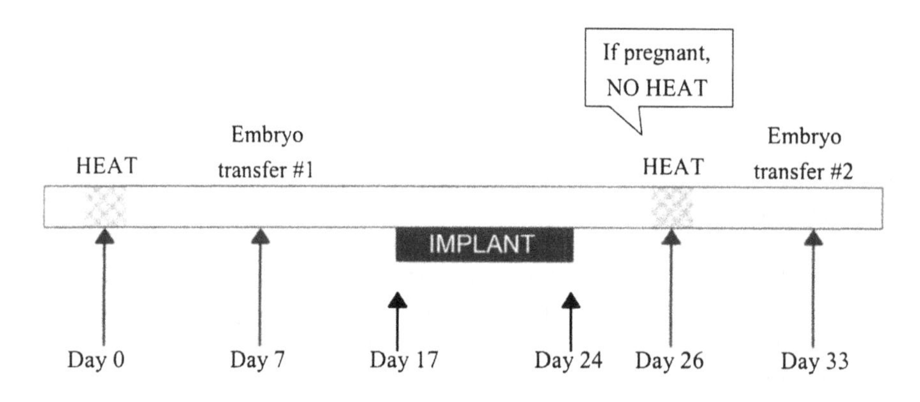

FIGURE 10.2 Scheme for resynchronizing estrus in embryo transfer recipients.

One hundred sixteen cows fitted with norgestomet implants 10 days after receiving an embryo were determined to be nonpregnant at the time of implant removal. Sixty-nine percent of the eligible cows were detected in estrus within 4 days following implant removal. Most cows that responded to the resynchronization were in estrus between 36 and 60 hours after implant removal.

Pregnancy rate per transfer and pregnancy rate per recipient following resynchronization of estrus and transfer of a second embryo (Table 10.9) were similar to those recorded in the same herd following the initial synchronization (Table 10.8). Hence, the resynchronization treatment used in this experiment enabled the immediate reutilization of nonpregnant recipients for the transfer of a second embryo.

Cost of Embryo Collection and Transfer

The cost of collecting and freezing embryos varies depending on the number of donors being collected and the fee schedule of the ET practitioner. The estimated costs presented in Table 10.10 represent the cost per frozen embryo ranging from a high-volume herd/moderately priced service to a low-volume herd/higher-priced service. The average number of embryos collected per donor (seven) is applicable to beef cows. It would be lower for beef heifers or Holstein cows. Note that while the average number of embryos collected is seven per donor, the range in embryos collected per donor (0 to >30) is the most variable component of an ET program.

TABLE 10.10
Range of Estimated Costs of Producing Frozen Bovine Embryos[a]

Item	Low-Volume[b] Estimated Cost	High-Volume[c] Estimated Cost
Collection fee/donor (drugs, flushing donors)	$300.00	$200.00
Semen (3 units @ $30/unit)	$90.00	$90.00
Freezing fee	$350.00	$200.00
Total	$740.00	$490.00
Average embryos/collection	7	7
Estimated cost/embryo	$105.71	$70.00

[a] Costs do not include farm labor, overhead, or donor maintenance.
[b] Estimated cost for collection of several donors, moderately priced service.
[c] Estimated cost for collection of single donor, higher-priced service.

The total cost per ET calf raised to weaning (7 to 10 mo) will depend on the cost of performing transfers, the pregnancy rate associated with ET, and the cost of maintaining the recipient cow and her calf. Obviously, those costs are highly variable, but the range can be estimated using a standard ET pregnancy rate and either home-raised recipient costs or the fees for preweaning development of ET calves in a contract recipient herd (see Table 10.7).

In a study comparing embryo freezing methods, we transferred 3531 frozen/thawed embryos prepared by 17 different ET practitioners. The overall pregnancy rate was 58.8%. For companies preparing more than 25 embryos, the pregnancy rate ranged from 49.0 to 72.5%. If the average pregnancy rate is combined with high or low embryo costs and fees for calf development in a contract recipient herd, the cost per embryo calf weaned will range from $1019 to $1080 (Table 10.11). Conversely, if recipients are home-raised with an 18-mo recipient cost from $400 to $650 and a transfer cost ranging from $25 to $50/transfer, the range in cost per embryo calf weaned will be $553 to $915. Note that the estimated cost using home-raised recipients does not include labor.

CONCLUSION

Embryo transfer has been used successfully by purebred breeders in the beef industry to produce progeny that are more marketable to other breeders or to AI studs. The use of ET has also enabled rapid expansion of purebred herds or establishment of high-volume marketing programs. In most cases, the cost of producing a purebred, weaned ET calf ($550 to $1100) has been covered by the marketability of breeding-age bulls or females. Conversely, other than the transfer of purebred embryos in contract recipient herds, the application of ET in the commercial industry has been nonexistent. Furthermore, the labor and cost of ET are likely to preclude its future application in commercial herds.

TABLE 10.11
Range of Estimated Cost of Developing Calves by Embryo Transfer with Varying Embryo Costs and Recipient Programs

Embryo Cost[a] Recipients	↓ Embryo $ Contract Recipient	↑ Embryo $ Contract Recipient	↓ Embryo $ Home-Raised Low Cost	↑ Embryo $ Home-Raised High Cost
Cost/frozen embryo	$70.00	$105.71	$70.00	$105.71
Pregnancy rate	58.8%	58.8%	58.8%	58.8%
Embryo cost/pregnancy	$119.04	$179.78	$119.04	$179.78
Development cost/calf[b]	$900.00	$900.00	—	—
Recipient maintenance[c]	—	—	$400.00	$650.00
Transfer cost/pregnancy[d]	—	—	$34.07	$85.18
Total	$1019.04	$1079.78	$553.11	$914.96

[a] See Table 10.10 for explanation of embryo costs.
[b] Average cost of calf development in contract recipient herd; includes recipient maintenance (18 mo), estrous synchronization, embryo transfer, calf vaccinations, creep feeding (variable), and delivery (limited).
[c] Recipient maintenance (18 mo); includes estrous synchronization, calf vaccinations, creep feeding (variable).
[d] Transfer cost for high- ($50/transfer) or low-cost ($25/transfer) service divided by average pregnancy rate per transfer (.588).

The use of ET in purebred herds is likely to expand. Improvements in methods for superovulation, collection and freezing of embryos from donors, or in estrous synchronization of donors and recipients will enable that expansion. The overall impact of ET in the beef industry is likely to be measured through the genetic impact of popular artificial insemination sires derived from ET or from the effects of the large volume of natural service bulls sold to commercial producers by breeders who use ET to increase supply.

REFERENCES

Bourdon, R.M. 1997. Biotechnology and animal breeding. In: *Understanding Animal Breeding*. pp. 405–406. Prentice Hall, Upper Saddle River, NJ.

Hasler, J.F. 1992. Current status and potential of embryo transfer and reproductive technology in dairy cattle. *J. Dairy Sci.* 75:2587.

Heape, W. 1891. Preliminary note on the transplantation and growth of mammalian ova within a uterine foster mother. *Proc. Roy. Soc.* (London) 48:457.

Sugie, T. 1965. Successful transfer of a fertilized bovine egg by nonsurgical techniques. *J. Reprod. Fertil.* 10:197.

Sugie, T., T. Soma, S. Fukumitsu, and K. Otsuki. 1972. Studies on ovum transfer in cattle, with special reference to collection of ova by means of nonsurgical techniques. *Bull. Natl. Inst. Anim. Ind.* (Ibaraki) 25:27.

Voelkel, S.A. and Y.X. Hu. 1992. Direct transfer of frozen thawed bovine embryos. *Theriogenology* 37:23.

Willett, E.L., W.G. Black, L.E. Casida, W.H. Stone, and P.J. Buckner. 1951. Successful transplantation of a fertilized bovine ovum. *Science* 113:247.

Wilmut, I. and L.E.A. Rowson. 1973. Experiments on the low temperature preservation of cow embryos. *Vet. Rec.* 92:686.

11 Embryo Transfer in Tropically Adapted Cattle in the Semitropics

Chad C. Chase, Jr.

CONTENTS

Over half the cattle in the world are found in the tropics (from the Tropic of Cancer [23.5°N latitude] to the Tropic of Capricorn [23.5°S latitude]). The semitropics (or subtropics, from 23.5° to 30°N and 23.5° to 30°S latitudes) of the United States include the Gulf Coast region of the southern states and all of Florida. Nearly 30% of the U.S. beef herd is maintained in this zone. Climate-mandated dependence on warm-season grasses and lack of significant grain production limits beef cattle producers in this region largely to the cow–calf segment of the industry.

Compatibility between beef cattle type (e.g., breed) and the environment is critical in harsh environmental zones such as the subtropics. In contrast to the abundance of beef cattle genotypes in the United States that are adapted to temperate climates, germ plasm resources with adaptation to warm climates, including the subtropics, are generally limited to the Zebu breeds (*Bos indicus*) and, within these, primarily to the American Brahman. The widespread assimilation of Brahman breeding into commercial beef herds throughout the warm regions of the United States attests to the economic value placed on tolerance to heat, diseases including internal and external parasites, and low-quality forage. Twenty-five percent of the cattle in the United States are estimated to have some percentage of Brahman breeding. The major niche for Brahman cattle has been in crossbreeding programs that combine the tropical adaptation of Brahman with the more desirable reproductive efficiency and carcass characteristics of temperate-adapted *Bos taurus* breeds.

At the Subtropical Agricultural Research Station (STARS), an overall research objective is to evaluate, improve, and preserve beef cattle germ plasm for adaptation to the subtropics of the United States. Toward meeting this objective, two major subobjectives are emphasized: 1) to evaluate and improve breeds existing in the United States, namely the Brahman, and 2) to characterize and evaluate tropically adapted breeds of *Bos taurus* that have potential for increasing production efficiency, especially reproductive efficiency, and potential for improving marketability.

Superovulation and embryo transfer have allowed the cattle industry to accelerate the production of genetically superior animals. These technologies have also proven useful in the safe and efficient movement of cattle germ plasm worldwide. At the STARS, these technologies have been utilized to further evaluate Brahman cattle as well as other tropically adapted breeds of beef cattle new to the United States. The objectives of this chapter are to summarize some of this research, emphasizing the use of these technologies in meeting an overall objective.

EFFECTS OF BREED OF RECIPIENT (i.e., SURROGATE) DAM

Selection of recipient dams is an important consideration to the success of an embryo transfer program. Recipient dams should be free from disease, structurally sound, in good body condition, and exhibiting regular estrous cycles. As part of the tropically adapted *Bos taurus* evaluation program, there is an interest in evaluating the Romosinuano, a criollo breed native to Colombia that is purported to exhibit high fertility in the tropics. Outside of Colombia there was an upgraded herd of Romosinuano at Centro Agronomico Tropical de Investigacion y Ensenanza (CATIE), Turrialba, Costa Rica. This herd was established in the 1950s using foundation Romosinuanos obtained from North Carolina State College. Through cooperative arrangements between CATIE and the University of Missouri, Romosinuano semen and embryos were collected at CATIE and imported into the United States in 1988, 1990, and 1992. In 1990, STARS received 77 of the imported Romosinuano embryos, and in 1992 we received another 38 embryos.

A study was conducted to determine the effect of breed of recipient dam (Angus vs. Brahman) on pregnancy and calving following transfer of Romosinuano embryos. In 1990, 75 embryos were transferred, and in 1992, 40 embryos were transferred (two from 1990 and 38 from 1992). In both years, Angus and Brahman recipient dams were used. Embryos were thawed (three-step method using glycerol) and transferred approximately 7 days after estrus. Breed of recipient dam did not affect the percentage of recipient dams pregnant or calving in either year (Table 11.1). Thirty-six Romosinuano calves were born in 1991 from the 75 embryos transferred in 1990 (48% calving), and 18 Romosinuano calves were born in 1993 from the 40 embryos transferred in 1992 (45% calving).

Additionally, a subset of these recipients was used to determine the effects of recipient breed (Angus vs. Brahman) and method of estrous synchronization on estrous response and the percentage of recipient dams pregnant and calving after embryo transfer (Chase et al., 1992a). Over a 3-year period, using imported Romosinuano embryos in 1990 and 1992, and Brahman embryos collected at the STARS in 1991, Angus ($n = 88$) and Brahman ($n = 87$) recipient cows (i.e., approximately 30 of each breed each year) were synchronized using either progestogen (Syncro-Mate-B®; Sanofi Animal Health, Inc., Merial, Iselin, NJ) or prostaglandin

TABLE 11.1
Effect of Breed of Recipient Dam on Pregnancy and Calving after Transfer of Romosinuano Embryos

		Breed of Recipient Dam	
Year	Item	Angus	Brahman
1990	No. of recipients	36	39
	No. pregnant	22	20
	Pregnancy, %	61	51
	No. calving	18	18
	Calving, %	50	46
1992	No. of recipients	20	20
	No. pregnant	10	9
	Pregnancy, %	50	45
	No. calving	10	8
	Calving, %	50	40

$F_{2\alpha}$ (Lutalyse®; The Upjohn Co., Kalamazoo, MI) treatments. The progestogen treatment included the administration of an ear implant (6 mg norgestomet), an injection (3 mg norgestomet and 5 mg estradiol valerate) on day 0, and removal of the implant on day 9. The prostaglandin treatment included injections of prostaglandin on days 0 (25 mg), 11 (12.5 mg), and 12 (12.5 mg). This sequence of prostaglandin administration previously had been shown to result in a higher percentage of pregnancies in artificially inseminated Red Brangus cows when compared to the standard double-injection sequence (25 mg of prostaglandin on days 0 and 11; Santos et al., 1988). Embryos were thawed (three-step method using glycerol) and transferred approximately 7 days after estrus. The overall estrous response, and percentage of recipient dams pregnant and calving did not differ ($P > 0.10$) between estrous synchronization treatments, and did not differ ($P > 0.10$) between breeds (Table 11.2). We concluded from these data that acceptable pregnancy and calving rates occurred with both progestogen and prostaglandin treatments and with both Angus and Brahman recipient dams.

The effect of breed of recipient dam on preweaning performance of Romosinuano calves born in 1991 and 1993 was also investigated (Chase et al., 1992b, 1994). In 1991, length of gestation was 5 days longer ($P > 0.05$) for Romosinuano calves from Brahman recipient dams than for Romosinuano calves from Angus recipient dams (Table 11.3). Romosinuano calves from Brahman recipient dams were 7.9 lb heavier ($P < 0.05$) at birth and 145 lb heavier ($P < 0.05$) at weaning (adjusted to 205 days) than Romosinuano calves from Angus recipient dams. These data suggested that milk production and/or milk composition may have affected weaning weights in 1991. Therefore, in 1993, we estimated 24-hour milk production and composition on days 28, 56, 84, 112, 140, and 169 after calving. However, unlike 1991, in 1993 there were no significant breed-of-recipient-dam effects on length of gestation, birth weight, or weaning weight. Furthermore, Angus cows had 3.08 lb/day greater ($P < 0.05$) milk production than Brahman cows, but similar milk fat production because Brahman cows

TABLE 11.2
Main Effects of Estrous Synchronization Treatment and Breed of Recipient Dam on Estrus and Pregnancy and Calving After Embryo Transfer

		Treatment		Breed of Recipient Dam	
Item	Year	Progestogen	Prostaglandin	Angus	Brahman
No. in estrus/No. treated (%)					
	1990	21/30 (70)	25/30 (83)	21/30 (70)	25/30 (83)
	1991	28/30 (93)	26/30 (87)	29/30 (97)	25/30 (83)
	1992	22/27 (81)	17/28 (61)	20/28 (71)	19/27 (70)
	All	71/87 (82)	68/88 (77)	70/88 (80)	69/87 (79)
No. pregnant/No. of recipients (%)					
	1990	12/21 (57)	18/24 (75)	14/21 (67)	16/24 (67)
	1991	13/28 (46)	10/25 (40)	14/29 (48)	9/24 (38)
	1992	9/21 (43)	8/16 (50)	9/19 (47)	8/18 (44)
	All	34/70 (48)	36/65 (55)	37/69 (54)	33/66 (50)
No. calving/No. of recipients (%)					
	1990	10/21 (48)	15/24 (62)	11/21 (52)	14/24 (58)
	1991	9/28 (32)	8/25 (32)	9/29 (31)	8/24 (33)
	1992	8/21 (38)	8/16 (50)	9/19 (47)	7/18 (39)
	All	27/70 (38)	31/65 (48)	29/69 (42)	29/66 (44)

had higher ($P > 0.05$) percentage of milk fat (Table 11.3). Differences in milk production between Angus and Brahman cows have been inconsistent in previous studies, but Brahman cows have been shown to have higher milk fat (Brown et al., 1993).

In order to objectively evaluate the Romosinuano breed, we recognized the importance of sampling pure Romosinuano germ plasm of Colombian origin. In 1996, 140 Romosinuano embryos collected in Venezuela produced from dams and sires of direct Colombian ancestry were imported. As with previous studies, there was an interest in breed-of-recipient-dam effects on estrous response, calving, and preweaning growth of Romosinuano embryo transfer calves (Chase et al., 1998a). Breeds of recipient dams used in this study included Angus and Senepol. The Senepol is a tropically adapted *Bos taurus* breed developed in the early 1900s in St. Croix, United States Virgin Islands, from crosses between the Red Poll and N'Dama (Hupp, 1981). Little published information was available on reproductive characteristics of Senepol cows. All of the Angus ($n = 88$) and Senepol ($n = 120$) cows were treated with progestogen (as previously described), and there was no difference in the percentage of Angus and Senepol cows observed in estrus (85 and 84%, respectively; Table 11.4). Romosinuano embryos were thawed (three-step method using glycerol) and transferred 7 days (±24 hours) after estrus to 67 Angus and 73 Senepol recipient dams. The percentage of pregnancy was similar between breeds of recipient dams and averaged 49% for Angus and 51% for Senepol. The percentage of recipient dams that calved also did not differ between breeds and averaged 48%

TABLE 11.3
Effect of Breed of Recipient on Dam Performance and Preweaning Performance of Romosinuano Embryo Transfer Calves

Year	Item	Breed of Recipient Dam	
		Angus	Brahman
1991	No. of cows	18	18
	Length of gestation,[a] days	282 ± 1.2	287 ± 1.1
	Birth weight,[a] lb	63.5 ± 2.25	71.4 ± 2.03
	Adj. 205-d weaning wt.,[a] lb	343.9 ± 13.23	489.4 ± 11.46
	Adj. wt./1000 lb of cow,[a] lb	368.2 ± 18.30	425.5 ± 16.09
1993	No. of cows	10	8
	Length of gestation, days	283 ± 1.3	284 ± 1.5
	Birth weight, lb	75.2 ± 2.27	69.7 ± 2.54
	Adj. 205-day weaning wt., lb	465.2 ± 7.50	449.7 ± 9.04
	Adj. wt./1000 lb of cow, lb	445.3 ± 14.11	436.5 ± 16.98
	Milk production,[a] lb/day	18.1 ± 0.82	15.0 ± 0.90
	Milk fat,[a] %	3.7 ± 0.20	5.0 ± 0.23
	Milk fat, lb/day	0.66 ± 0.033	0.75 ± 0.038
	Milk protein, %	3.0 ± 0.06	3.1 ± 0.07
	Milk protein, lb/day	0.53 ± 0.022	0.46 ± 0.027

[a] Effect of breed ($P < 0.05$).

TABLE 11.4
Effect of Breed of Recipient Dam on Estrous Synchronization, Pregnancy and Dam Performance, and Preweaning Performance of Romosinuano Embryo Transfer Calves from Venezuela

Item	Breed of Recipient Dam	
	Angus	Senepol
No. of cows	88	120
No. in estrus	75	101
Synchronized, %	85	84
No. of recipients	67	73
No. pregnant	33	37
Pregnancy, %	49	51
No. of cows calving	32	37
Length of gestation,[a] days	289 ± 1.1	293 ± 0.8
Birth weight,[a] lb	72.3 ± 1.68	78.7 ± 1.21
No. of calves weaned	30	37
Adj. 205-day weaning wt., lb	451.9 ± 7.94	460.8 ± 7.05

[a] Effect of breed ($P < 0.05$).

for Angus and 51% for Senepol. Length of gestation was 4 days longer ($P < 0.05$) and birth weights were 6.4 lb heavier ($P < 0.05$) for Romosinuano calves from Senepol than from Angus recipient dams. Weaning weights (adjusted to 205 days), however, were not affected by breed of recipient dam.

In summary, these studies demonstrated that the use of available estrous synchronization treatments resulted in similar and acceptable estrous responses, pregnancies, and calvings between Angus and Brahman recipient cows and between Angus and Senepol recipient cows. Significant breed of recipient dam effects are likely to occur for length of gestation and birth weight; however, breed-of-recipient-dam effects on weaning weight were inconsistent (i.e., only observed in 1 of 3 years).

EFFECTS OF BREED OF DONOR

Previous research has documented differences in reproductive characteristics between *Bos indicus* (primarily Brahman and Brahman-derivative breeds) and *Bos taurus* cattle (for reviews, see Randel, 1984; 1994). We have also investigated some reproductive characteristics between *Bos indicus* (Brahman) and *Bos taurus* breeds that may have implications for the successful control of the estrous cycle and superovulation in tropically adapted cattle. More recently, to determine breed differences in ovarian function, daily rectal ultrasonography of the ovaries was conducted on Angus, Brahman, and Senepol cows during an estrous cycle in summer (Alvarez et al., 2000). Numbers of small (2 to 5 mm; SE = ±4), medium (6 to 8 mm; SE = ±0.7), and large (9 mm; SE = ±0.2) follicles were greatest for Brahman (39, 5.0, and 1.6, respectively), least for Angus (21, 2.3, and 0.9, respectively), and intermediate for Senepol (33, 3.9, and 1.2, respectively) when averaged over an estrous cycle. Angus and Brahman cows had mostly two waves of follicular development during an estrous cycle (73 and 56%, respectively), whereas Senepol cows had mostly three waves of follicular development during an estrous cycle (70%); however, these breed differences were not statistically significant (Table 11.5). Length of an estrous cycle (SE = ±0.6 days) did not differ among Angus (19.5 days), Brahman (19.7 days),

TABLE 11.5
Number of Waves of Ovarian Follicular Development and Length of the Estrous Cycle in Angus, Brahman, and Senepol Cows as Determined by Rectal Ultrasonography

		2-Wave Cycles		3-Wave Cycles	
Breed	Number of Cows	No. (%)	Cycle Length Days	No. (%)	Cycle Length Days
Angus	11	8 (72.7)	18.3 ± 0.6	3 (27.3)	20.7 ± 0.9
Brahman	9	5 (55.6)	18.6 ± 0.7	4 (44.4)	20.8 ± 0.8
Senepol	10	3 (30.0)	19.3 ± 0.9	7 (70.0)	21.4 ± 0.6

From Alvarez et al. (2000). With permission.

TABLE 11.6
Effect of Breed on Number of Follicles Observed at Ovariectomy Following a 5-Day Superovulation Regimen

	Breed	
Follicle Size	**Angus**	**Brahman**
Small (1.0 to 3.9 mm)	8.0 ± 1.8	12.0 ± 1.8
Medium (4.0 to 7.9 mm)[a]	7.5 ± 2.6	19.4 ± 2.5
Large (≥8.0 mm)	14.9 ± 4.2	29.1 ± 4.2
Total[a]	30.5 ± 5.6	60.5 ± 5.5

[a] Effect of breed ($P < 0.01$).

From Simpson et al. (1994). With permission.

and Senepol (20.4 days) cows. Length of an estrous cycle was longer ($P < 0.01$) in cows with three waves of follicular development (20.9 ± 0.47 days) than in cows with two waves of follicular development (18.7 ± 0.44 days) during an estrous cycle.

The effect of breed on ovarian function following superovulation was also investigated (Simpson et al., 1994). Estrus was synchronized using progestogen (as previously described) and approximately 10 days after estrus, cows were subjected to a 5-day superovulation regimen (days 0 to 4) using a total of 40 mg follicle-stimulating hormone (FSH; Schering–Plough Animal Health Corp., Kenilworth, NJ). Injections were given twice daily with a dose sequence of 5 mg per injection on days 0 and 1, 3.75 mg per injection on days 2 and 3, and 2.5 mg per injection on day 4. On day 3, cows were administered prostaglandin $F_{2\alpha}$ (25 mg in morning and 15 mg in evening). All cows were ovariectomized on day 5, 12 to 24 hours prior to expected time of ovulation. Brahman cows had greater ($P < 0.01$) numbers of medium and total follicles than Angus cows (Table 11.6). These data suggested that the Brahman cows responded to the superovulation regimen as well as the Angus cows (i.e., numbers of large follicles at ovariectomy).

Embryos were collected using the same synchronization and superovulation regimen of Simpson et al. (1994), but direct comparisons were not made among or between breeds. Initially, embryos were collected from Line 4 Hereford cows from a long-term genotype × environment interaction study conducted at STARS. These Line 4 Hereford cows originated from Line 1 Hereford cows and Line 1 Hereford bulls from Montana but were bred and raised in Florida. The Hereford herd was being dispersed at the STARS and our objective was to preserve semen and embryos from the Line 4 Herefords prior to the dispersal of the herd. In 1990, 14 Line 4 Hereford cows were subjected to the synchronization and superovulation regimen. Because fertile Line 4 Hereford bulls were available, donors were bred by natural service after observed estrus. However, by 48 hours following the first prostaglandin injection, only six cows had been observed in estrus and were exposed and serviced by bulls. Therefore, all cows were exposed to bulls at 50 to 52, 54 to 56, 58 to 60, and 73 to 75 hours after the first injection of prostaglandin. Seven of the 14 cows were observed to be serviced by bulls, and these seven cows were flushed (nonsurgically) approximately 7 days later. A total of 25 quality embryos (i.e., freezable embryos) were collected and frozen (in glycerol) from five of the seven donor cows (3.6 embryos per donor flushed or 1.8 embryos per superovulated donor). In 1992, 10 Line 4 Hereford cows were subjected to the synchronization and superovulation regimen, but for all further studies animals were bred by artificial insemination (AI) three times (48, 60, and 72 hours following the first prostaglandin injection; i.e., twice on day 5 and once on day 6; one straw of semen per AI) before embryo collection. A total of 24 quality embryos were collected and frozen (in glycerol) from six of the 10 donor cows (2.4 embryos per superovulated donor). This was repeated using the same 10 donor cows approximately 3 months later, and 38 embryos were collected and frozen (in glycerol) from seven of the 10 donor cows (3.8 embryos per superovulated donor). In 1993, eight Line 4 Hereford cows were subjected to the synchronization and superovulation regimen and bred by AI. A total of seven embryos were collected and frozen (in glycerol) from two of the eight donor cows (0.9 embryos per superovulated donor). These eight donor cows were all flushed as part of a training program, and this may explain the poor results. In 1995, 16 Line 4 Hereford cows were subjected to the synchronization and superovulation regimen and bred by AI. A total of 24 embryos were collected and frozen (in glycerol) from nine of the 16 donor cows (1.5 embryos per superovulated donor). In summary, a total of 118 Line 4 Hereford embryos were collected and frozen from 29 of 58 superovulated donor cows (2.0 embryos per superovulated donor). These results may be influenced by the Line 4 Herefords not being adapted to the subtropical environment and also that these cattle are inbred. Although donor cows can be bred by bulls using natural service, it was deemed more efficient to use AI. Perhaps some donor cows may not exhibit a strong estrus or be serviced by bulls but could potentially yield quality embryos if bred by AI. Use of AI, however, requires the use of good quality semen. It is also essential that highly skilled embryo transfer specialists perform the embryo collections and processing of the embryos to ensure optimal results.

In another study, miniature Brahman cattle were produced to study the effects of this anomaly on the regulation of growth (Hammond et al., 1998). Three miniature Brahman cows and five normal-stature Brahman cows believed to be heterozygous for the miniature gene were synchronized, superovulated, bred by AI using semen from miniature sires, and embryos were collected 7 days after AI. A total of 49 quality embryos (i.e., suitable for fresh transfer or freezing in glycerol) were collected from five of the cows. Two of these embryos were collected from one of the three miniature Brahman cows, and the remaining 47 embryos were collected from four of the five normal-stature Brahman cows. These same eight cows were again subjected to the synchronization regimen and approximately 12 days after estrus, cows were subjected to a 3-day superovulation regimen (days 0 to 2) using a total of 24 mg recombinant bovine FSH (r-bFSH; Granada Genetics, Texas). The r-bFSH product used in this study was not commercially available. Injections were given once daily (8 mg per injection). Prostaglandin $F_{2\alpha}$ (one injection, 25 mg) was given on day 2, and cows were bred by AI three times (48, 60, and 72 hours following prostaglandin; i.e., twice on day 4 and once on day 5). A total of 14 quality embryos were collected. Three of these embryos were collected from one of the three miniature cows, and the remaining 11 embryos were collected from three of the five normal-stature Brahman cows. In summary, using FSH, an average of six embryos were collected from eight Brahman superovulated donors, whereas using r-bFSH an average of 1.8 embryos were collected from eight Brahman superovulated donors. Following the transfer of these 64 embryos to Angus and Brahman recipients, 28 of 64 (44%) were pregnant and 22 of 64 calved (34%; Table 11.2 data for 1991). The miniature Brahman dams and the embryos produced from the matings were highly inbred and may have affected the results of this study.

There are differences in ovarian characteristics between miniature and normal Brahman cows (Chase et al., 1998b). Furthermore, there is a high death loss at birth of these miniature Brahman calves under natural conditions (i.e., when miniature Brahman cows are bred to miniature Brahman bulls) as well as when miniature Brahman embryos were transferred to normal Angus and Brahman recipient dams (Hammond et al., 1998). Thus, the lower-than-expected pregnancy and calving rates observed in this study may be a reflection of increased embryonic and fetal losses. Subsequent to this study, five miniature Brahman cows were subjected to the synchronization protocol and, at approximately 10 days after estrus, a 4-day superovulation regimen (days 0 to 3) using a total of 220 mg FSH (FSH-V; Folltropin-V®, Vetrapharm Canada, Inc., London, Ontario). Injections were given twice daily with a dose sequence of 30 mg per injection on days 0 to 2 and 20 mg per injection on day 3. On day 3, cows were administered prostaglandin $F_{2\alpha}$ (25 mg in morning and 15 mg in evening). Cows were bred by AI three times (twice on day 5 and once on day 6), and embryos were collected 7 days after AI. The reason for using FSH-V rather than FSH was that FSH was no longer marketed in the United States. A total of two embryos were collected and frozen from two of the five miniature Brahman donor cows.

As previously mentioned, in order to objectively evaluate the Romosinuano breed, we recognized the importance of sampling pure Romosinuano germ plasm of Colombian origin. Initial importations of Romosinuano embryos were from Costa Rica, a country which is recognized as free of foot-and-mouth disease and thus embryo importations were not restricted by health and sanitary regulations pertaining to foot-and-mouth disease. In 1991, the U.S. Animal and Plant Health Inspection Service (APHIS, 1991) published regulations that allowed the importation of washed embryos into the United States from countries with endemic foot-and-mouth disease and rinderpest. Working under these regulations, in cooperation with the Central University of Venezuela–Maracay and Romosinuano producers in Venezuela, we collected Romosinuano embryos in Venezuela in 1995. Forty-four donor cows and 14 bulls were quarantined and used. Semen was collected and frozen from the bulls to artificially inseminate the donor cows. The 44 Romosinuano donor cows were subjected to the synchronization and superovulation (FSH) regimen and bred by AI. A total of 143 embryos were collected and frozen (in glycerol) from 23 of the 44 donor cows (3.2 embryos per superovulated donor). Following health testing conducted by APHIS, 140 embryos from 22 cows bred to 12 bulls were imported to the STARS in February 1996 (Chase, 1996). These 140 Romosinuano embryos were thawed (three-step method using glycerol) and transferred into Angus and Senepol recipient dams, and 69 live calves were born (49% calving; Table 11.4). These results support the use of embryo transfer for the movement of cattle embryos on a worldwide basis. To our knowledge, this was the first importation of cattle embryos into the United States from a country with endemic foot-and-mouth disease.

Following the evaluation of Senepol at the STARS, semen and embryos were collected and frozen prior to the dispersal of the Senepol herd. This allows us the flexibility to be able to regenerate Senepol germ plasm if there is a research need in the future. Seventeen Senepol cows were subjected to the synchronization and superovulation (FSH-V) regimen and were bred by AI. A total of 50 embryos were collected and frozen (in ethylene glycol for direct transfer) from 10 of the 17 Senepol donor cows (2.9 embryos per superovulated donor). Subsequently, this was repeated on 21 Senepol donor cows (six of which had previously been used), and a total of 58 embryos were collected and frozen (in ethylene glycol) from 15 of the 21 donor cows (2.8 embryos per superovulated donor). This was the first time we used ethylene glycol in the freezing process. The primary advantage to the use of ethylene glycol is in the thaw process, which allows direct transfer. For direct transfer, a thawing sequence is applied to the straw containing the embryo, and then the embryo is directly transferred from the straw to the recipient dam much like as in AI. Currently, freezing in ethylene glycol and direct transfer are commonly practiced by embryo collection and transfer specialists.

SUMMARY

Superovulation and embryo transfer techniques can be used successfully to accelerate the production of desirable animals. These technologies are also useful for the safe and efficient movement of cattle genetics on a worldwide basis. It is important that objectives and goals be set for a particular situation (e.g., to collect and market embryos or to collect and transfer embryos on site). In addition to the importance of selecting the donor cows and sires used to produce the embryos, it is also important to consider the recipient dams. In our studies conducted in the subtropics of Florida, using available estrous synchronization protocols, similar calving rates were observed after the transfer of embryos to Angus and Brahman recipient dams as well as to Angus and Senepol recipient dams. Although breed of recipient dam is likely to significantly influence length of gestation and birth weight, the effect of the breed of recipient dam on weaning weight was inconsistent. Reproductive characteristics do differ between tropical and temperate breeds of cattle and must be considered in designing synchronization and superovulation protocols. However, embryos can be successfully recovered and transferred from tropically adapted cattle including the Brahman, particularly when conducted in the tropics or subtropics. Tropically adapted *Bos taurus* breeds, such as the Senepol and Romosinuano appear to be as heat-tolerant as Brahman and may offer some alternative traits for beef cattle producers. In the future, new estrous synchronization and superovulation regimens will undoubtedly be designed for better control of the estrous cycle and superovulation. These new products and regimens will need to be examined using tropically adapted cattle in warm regions of the United States to ensure the efficient propagation of superior genetics to a large segment of the U.S. beef cattle population and producers.

REFERENCES

Alvarez, P., L. J. Spicer, C. C. Chase, Jr., M. E. Payton, T. D. Hamilton, R. E. Stewart, A. C. Hammond, T. A. Olson, and R. P. Wettemann. 2000. Ovarian and endocrine characteristics during an estrous cycle in Angus, Brahman, and Senepol cows in a subtropical environment. *J. Anim. Sci.* 78:1291.

APHIS. 1991. Importation of cattle embryos from countries where rinderpest or foot-and-mouth disease exists. *Federal Register* 56:55804.

Brown, M. A., L. M. Tharel, A. H. Brown, Jr., W. G. Jackson, and J. R. Miesner. 1993. Milk production in Brahman and Angus cows on endophyte-infected fescue and common Bermudagrass. *J. Anim. Sci.* 71:1117.

Chase, C. C., Jr. 1996. The whole animal approach to the study of bovine reproduction. *J. Anim. Sci.* 74 (Suppl. 1):235.

Chase, C. C., Jr., A. C. Hammond, T. A. Olson, J. L. Griffin, C. N. Murphy, D. W. Vogt, and A. Tewolde. 1994. Effect of recipient breed on milk production and performance traits of Romosinuano embryo transfer calves. *J. Anim. Sci.* 72 (Suppl. 1):87.

Chase, C. C., Jr., A. C. Hammond, T. A. Olson, C. N. Murphy, and J. L. Griffin. 1998a. Effect of recipient breed (Angus and Senepol) on pregnancy and performance traits of Romosinuano embryo transfer calves of Colombian origin. *J. Anim. Sci.* 76 (Suppl. 1):216.

Chase, C. C., Jr., C. J. Kirby, A. C. Hammond, T. A. Olson, and M. C. Lucy. 1998b. Patterns of ovarian growth and development in cattle with a growth hormone receptor deficiency. *J. Anim. Sci.* 76:212.

Chase, C. C. Jr., T. A. Olson, A. C. Hammond, J. L. Griffin, and M. J. Fields. 1992a. Estrous synchronization and pregnancy in Angus and Brahman recipient cows. *J. Anim. Sci.* 70 (Suppl. 1):278.

Chase, C. C., Jr., T. A. Olson, A. C. Hammond, J. L. Griffin, D. W. Vogt, A. Tewolde, and C. N. Murphy. 1992b. Effect of recipient breed on pregnancy and performance traits of Romosinuano embryo transfer calves. *J. Anim. Sci.* 70 (Suppl. 1):25.

Hammond, A. C., T. H. Elsasser, M. C. Lucy, C. C. Chase, Jr., and T. A. Olson. 1998. Dwarf and miniature models in cattle. In: J. W. Blum, T. Elsasser, and P. Guilloteau (Eds.). *Proceedings, Symposium on Growth in Ruminants: Basic Aspects, Theory and Practice for the Future.* pp.188. Univ. Berne, School Vet. Med., Berne, Switzerland.

Hupp, H. D. 1981. History and development of Senepol cattle. College of the Virgin Islands, *Agric. Exp. Stn. Rep.* 11. St. Croix.

Randel, R. D. 1984. Seasonal effects on female reproductive functions in the bovine (Indian breeds). *Theriogenology* 21:170.

Randel, R. D. 1994. Unique reproductive traits of Brahman and Brahman based cows. In: M. J. Fields and R. S. Sand (Eds.) *Factors Affecting Calf Crop.* pp. 23. CRC Press, Boca Raton.

Santos, E. A., A. C. Warnick, J. R. Chenault, D. L. Wakeman, and M. J. Fields. 1988. A novel approach for prostaglandin $F_{2\alpha}$ estrous synchronization in beef cattle. *Proc. 11th Int. Congr. Anim. Reprod. Artifil. Insem.,* Dublin, Ireland. 4:459.

Simpson, R. B., C. C. Chase, Jr., L. J. Spicer, R. K. Vernon, A. C. Hammond, and D. O. Rae. 1994. Effect of exogenous insulin on plasma and follicular insulin-like growth factor I, insulin-like growth factor binding protein activity, follicular oestradiol and progesterone, and follicular growth in superovulated Angus and Brahman cows. *J. Reprod. Fertil.* 102:483.

12 *In Vitro* Fertilization to Improve Cattle Production

J. J. Rutledge

CONTENTS

In vitro fertilization (IVF) of bovine oocytes is a routine procedure in hundreds of laboratories around the world (Gordon, 1994), and in the space of a decade or so has grown from an experimental science to a mature technology ripe for exploitation. This does not imply that there are few scientific questions left to answer about or with IVF; rather, it should be taken as a compliment to the researchers. It is the aim of this chapter to explore possible applications of IVF in beef production. Addressed are some applications in dairying elsewhere (Rutledge, 1996; 1998). Herein lies a broad definition of IVF, much like that of Trounson and Gardner (1999) in that IVF includes sister technologies, which, when packaged together and applied to some production process, exhibit synergism (Rutledge and Siedel, 1983). The opening sentence might be taken to imply that this area of reproductive biology is relatively new; this is not the case, as evidenced by Table 12.1.

PUREBRED PRODUCTION

Considerable use of IVF is made in the purebred industry, with several clinics offering this service (Looney et al., 1994; Hasler et al., 1995). Characteristic clients are those having valuable but clinically infertile cows, often geriatric. Typically, the cows had failed or ceased to be productive in conventional superovulation and embryo transfer programs. For consideration for IVF, both groups used only those ova having at least one layer of compact cumulus cells, and there were an average of about 4.5 usable ova per collection. Cows were scheduled for weekly collections. Looney et al. (1994) transferred 813 embryos from 6344 oocytes recovered and obtained a 40% pregnancy rate.

0-8493-1117-9/02/$0.00+$1.50

TABLE 12.1
Some Watershed Events in Development of *In Vitro* Produced Embryos of Mammals

Person (date)	Events
de Graff (1672)	Discovered and described the antral follicle*
van Leeuwenhoek (1677)	Discovered and described sperm in semen*
Spallanzani (1780)	Hypothesized that sperm were fertilizing agent*
von Baer (1828)	Discovery of mammalian ova
Heape (1891)	Embryo transfer in rabbit
Pincus and Enzmann (1935)	*In vitro* maturation of mammalian oocyte (rabbit)
Willett et al. (1951)	Birth of calf from embryo transfer
Chang (1951); Austin (1951)	Independently discovered and described sperm capacitation
McLaren and Biggers (1958)	Culture to blastocyst with transfer and live births (mouse)
Chang (1959)	*In vitro* fertilization with transfer and live births (rabbits)
Brackett et al. (1982)	First calf born from *in vitro* fertilization
Hanada et al. (1986)	First calf born from *in vitro* matured ovum
Sirard et al. (1988)	First calf born from *in vitro* matured ovum, fertilized *in vitro* and cultured *in vitro*

* Cited in Senger (1999).

TABLE 12.2
Pregnancy Rates of *In Vivo* and *In Vitro* Produced Day-7 Embryos

Embryo	No. Transfer	Grade 1 % Pregnant	Grade 2 % Pregnant
IVF—fresh	1884	59	45
In vivo—fresh	320	76	65
IVF—frozen	67	42	—
In vivo—frozen	325	67	—

Adapted from Hasler et al. (1995).

Hasler et al. (1995) transferred both fresh and frozen IVF and *in vivo* day 7 embryos, which allows comparison (Table 12.2). Clearly, the IVF embryo is inferior to the *in vivo* embryo. Also note that there appears to be a greater susceptibility to insult as the drop in pregnancy rate due to freezing/thawing was about twofold that of the *in vivo* embryo. Likewise, there was a greater drop in pregnancy rate with IVF embryos with respect to grade. But IVF has found a distinct niche in extending the reproductive life of valuable cows. Cost considerations, however, restrict its use to a small proportion of purebreds.

Other opportunities exist for use of IVF in the purebred industry. For example, IVF would allow direct selection for carcass traits in slaughter heifers. To accomplish this, one could harvest ovaries at slaughter, keeping identity intact, and evaluate the carcasses for slaughter traits. If the evaluation could be completed and selection of superior animals made in less than 12 hours, ova from only the selected ovaries would need harvest and further processing. Most carcass traits are at least moderately heritable, and direct selection would be efficient. There are many circumstances where purebreeding is desirable, but the maintenance of intact pedigrees is unnecessary. For example, a rancher might cull a large number of females at one time intending to collect ovaries and use the resulting embryos for export or other uses where pedigrees are unimportant. Because of the bulk processing associated with IVF, very inexpensive but high-quality embryos could be made.

Most of the costs of beef calf production are in the maintenance of the brood cow herd (Dickerson, 1978). Recognizing this, Taylor et al. (1985) proposed producing calves without a brood cow herd. To accomplish this, the producer impregnates yearling heifers with female embryos and at parturition places the cow–calf pair in a feedlot for finishing of the one-calved heifer. Early weaning coupled with high-energy feed yields a slaughter heiferette at about 28 to 30 months of age. Her calf (or calves) are impregnated as yearlings to continue the cycle, etc. Biological efficiency (grams lean tissue per MJ metabolizable energy) of the single-calving heifer system was projected to be 24.6% assuming a reproductive rate of 0.75 and slaughter at 0.8 mature weight (Taylor et al., 1985). Bourdon and Brinks (1987), using a different measure of efficiency, obtained a predicted increase of 11%. Hermesmeyer et al. (2000) evaluated feedlot performance of heifer–calf pairs and concluded that the heifers can be finished for slaughter at 30 months, yielding a USDA choice carcass even though they are lactating.

Another use of IVF in purebreeding or straightbreeding systems would be with the Twinner breed, which is currently organizing its herdbook (B. L. Kirkpatrick, personal communication). Much of the foundation heredity of the Twinner breed comes from the selection project for twinning conducted at the Meat Animal Research Center in Nebraska, where up to 40% of the cows twin. Owners of Twinner cattle are thus committed to providing the extra husbandry, nutrition, and other support required by the twin-bearing cow. One could palpate cows 7 days post-estrus, determine those which had a single ovulation and augment the suspected pregnancy by implanting an IVF embryo on the side contralateral to the corpus luteum. Later, at 70 days of gestation, ultrasound technology could be used to determine which cows are carrying twins. At Wisconsin (Agca et al., 1998), a production system was investigated using biopsied, sexed, and frozen IVF embryos and Holstein × Hereford heifers as dams. Artificial insemination alone gave a 65% pregnancy rate in 31 heifers. Artificial insemination plus an embryo transfer yielded a 58% pregnancy rate in 36 heifers, but 11 carried twins so that 21 heifers carried a total of 32 calves. Use of fresh IVF embryos rather than biopsied and frozen IVF embryos should improve the rate of calf production. In a similar study conducted in Japan with mature cow recipients and transfer of two IVF embryos to the opposite side subsequent to artificial insemination, Suzuki et al. (1994) obtained a 95.7% calving rate.

Another use of IVF embryos in purebreeding, which has not been investigated to my knowledge, might be as a "supporter embryo" for high-value embryos which are in some way compromised such that the expected development to live calf is low. For example, included as treatments in Agca et al. (1998) were single (transfer to same side as corpus luteum) and twin (bilateral) transfers of IVF embryos. In the former, eight calves resulted from 40 transfers (20%), while in the latter, 27 of 72

(37.5%) transfers resulted in calves born. These two rates differ ($P = 0.06$). Inarguably, the IVF biopsied, frozen, and singly transferred embryo is compromised. However, if co-transferred with another embryo, even one compromised by biopsy and vitrification, the probability of a calf birth from such an embryo is nearly doubled. McMillan et al. (1994) found evidence for nonindependence in embryo survival in twin fresh transfers of IVF embryos. This use of IVF embryos needs further investigation, and an efficient sexing mechanism for the supporter embryo, which would avoid producing a freemartin from the valuable embryo, would be desirable.

CROSSBRED PRODUCTION

Broadly there are two kinds of crossbreeding systems: rotational and terminal. In rotational systems, two or three breeds of sires are used sequentially over generations, but since cow generations overlap, usually sires of each breed need to be in use at any one time. This leads to difficulties in administration of the crossbreeding program. Further, it is difficult to take advantage of complementarity since breeds used must be somewhat similar in size. In contrast, in terminal systems there are distinct sire and dam lines with up to a 400-lb difference in the mature weight of cows of the lines (Notter, 1977). The systems also differ in purebred (or genetic) inputs. The rotational system requires only input of bulls, while terminal systems require not only the terminal sire breed but also a source of cows. There are systems, so-called roto-terminal systems, where the crossbred cows are produced in a rotational system and those cows not needed for replacements are bred to a terminal breed of bull. To investigate options, lets look briefly at crossbreeding theory (Figure 12.1). The geometry of Figure 12.1 (Cunningham, 1987), which is based on a very general theory, has been found to fit or explain much beef cattle data (Gregory et al., 1995), although there needs to be more work done on *Bos indicus* crosses and particularly the *inter se* generations of such crosses. The first cross of purebreds is the F1, and with positive heterosis its value will be greater than the parental average. The cross of two F1s yields the F2, with decreased expected performance owing to recurrence of homozygotes, hence less heterozygosity. Generally, the performance of any advanced cross, i.e., the *inter se* mating generations, lies on a smooth curve halfway between the "roof" and the straight line connecting the purebreds. In a two-breed rotation, at equilibrium, the expected performance oscillates on the roof between the points labeled C1 and C2. A key point of this theory is that expected performance of all composites, such as the F2 and later generations, is less than that of the initial cross. As more breeds enter the initial cross, this effect is dampened (Gregory et al., 1995).

In the southern U.S., some amount (approximately one half) of *Bos indicus* breeding in the cow herd has been found advantageous. Such cows, particularly if they are F1, are efficient producers of beef (Koger et al., 1962; Turner et al., 1968), producing on average about 21 more calves weaned per 100 cows exposed than either the *Bos taurus* or *Bos indicus* purebred. Strictly speaking, these two are not true species in the taxonomic sense (Groves, 1981); however, they have been genetically isolated as straightbreds for 200,000 to one million years (Loftus et al., 1994)

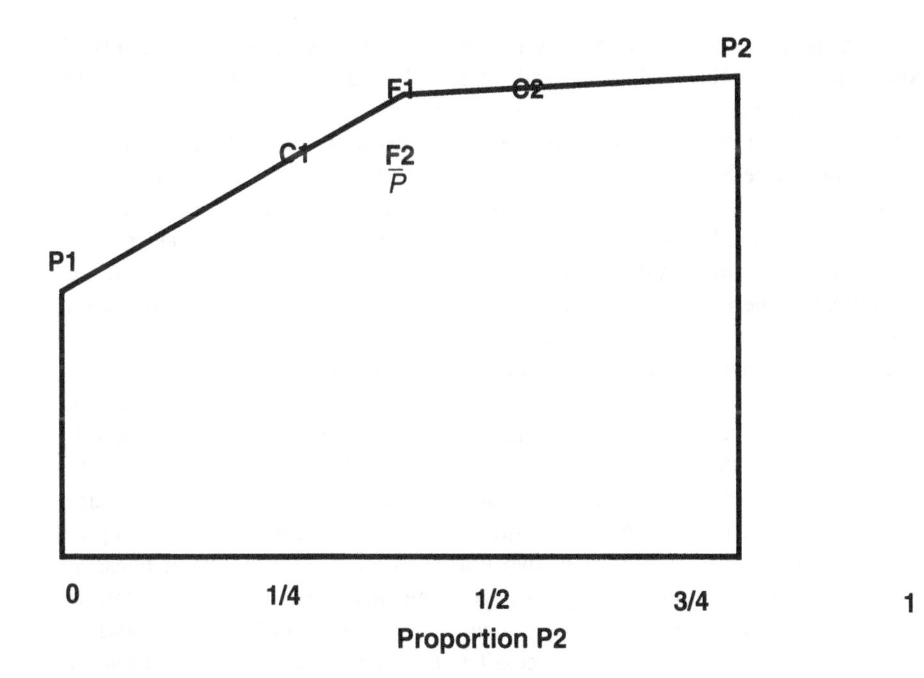

FIGURE 12.1 Cunningham's Greek Temple Model. In the absence of epistasis, expected values of all intergrades between the two parental breeds (P1 and P2) lie on the roof of the peak which is determined by the F1. F2 and subsequent *inter se* generations have expected value intermediate between the F1 and parental mean $\bar{P}$. A two-breed rotational system would oscillate between the points C1 and C2, depending upon sire breed.

and exhibit some of the attributes of species. Were the wild forms still available, taxonomists would designate them semi-species (Mayer, 1940).

A primary problem in exploiting the F1 superiority in cattle is insufficient reproductive excess to efficiently produce and maintain an F1 brood cow herd. This can be illustrated using data from Turner et al. (1968). Straightbred cows (Angus, Hereford, or Brahman) all averaged about 61 calves weaned per 100 cows exposed (W = 0.61). Assuming a 15% replacement rate (R), we can determine the number (N) of straightbred cows necessary to maintain an F1 herd of, say 1000, mature females. To sustain the system, 150 F1 heifers plus 0.15 N straightbred heifers are needed annually. Production of heifers by the straightbred herd is 0.61 N/2, where the divisor reflects an equal sex ratio. Equating production to need, we obtain a single equation in N: $WN/2 = R(1000 + N)$, generally, or for our example, N = 968 straightbred cows are required to sustain the system. Thus, about half the brood cow resources would need to be allotted to the relatively unproductive straightbreds (Koger, 1980). On the other hand, assuming a source of gametes and exploitation of IVF technology, the entire brood cow herd could be the relatively productive F1. To effect this, one would implant sufficient F1 cows with F1 IVF embryos to obtain replacement heifers; the rest of the herd would be mated to a terminal sire breed.

Of the terminal cross system using F1 dams, that great student of southern beef production Marvin Koger (Koger, 1980) wrote: "Insofar as F1 dams are available for generation three-breed crosses, this system of crossbreeding is highly effective. Certain features of the system, however, limit its utility to the production of relatively few terminal three-breed crosses. The system has three principal disadvantages: 1) the system involves three phases, including (a) the production of straightbred dams, (b) the production of F1 dams, and (c) production of the three-breed terminal crosses; 2) considering the total system, approximately one half of the dams involved are straightbreds, whereas the principal objective of crossbreeding is to capitalize on hybrid dams; 3) acquisition of adequate numbers of replacement F1 females is a difficult undertaking. Few F1 females are available at affordable prices."

All of what Professor Koger wrote was true in 1980. Two decades later, however, additional options in cattle production are available from IVF technology. Hundreds of thousands of high-quality *Bos taurus* cows are slaughtered annually in the U.S. from which ova could be collected. This completely removes the necessity of maintaining straightbred cows. In this regard, there should be little concern if the "straightbred" slaughter cow were a cross of two European breeds, for the major boost of heterosis likely comes from heterozygosity generated in the cross between *Bos taurus* and *Bos indicus* genomes. Very large numbers of replacement F1 females could be made using IVF at very reasonable prices. The *Bos indicus* bull has a reputation for rather low fertility in natural matings (Cartwright, 1980). Since a single straw of semen might be used on several hundred ova, fertility of the straightbred *Bos indicus* bull in natural matings is of no concern.

CONCLUSIONS

The technologies collectively known as IVF have matured greatly in the last two decades, but their emergence was made possible only from a large store of basic research in reproductive biology, much of which probably seemed to the practical man to have been motivated by idle curiosity. It is a lesson difficult to learn that only in retrospect can the importance of some bit of scientific work be accurately judged. The mature IVF technology can and should be exploited by cattle producers in a variety of ways. Herein a sampler of uses has been presented, but innovative minds having IVF in their tool kit of problem solvers will ultimately demonstrate the full capabilities of IVF when applied to cattle production.

REFERENCES

Agca, Y., R.L. Monson, D.L. Northey, D.E. Peschel, D.M. Schaefer, and J.J. Rutledge. 1998. Normal calves from transfer of biopsied, sexed and vitrified IVP bovine embryos. *Theriogenology* 50:129.

Austin, C.R. 1951. Observations of the penetration of the sperm into the mammalian egg. *Aust J. Sci. Res.* 4:581.

Baer, Karl Ernst von. 1828. Über Entwickelungsgeschichte der Thiere: Beobachtung und Reflexion. Königsberg.

Bourdon, R.M. and J.S. Brinks. 1987. Simulated efficiency of range beef production III. Culling strategies and nontraditional management systems. *J. Anim. Sci.* 65:963.

Brackett, B.G., D. Bousquet, M.L. Boice, W.J. Donawick,. J.F. Evans, and M.A. Dressel. 1982. Normal development following in vitro fertilization in the cow. *Biol. Reprod.* 27:147.

Cartwright, T.C. 1980. Prognosis of Zebu cattle: research and application. *J. Anim Sci.* 50:1221.

Chang, M.C. 1951. Fertilizing capacity of spermatozoa deposited in the fallopian tubes. *Nature* 168:697.

Chang, M.C. 1959. Fertilization of rabbit ova *in vitro. Nature* 184:466.

Cunningham, E.P. 1987. Crossbreeding—the Greek temple model. *J. Anim. Breed. Genet.* 104:2.

Dickerson, G.E. 1978. Animal size and efficiency: basic concepts. *Anim. Prod.* 27:367.

Gordon, I. 1994. *Laboratory Production of Cattle Embryos.* CAB Inter., Wallingford, UK.

Gregory, K.E., L.V. Cundiff, and R.M. Koch. 1995. Composite breeds to use heterosis and breed differences to improve efficiency of beef production. Memo. Roman L. Hruska U.S. Meat Anim. Res. Ctr., Clay Center, NE.

Groves, C.P. 1981. Systematic relationships in the Bovine (Artiodactyla, Bovidae). *Zool. Syst Evolut.* 19:264.

Hanada, A., U. Enya, and T. Suzuki. 1986. Birth of calves by non-surgical transfer of *in vitro* fertilized embryos obtained from oocytes matured in vitro. *Jpn. J. Anim. Reprod.* 32:208.

Hasler, J.F., W.B. Henderson, P.J. Hurtgen, Z.Q. Jin, A.D. McCauley, S.A. Mower, B. Neely, L.S. Shuey, J.E. Stokes, and S.A. Trimmer. 1995. Production, freezing and transfer of bovine IVF embryos and subsequent calving results. *Theriogenology* 43:141.

Heape, W. 1891. Preliminary note on the transplantation and growth of mammalian ova with a uterine foster-mother. *Proc. R. Soc.* 48:457.

Hermesmeyer. G.N., L.L. Berger, D.B. Faulkner, D.J. Kesler, T.G. Nash, and D.M. Schaefer. 2000. Effects of prenatal androgenization and implantation on the performance and carcass composition of lactating heifers in the single-calf heifer system. *Prof. Anim. Sci.* 15:173.

Koger, M., W.L. Reynolds, W.G. Kirk, F. M. Peacock, and A.C. Warnick. 1962. Reproductive performance of crossbred and straightbred cattle on different pasture programs in Florida. *J. Anim. Sci.* 21:14.

Koger, M. 1980. Effective crossbreeding systems utilizing Zebu cattle. *J. Anim. Sci.* 50:1215.

Loftus, R.T., D. E. MacHugh, D.G. Bradley, P.M. Sharp, and P. Cunningham. 1994. Evidence for two independent domestications of cattle. *Proc. Nat. Acad. Sci.* 91:2757.

Looney, C.R. B.R. Lindsey, C.L. Gonseth, and D.L. Johnson. 1994. Commercial aspects of oocyte retrieval and *in vitro* fertilization (IVF) for embryo production in problem cows. *Theriogenology* 41:67.

Mayer, E. 1940. Speciation phenomena in birds. *Amer. Nat.* 74:249.

McLaren, A. and J.D. Biggers. 1958. Successful development and birth of mice cultivated *in vitro* as early embryos. *Nature* 182:877.

McMillan, W.H., P.A. Pugh, A.J. Peterson, and K.L. Macmillan. 1994. Evidence for non-independent embryo survival following the twin-transfer of fresh but not frozen-thawed in vitro produced bovine embryos. *Theriogenology* 41:254.

Notter, D.R. 1977. Simulated efficiency of beef production for a cow–calf–feedlot management system. Ph. D. thesis, Univ. Nebraska, Lincoln, NE.

Pincus, G. and E.V. Enzmann. 1935. The comparative behavior of mammalian eggs *in vivo* and *in vitro. J. Exp. Med.* 62:655.

Rutledge, J.J. 1996. Cattle breeding systems enabled by *in vitro* embryo production. *Embryo Trans. Newsletter* 15:14.

Rutledge, J.J. 1998. Applications of *in vitro* methodology in tropical dairying. In: *Proc. Seventeenth Tech. Conf. AI and Reprod. Nat. Assn. Anim. Brds.* Sept. 25–26 Madison, WI.

Rutledge, J.J. and G.E. Siedel, Jr. 1983. Genetic engineering and animal production. *J. Anim Sci.* 57:265.

Senger, P.L. 1999. *Pathways to Pregnancy and Parturition.* Current Conceptions, Inc., Pullman, WA.

Sirard, M.A., M.L. Leibfried-Rutledge, J.J. Parrish, C.W. Ware, and N.L. First. 1988. The culture of bovine oocytes to obtain developmentally competent embryos. *Bio. Reprod.* 39:546.

Suzuki, O., M. Geshi, M. Yonai, and M. Sakaguchi. 1994. Effects of method of embryo production and transfer on pregnancy rate, embryo survival rate, abortion and calf production in beef cows. *Theriogenology* 41:309.

Taylor, St. C.S., A.J. Moore, R.B. Thiessen, and C.M. Bailey. 1985. Food efficiency in traditional and sex controlled systems of beef production. *Anim. Prod.* 40:401.

Trounson, A.O. and D.K. Gardner, Eds. 1999. *Handbook of In Vitro Fertilization,* 2nd ed., CRC Press, Boca Raton, FL.

Turner, J.W., B.R. Farthing, and G.L. Robertson. 1968. Heterosis in reproductive performance of beef cows. *J. Anim. Sci.* 27:336.

Willett, E.L., W.G. Black, L.E. Casida, W.H. Stone, and P.J. Buckner. 1951. Successful transplantation of a fertilized bovine ovum. *Science* 113:247.

13 National Bovine Genomics Projects: Present Status, Future Directions, and Why They Are Important

R. Michael Roberts

CONTENTS

This chapter describes current federally funded efforts to provide a detailed genetic map of the bovine genome. It traces the history of cattle genetics and indicates how a formalized genetic map of the bovine genome is being created and how such efforts are likely to benefit the beef industry. The author predicts that an effort to provide the entire nucleotide sequence of all 30 chromosomes of the bovine genome is unlikely to happen for many years because of expense. However, a bovine genome project will be the beneficiary of the Human Genome Initiative, since the great majority of genes are well conserved, both in sequence and in the order in which they are arranged on chromosomes, between humans and cattle. Nevertheless, there

0-8493-1117-9/02/$0.00+$1.50

is a national need to invest in genomics research so that the information encrypted in the genomes of agriculturally important species can be of value to producers.

The USDA supports research on animal genomes through both its intramural (ARS; Agricultural Research Service) and its extramural (CSREES; Cooperative States Research, Education and Extension) programs. The ARS (www.ars.usda.gov) plays a major role in national efforts to improve agricultural productivity by genetic selection. It is particularly prominent in developing genetic maps and in identifying and maintaining genetic resources. CSREES (www.reeusda.gov) provides funding to scientists throughout the U.S. in the form of grants-in-aid. There are so-called special grants. Some of these are awarded in a competitive, peer-reviewed manner, while others are earmarked to specific institutions by congressional directive. The major arm of CSREES involved in competitive grants, however, has been the NRI (National Research Initiative). In 1999, the NRI awarded approximately $6 million for grants for animal genetic research, most of it going to universities associated with the Land Grant System. Approximately half of this amount was in the form of larger-than-usual grants to foster the development of reagents and methods that would accelerate research in food animal genomics (a term that includes all aspects of research on the genome). The data and reagents so generated are to be made immediately available to the scientific community. Other, generally smaller, grants will permit the genetic selection of animals with superior production traits and disease resistance. In the years 2000 and beyond, the responsibility for much of the genomics effort in cattle and other food animals was shifted from the NRI to a sister competitive program, IFAFS (Initiative for Future Agriculture and Food Systems; Program 10) (www.reeusda.gov/ifafs/). The USDA, through the NRI, also emphasizes research in so-called functional genomics, including the identification, isolation, and characterization of genes that are likely to be of significance to the agricultural mission, such as ones involved in growth, reproduction, and susceptibility to disease. Grants have been awarded for studying how genes are regulated and organized in the chromosome and for producing transgenic and cloned animals.

In the sections that follow, a justification is given for the present investment in genomics research on livestock by the USDA and for why more funds are needed if the United States is to remain competitive. I shall show how the genomics revolution promises to provide agricultural products more efficiently and safely, and improve the health of animals. I also suggest that these endeavors, though expensive, will be cost-efficient as a beneficiary of the Human Genome Project. Before proceeding with these topics, however, a brief description of the history of genome research and some background to gene mapping and other procedures that are the core of present research efforts will be provided.

A BRIEF HISTORY OF GENETICS AND ITS IMPACT ON ANIMAL AGRICULTURE

The first mammalian progenitor species probably emerged from its reptilian relatives about 200 million years ago. Its members were most likely small and insignificant, as were all mammals until the sudden disappearance of the dominant dinosaurs about 65 million years ago. The very early mammals likely had genomes smaller in both

DNA content and gene number than today's species, but their chromosomes undoubtedly contained the recognizable precursors (orthologs) of just about all the genes of modern-day species, including the human and bovine. Today, the eutherian or placental mammals are comprised of 20 orders and about 4500 species, but, as we shall see, the genome organizations of these various animals are remarkably similar, despite differences in chromosome number. For example, the genomes of the human and the chimpanzee appear to be greater than 99% identical in terms of the sequences of the more than 3300 million bases in their genomes (Crouau-Roy et al. 1996), yet these two closely related species are clearly distinguishable in appearance, habits, and other features. Somehow, the genetic blueprint of each species has been subtly modified, to provide the two with the abilities to survive and reproduce in their respective environments.

The practice of genetics is as ancient as agriculture itself, which began independently in various parts of the world (Diamond, 1999). The most ancient agricultural culture, with origins about 30,000 years ago, is the one that began in the so-called fertile triangle within an area that now encompasses Iraq and other areas of the Middle East. It is from this region that the majority of the 250 or so crop plants planted in the United States originated. Cattle, sheep, and goats were probably domesticated in this region as well, but less than 10,000 years ago. Farmers selected plants and animals with the characteristics they considered desirable. In the case of cattle, which are derived from several subspecies of the now extinct auroch, attention was undoubtedly paid to how fast the animals grew, how quickly they matured, and to their breeding habits and manageability. Although originally domesticated for meat and fiber, the emerging *Bos taurus* and *indicus* species became sources of milk, which ultimately led to the selection of the dairy breeds of today. Experimental breeding was also practiced, as is evident from the discovery of *Bos indicus* genes mixed with *taurus* genes in African breeds of cattle (MacHugh et al., 1997; Nijman et al., 1999). This agrarian practice of judicious breeding followed by selection was pragmatic and intuitive, and based on the recognition that many of the characteristics that were observed in an animal (its phenotype) were transmitted to its different offspring, although not always in a predictable manner.

It was not until the rediscovery of Mendel's experiments on peas and the explosion of genetic studies at the beginning of the 20th century, particularly on model organisms such as the fruit fly, *Drosophila*, that genes could be conceptualized as predictable, physical entities, even though their chemical composition remained enigmatic. It became clear that single genes could exist in different forms, known as alleles, and that a gene could be defined by its position or locus relative to other genes. In this manner, linkage groups and simple genetic maps were established, particularly in species where mutations with precise phenotypes could be created at will and crosses easily conducted. Cattle were not among the species easily amenable to this kind of study, but the realization that there were genes and that these genes obeyed certain rules allowed breeding programs to become less empirical than in preceding centuries. That genes encoded proteins was deduced well before it was confirmed that genes were made up of DNA, a concept that emerged only slowly, because it was a mystery as to how a substance varying only in two purine and two pyrimidine components could be the genetic material. The Watson–Crick model and the flowering of molecular biology over the next three decades defined the genetic code and the physical nature of genes. The work also provided a biochemical basis for many simple genetic diseases, i.e., ones involving mutations within single genes, and a better understanding of mutational processes and genetic recombination. By 1985, it was possible to discuss seriously the possibility of defining the entire blueprint of a eukaryotic organism by the nucleotide sequence of the DNA contained in its nuclei. As a consequence, the sequence of the human genome is completed, and the sequences for several other organisms are already on hand. In the narrative that follows, I shall attempt to explain how data from these projects will influence the course of animal agriculture over the first quarter of the 21st century.

WHAT IS A GENETIC MAP AND HOW IS IT HELPFUL?

LINKAGE MAPS

A gene exists at a specific location or locus on the chromosome (Figure 13.1). As discussed above, genetic maps can be created by estimating recombination frequencies among loci, but the density of markers so produced tends to be very low in species where such crosses are laborious to perform. It was far more successful for mapping the genomes of plants and mice where large numbers of progeny can be obtained quickly. In 1980, however, a new type of genetic marker was discovered, known as a restriction fragment polymorphism (RFLP) (Botstein et al., 1980). In this procedure, restriction enzymes, which cut DNA and hence specific genes at particular sequences of nucleotides, are used to detect base changes, deletions, and inversions in the genes of related and unrelated individuals. Inheritance patterns of the DNA banding patterns, and hence individual gene variants, can be determined, thereby allowing different loci to be mapped relative to one another in large numbers of animals. Such changes are, however, relatively rare and difficult to find and were

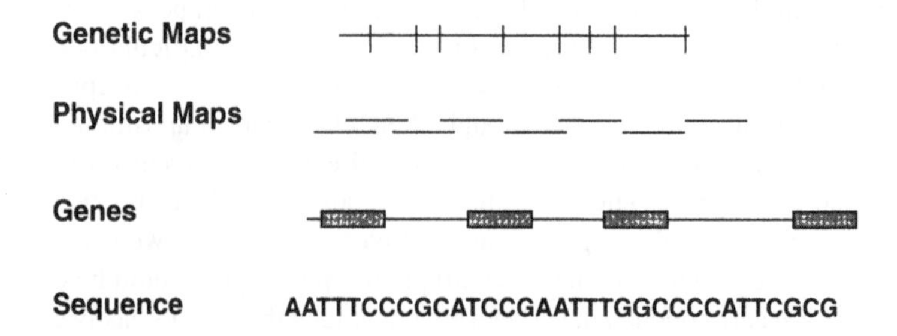

FIGURE 13.1 Mapping and organization of the genome of a typical mammal. In line 1, loci for genes and genetic markers, such as STPs, are arranged in linkage groups in an order that reflects recombination frequencies. In line 2, the DNA has been broken into identifiable, linearly arranged pieces, e.g., BACs, to create a physical map of the genome. Line 3 shows the physical placement of the genes themselves within a small piece of DNA. Line 4 illustrates the ultimate physical map, the nucleotide sequence itself.

more useful for forensic purposes, e.g., for analyzing blood or semen samples in criminal investigations, than for the livestock industry.

More recently other sources of DNA variation between individuals have been exploited to create sophisticated pedigrees and denser genetic maps. For example, microsatellites consist of inheritable sequences of di-, tri- and tetranucleotide repeats, occurring at well-defined positions throughout the genome. These short, tandem repeats (STRs) are highly polymorphic, varying in length according to the number of repeats present, but are flanked by less variable DNA sequences. A DNA amplification procedure, known as the polymerase chain reaction (PCR), can be used to measure the lengths of such microsatellites. Conserved sequences on either side of the repetitive sequence are used to prime the amplification reaction, and the sizes of the microsatellite amplicons so generated are measured by comparing their mobilities on electrophoretic gels. The inheritance of many such markers can be followed after experimental crosses. Consequently microsatellites have become the most common marker on genetic maps of livestock.

A further sophistication in mapping has been to follow single nucleotide polymorphisms (or SNPs) that often distinguish alleles of a single gene. The frequencies of such mutational differences depend on the species. In the human, they are relatively rare and are found every 500 to 1000 bases along the DNA. The methods for their detection, all of which depend in one form or another on the PCR (Landergren et al., 1999), are beyond the scope of this review. Moreover, such markers have not yet been heavily exploited for livestock. An advantage of the STR and SNP procedures is that large numbers of individuals can be compared for differences at many sites by using automated procedures. The SNP method also has enormous potential utility for screening herds and flocks for deleterious alleles caused by single base mutations. Linkage maps have been constructed for cattle, sheep, pigs, and deer by making use of a variety of such polymorphic loci. The goal for each species is to create a dense, evenly ordered map for every chromosome that incorporates several thousand markers anchored by the accurate placement of specific known genes. The information presently available can be accessed at a variety of sites on the World Wide Web (Montgomery, 2000). For all the above species, map densities are well below what is optimal. In contrast, maps for at least a dozen plant species have been constructed at such a density that it is possible to locate genes responsible for economically important traits and to apply marker-assisted selection procedures (see section below). These comparative maps also allow the positions of genes in poorly mapped species to be predicted.

Physical Maps

The genome of a cow consists of 30 pairs of haploid chromosomes. Its haploid complement is comprised of about 3000 million bases, not very different from that of the human, pig, sheep, or goat (Table 13.1). Like the human, there are probably between 30,000 and 40,000 genes, and these genes are most likely contained in less than 5% of the DNA. The rest is so-called junk DNA whose function is unknown, but may have accumulated inadvertently over evolutionary time. As we shall see, the massive amounts of gene-poor DNA, often containing difficult-to-distinguish segments of repetitive material, create major problems for the gene sequencer but

TABLE 13.1
Comparisons of Sizes of Mammalian Genomes and of Current Genome Maps

Species	Order	Haploid Chromosome Number	Number of Mapped Genes	Number of Mapped Microsatellites	Genome Length (in million bases)
Human	Primate	23	>30,000	~8000	3300
Mouse	Rodentia	20	~7000	~7500	1450
Cow	Artiodactyla	30	~500	~1300	2990–3532
Goat	Artiodactyla	30	257	307	3100
Sheep	Artiodactyla	27	254	504	3063
Pig	Artiodactyla	19	369	~1000	2300–2500
Horse	Perissodactyla	32	53	309	2000

Data from O'Brien et al. (1999).

must be accounted for in accurate maps of the genome. Physical mapping involves the assignment of genes to chromosomal segments (Figure 13.1).

Until about 25 years ago it was only possible to infer the location of a gene on a particular chromosome through pedigree analysis. At about that time, other techniques, such as *in situ* hybridization, were developed, which allowed chromosomal assignments to be made directly. More recently, radiation hybrid panels have been used to map linked genes to a relatively small segment of a chromosome (Yang and Womach, 1998). Again, however, a description of this technique is beyond the scope of this chapter.

A more detailed physical map can be obtained by narrowing the location of genes or genetic markers to stretches of DNA no longer than a few hundred kilobases. Genomic DNA is broken up by partial restriction enzyme digestion, and each small piece propagated in the form of artificial chromosomes in yeast, bacteria, or bacteriophage. Bacterial artificial chromosome (BAC) libraries are now available for several livestock species. A library consists of millions of bacteria, such as *Escherichia coli*, with each bacterial cell carrying a single BAC of about 150,000 nucleotide pairs. These libraries are large enough that they represent over a tenfold coverage of an entire genome. Each BAC can be identified in a variety of ways, such as by the pattern in which it is cut by restriction enzymes, by partial nucleotide sequencing, or by markers to which they hybridize. Most of them overlap other BAC, thereby providing a series of "contigs" across the genome. Genes and other genetic markers, such as STRs or expressed sequence tags (ESTs), can be assigned to such BACs, thereby physically placing markers at specific sites in the genome. If necessary, but at considerable cost, the entire BAC can be sequenced. Such a physical map, consisting of overlapping artificial chromosomes, is shown in Figure 13.1, line 2.

Genome Sequence

The ultimate map of the genome of an organism is the complete nucleotide sequence of its DNA (Figure 13.1). Full genome sequences are available for more than 50 bacteria, one plant (*Arabidopsis thaliani*), baker's yeast (*Saccharomyces cerevieae*), the nematode *Caenorhabditis elegans*, and *Drosophila melanogaster*. Those for a second plant species, rice, the human, and mouse are essentially completed but are incompletely annotated, i.e., the genes have not all been classified as to likely function and relation to other genes in terms of their sequences. The mouse is currently being targeted. Two approaches have been used for sequencing the larger genomes. The first, which has been successful for *Arabidopsis* and for much of the publicly funded effort to sequence the human genome, has involved a careful sequencing of overlapping segments of DNA, mainly in the form of BACs. Human chromosome 22 was analyzed in this manner (Dunham et al., 1999). An almost complete sequence was provided, although several gaps still remained of regions that either had defied sequencing efforts or had resisted cloning into BACs. The alternative approach is that of so-called "shotgun" sequencing, which was used successfully for *Drosophila* and also to complete the human and mouse sequences quickly. Here the entire genome is broken into relatively small pieces by restriction enzymes. These pieces are subcloned in bacterial plasmids and sequenced. Sufficient random sequencing is carried out to cover the entire genome numerous times, and the complete data set then assembled by computer with a program that utilizes overlaps to piece the data together into groupings that represent each chromosome. Although the latter procedure has the virtue of being fast, it is probably more error prone and fails in regions where there are multiple repeats. "Difficult" regions are, of course, missed, and enormous sequencing and computing power is needed for success. A company such as Celera, which specializes in sequencing, could, in theory, generate the unassembled, unannotated nucleotide sequence for *Bos taurus* in a month or less if it turned over its entire sequencing capacity to that task.

It is a relatively simple matter to analyze this great mass of assembled sequence data for the location of genes that have been previously identified from cloned copies of their expressed mRNA (so-called expressed sequence tags, or ESTs). On the other hand, there are estimated to be over 30,000 human genes, and EST data are available for only about half this number. Although the missing genes have certain characteristics, for example open-reading frames of coded sequence, these are usually broken up by introns, and contemporary computer programs are not sophisticated enough to recognize every gene submerged in a sea of junk sequence. In addition, less than 10,000 human genes have actual names and known functions. Even with a complete nucleotide sequence of each chromosome, it will be many years before a completely annotated gene map will be available.

THE BENEFITS OF GENE MAPS TO THE BEEF INDUSTRY

Most genetic traits in cattle that are of interest to the beef cattle industry are complex. They are under the control of many genes, each of which contributes only partially to the observed phenotype. Examples of complex traits include disease resistance, lactation, rate of gain, meat tenderness, twinning, timing of puberty, and other reproductive characteristics. The genes responsible for these traits have been termed *quantitative trait loci* (QTLs) (Montgomery, 2000). Those of economic relevance are often known as economic trait loci (ETLs). The ability to track QTLs has considerable value if livestock pedigrees are to be improved, but this task is no simple matter in an outbred population. A typical strategy for identifying loci is described by (Heyen et al., 1999). These workers employed a resource family of more than 1000 cattle and established linkage patterns between the phenotype under study (for example, milk yield or milk fat content) and markers on a genetic map. A complex statistical approach then allowed them to establish approximate chromosomal positions for several QTLs. The method remains far too crude to identify the actual genes, but it does provide markers that segregate with the trait, thereby allowing marker-assisted selection to be practiced.

Simple genetic traits are likely to be dealt with more easily (particularly if the phenotype is readily scored), for example, the presence of horns or double muscling. The latter trait was mapped to bovine chromosome 2 by a mapping approach similar to that described above and shown to be due to a mutation in a gene for the muscle protein myostatin (Grobet et al., 1997). Such identification would not have been possible, however, if a double muscling phenotype had not been earlier described in a mouse in which the myostatin gene had been deleted experimentally. Indeed QTL identification in cattle is unlikely to advance at a worthwhile pace without making use of comparative mapping and the use of data from the human and mouse genome projects.

COMPARATIVE MAPPING AND THE VALUE OF WHOLE GENOME SEQUENCE FROM THE HUMAN AND MOUSE

As stated earlier, the genomes of mammals all contain a roughly similar number of genes. As species have evolved, exchanges of chromosomal segments have been relatively infrequent, so that a considerable degree of synteny (chromosomal conservation) is evident, even across different orders of mammals. Particularly good preservation of large stretches of chromosome is evident between humans and cattle so that orthologous genes, i.e., ones that have the same function and evolutionary origin, often map to homologous stretches of chromosome. It has been calculated that as few as 27 major chromosomal rearrangements distinguish *Bos taurus* and the human (O'Brien et al., 1999). Even minor rearrangements involving inversions and translocations seem limited in number. As a result, the linear orders of many large groups of genes are maintained across the two species. Clearly, the ability to extrapolate from the human and mouse, where the positions and nucleotide sequences of most genes will soon be known, to cattle and then to related species, such as sheep, will be of enormous benefit. For example, once the rough map position of a QTL has been established, it should be possible to focus on all the genes in that region and select the one that seems the most reasonable candidate for controlling the trait. Cattle can then be selected on the basis of the most favorable allele of that gene, e.g., the one associated with best growth rate or resistance to a particular disease. In such a manner, the optimal combination of genes that together provide the superior phenotype might be selected as a combination to produce an elite animal.

Beef cattle genomics will thus be cost efficient as a beneficiary of the Human Genome Project. That effort, which conservatively cost $2 billion, is unlikely to be repeated for *Bos taurus* unless sequencing costs are reduced dramatically. Even at present-day rates of approximately $0.25 per base of annotated and aligned sequence, it would cost over $500 million, well beyond the budget scope of the USDA, to accomplish the task. The failure to provide a complete bovine sequence means that genes present in cattle and not in the human, as well as many genes that have been modified or placed at unexpected positions in the bovine genome may not be discovered any time soon. Since these genes are likely to include the very ones that have allowed species adaptation and divergence, much useful information will remain missing. Unfortunately, agriculture is likely to remain the poor stepchild of biomedical science. The first priority of any federally funded animal genome project will not, therefore, be whole genome sequencing but will remain the provision of high-density maps, each linked to the sequenced genome of the human and eventually the mouse.

WHAT ARE THE LIKELY OUTCOMES OF A FEDERALLY FUNDED ANIMAL GENOME INITIATIVE?

A major mission of research supported by the USDA is to provide nutritionally sound, safe products more efficiently. The comparative gene mapping approaches outlined previously will allow elite animals to be identified at birth or even *in utero*, thereby providing needed accuracy in selection of livestock for breeding programs or even for cloning (Kappes, 1999). The nation's food supply will be made safer by allowing animals and animal products to be tracked from farm to table. Selection of animals with increased resistance to disease and exhibiting better food utilization will lower the need for antibiotics and for hormones. Alleles that are associated with ill health will be eliminated from the population, and relatively inexpensive tests will become available for potentially undesirable as well as desirable traits.

There will be novel discoveries. Genes controlling traits and developmental steps unique to cattle, for example, will be discovered. Unexpected patterns of inheritance will be revealed. The Calipyge gene, which increases muscle mass in sheep, for example, is transmitted in an unusual manner (Crockett et al., 1996).

WHY IS MORE FEDERAL SUPPORT NEEDED FOR ANIMAL GENOME STUDIES?

A failure to invest in animal genomics reduces the efficiency of agricultural production generally. The majority of corn and soybeans produced in the United States does not go directly into human diets, but instead is "processed" to milk, eggs, and meat. It makes no sense to reap the benefits of genomics through crop improvements alone when further gains can be made in animal production. To be frank, would the American public prefer cheaper broccoli or cheaper hamburger? In addition, economic security strongly depends on global food security. Increasing genetic capacity for milk and meat production will be required to meet projected demand for food as the world population soars.

There are other reasons why the USDA needs to fund genomics research. One is to maintain U.S. competitiveness in world markets. There is already good reason to believe that the U.S. is no longer ahead of the competition. Significant challenges to

U.S. dominance comes from the European Community (in marker-assisted selection of cattle, for example) and from aggressive programs in New Zealand, Australia, and Japan. Such challenges underscore the importance of developing and maintaining a strong, national infrastructure for gene discovery. A second reason for federal investment in the area is that there is a need to maintain as much information as possible on genes in the public domain. Industry is currently accumulating data without releasing its findings. Such trade secrets tend to impede scientific progress and discovery, and render scientists who make use of the secret data the lackeys of big business.

WHAT ADDITIONAL STEPS ARE NEEDED IF THE GENOME REVOLUTION IS TO BE SUCCESSFUL?

There needs to be investment in flocks and herds in which interesting traits are segregating so that such loci can be placed in the context of the high-density maps being constructed as above. Such resources are presently scarce. There should be intensive analysis of ESTs so that as many of the expressed genes as possible can be placed on the high-density maps described above. In this regard, it may be preferable to concentrate activities on specific organs of interest (e.g., muscle, mammary gland, uterus, ovary, and lymphoid tissue). Partial sequencing of some regions of the genome, such as ones dealing with disease resistance or with features found in cattle and not humans, will also have real value. Such efforts will become cheaper as techniques continue to improve.

There is also an acute shortage of personnel working in agricultural genomics. Individuals trained in the necessary statistical methods and in developing the informatics and computer software necessary for analyzing sequence information and extracting data from comparative maps are in particularly short supply. Such individuals will be keys to unlocking the secrets and hidden potential of genomes.

REFERENCES

Botstein, D., R.L. White, M. Skolnick, and R.W. Davis. 1980. Construction of a genetic linkage map in man using restriction length polymorphisms. *Am. J. Human Genet.* 32:314.

Crockett N.E., S.P. Jackson, T.L. Shay, F. Farnir, S. Berghmans, G. Snowder, D.M. Nielsen, and M. Georges. 1996. Polar overdominance at the ovine callipyge locus. *Science* 273:236.

Crouau-Roy, B., S. Service, M. Slatkin, and N. Freimer. 1996. A fine-scale comparison of the human and chimpanzee genome linkage, linkage disequilibrium and sequence analysis. *Hum. Mol. Genet.* 5:1131.

Diamond, J. 1999. *Guns, Germs and Steel: The Fates of Human Societies.* W.W. Norton and Co. New York.

Dunham, I. et al. 1999. The DNA sequence of human chromosome 22. *Nature* 402:489.

Grobet, L., L.J. Martin, D. Poncelet, D. Pirottin, B. Brouwers, J. Riquet, A. Schoeberlein, S. Dunner, F. Menissier, J. Massabanda, R. Fries, R. Hanset, and M. Georges. 1997. A deletion in the bovine myostatin gene causes the double-muscled phenotype in cattle. *Nat. Genet.* 17:71.

Heyen, D.W., J.I. Weller, M. Ron, M. Band, J.E. Beever, E. Feldmesser, Y. Da, G.R. Wiggans, P.M. VanRuden, and H.A. Lewin. 1999. A genome scan for QTL influencing milk production and health traits in dairy cattle. *Physiol. Genomics* 1:165.

Kappes, S.M. 1999. Utilization of gene mapping information in livestock animals. *Theriogenology* 51:135.

Landergren, U., M. Nilsson, and P.Y. Kwok. 1999. Reading bits of genetic information: methods for single-nucleotide polymorphisms. *Genome Res.* 8:769.

MacHugh, D.E., M.D. Shriver, R.T. Loftus, P. Cunningham, and D.G. Bradley. 1997. Microsatellite DNA variation and the evolution, domestication phyleogeography of taurine and zebu cattle (*Bos taurus* and *Bos indicus*). *Genetics* 146:1071.

Montgomery, G.W. 2000. Genome mapping in ruminants and map locations for genes influencing reproduction. *Rev. Reprod.* 5:25.

Nijman, I.J., D.G. Bradley, O. Hanotte, M. Otsen, and J.A. Lenstra. 1999. Satellite DNA polymorphisms and AFLP correlate with *Bos indicus–taurus* hybridization. *Anim. Genet.* 30:265.

O'Brien, S.J., M. Menotti-Raymond, W.J. Murphy, W.G. Nash, J. Wienberg, R. Stanyon, N.G. Copeland, N.A. Jenkins, J.E. Womack, and J.A. Graves. 1999. The promise of comparative genomics in mammals. *Science* 286:458.

Yang, Y.P. and J.E. Womach. 1998. Parallel radiation hybrid mapping: a powerful tool for high-resolution genomic comparison. *Genome Res.* 8:731.

14 Genetic Technologies in Cow–Calf Operations

S. K. DeNise and J. F. Medrano

CONTENTS

Advances in molecular genetic technology and understanding of the bovine genome have led to the development of tools that can be used to enhance profitability of cow–calf enterprises. Animals can be tracked from birth to harvest by comparing DNA; progeny can be accurately and quickly matched to their parents; and inheritance of favorable/unfavorable characteristics can be determined prior to expression of the specific trait. We have just begun to understand the potential of the technology to impact production at the cow–calf level. The synergy of traditional selection methods and new technologies will provide the seedstock industry new tools to enhance genetic improvement. The next generation of genomic studies will provide new biologicals and diagnostics to enhance performance of individual animals and improve production efficiency in ways we can only imagine today.

The phenotype of an animal is the result of complex reactions, orchestrated by genes, that take place in each individual cell. These phenotypes at the cellular level combine and interact to create the performance we measure in the animal. Significant strides have been made in recent years in our understanding of the molecular nature of these complex processes, and the field of genomics has appeared central to this line of study. Genomics refers to the study of an organism's complete set of genes and chromosomes, i.e., its genome. Genomics has evolved to encompass not only structural aspects of a genome leading to its complete sequencing, but also to functional genomics and proteomics. This field is moving at a rapid pace, unlocking secrets of the information coded in DNA, transcribed to RNA, and expressed as a protein product.

0-8493-1117-9/02/$0.00+$1.50

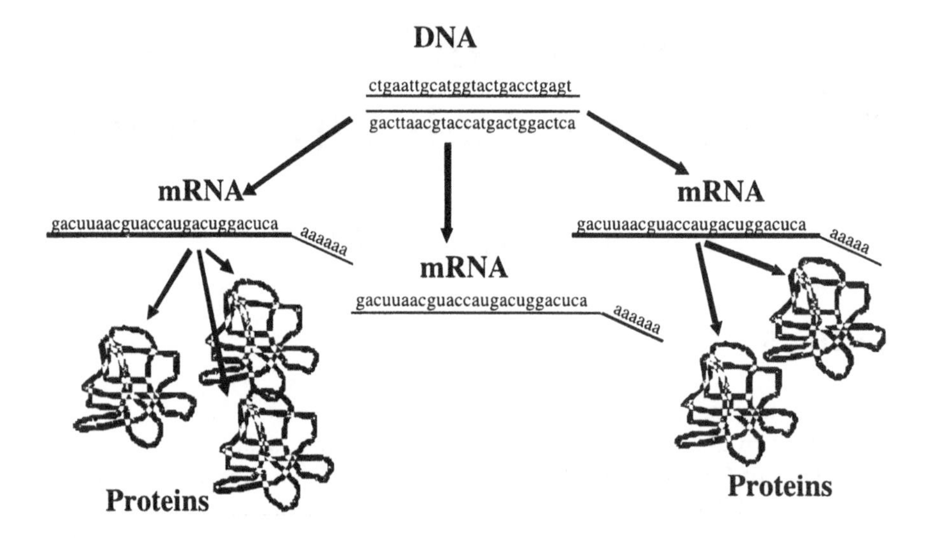

FIGURE 14.1 Diagram showing transfer of information from DNA to protein. Chromosomal DNA is transcribed to mRNA in the nucleus of cells, and mRNA is translated in the cytosol to produce proteins. All of these steps are critical for protein synthesis from the information stored in DNA, multiple copies of mRNA are transcribed, but some are degraded prior to translation to a protein. Multiple copies of a protein can be translated from each mRNA.

The phenotype at the cellular level relies on gene expression from the template of DNA at specific times and in specific tissues creating mRNA. mRNA is translated into proteins that interact to carry out the function of the cell. Figure 14.1 depicts the biological relationship between each of these areas. Simply knowing the gene structure of the DNA gives no indication of how many transcripts of mRNA will be made of a specific gene. Most genes occur in a single copy, but many copies of mRNA can be made within a short period of time. The number of transcripts will correlate to the amount of protein present in the cell; but that correlation is not perfect. Many mRNA molecules are degraded prior to reaching the point of protein synthesis, and each mRNA is capable of serving as a template for making multiple copies of the protein.

The three different areas of genomics represent the phases of this evolving field. *Structural genomics* is the study of DNA sequences at the chromosome level and first level transcripts. These sequences can be studied to compare sequence differences that create genetic variation through different alleles (e.g., the red and black coat color alleles, or horned and polled alleles). This structural information can also be used to compare genomes of different species providing insight into the function of genes coding for proteins and other important sequences. *Functional genomics* is a relatively new area of research with the goal of elucidating the functions of many genes by studying differences among developmental and life-cycle patterns of transcription in specific tissues compared to controls (Hieter and Boguski, 1997). For example, functional genomics' strategies are used to study the pattern of changes in a normal cell as it becomes a cancer cell. Scientists hope to diagnose the type of cancer at earlier stages or develop specific treatments targeted to the cancer type. In agriculture, this type of research is likely to be used to develop products that may enhance production, or for diagnosing potential metabolic diseases or infections long before the animal exhibits symptoms. The third area of study has been labeled *proteomics*, and focuses on how the transcript of RNA is remodeled and made into proteins and how the proteins interact to control specific functions within the organism. All three areas are important for providing tools and products to livestock producers that will enhance economic efficiency.

GENES AND MAPS

Cattle genome researchers have benefited greatly from the technologies developed and the information produced from the Human Genome Project. Even if the bovine genome is only partially sequenced, we will be able to capitalize on information acquired from human and other model organisms because of the similarity among animals in their coding sequences. Genes of great importance to the survival of an animal are well conserved from the simplest animals to the most complex. Thus, we can utilize genomic information generated from a number of organisms to help unlock the secrets of the bovine genome.

The human maps will be of the greatest utility because of the effort to sequence and map the genome in its entirety. Two human genome maps have been completed and recently published. The private effort of Celera Genomics (Venter et al., 2001) conservatively reported 26,383 coding genes, whereas the public effort (International Human Genome Sequencing Consortium, 2001) estimates 32,000 genes. Both of these estimates are less than would have been expected, given the diversity of protein products. Because of the degree of similarity between cattle and human genes and the comparable number of bases in both species, we can expect to find that cattle will have a similar number of genes as humans.

Already there are a number of integrated and annotated sequence databases and web-based maps (Table 14.1) that allow researchers to compare and search for genes, transcripts, and proteins identified in a number of organisms, not just cattle. The National Center for Biotechnology Information (NCBI) is the primary storehouse for genomic information for all species and has specific tools for cattle, human, and model organisms. Databases integrating all published information on expressed sequences can be found through UniGene (Boguski and Schuler, 1995; Schuler, 1997), the TIGR *Bos taurus* Gene Index, and the COMPASS database (Rebeiz and Lewin, 2000).

Unlocking the secrets within the three areas of structural and functional genomics and proteomics requires maps with increasing levels of complexity to summarize, interpret, and further understand the genome of an organism. The types of maps that will continue to be developed are genetic, physical, molecular, biochemical, and physiological (see Table 14.2 for a complete description). We currently have several genetic maps in cattle, and molecular maps are being developed as genome sequencing progresses. The remaining maps will evolve as we learn more about the bovine genome.

Genetic and molecular maps require DNA markers with variants in their sequence in order to assign location. There are primarily two types of marker sequences that have been used, microsatellites and single nucleotide polymorphisms (SNPs). Microsatellites have been the most common type of marker. They are composed of repetitive DNA sequences (two- to six-base repeats) that are interspersed throughout the genome

TABLE 14.1
Genomic Resources Located on the World Wide Web

Source	Website	Species	Information
National Center for Biotechnology (NCBI)	http://www3.ncbi.nlm.nih.gov	All	Genomic and cDNA sequence databases; annotations; links to protein databases; bovine UniGene database
TIGR *Bos taurus* Gene Index (BtGI)	http://www.tigr.org/tdb/btgi/	Bovine	Tentative *Bos taurus* consensus sequences (Tcs); expressed sequence tags (ESTs); mature *Bos taurus* transcripts (ETs)
UniGene Bovine Version	http://www.ncbi. nlm.nih.gov/UniGene/	Bovine	Compilation of consensus sequences
Compass University of Illinois	http://cagst.animal.uiuc.edu/genemap/	Bovine	Comparative mapping of expressed sequence tags
TAMU Bovine Genome Database	http://bos.cvm.tamu.edu/bovarkdb.html	Bovine	Human, bovine and mouse comparative maps, links to genetic resources
Bovmap Database, Institut National de Recherche Agronomique, Laboratoire de génétique biochimique, Jouy-en-Josas	http://locus.jouy.inra.fr/cgi-bin/lgbc/ mapping/bovmap/Bovmap/main.pl	Bovine	Genetic maps of bovine loci, sequence, annotations
USDA–ARS Cattle Genome Mapping Project	http://sol.marc.usda.gov/genome/cattle/ cattle.html	Bovine	Genetic maps of bovine genome
Bovine ARK–DB Roslin Institute, UK	http://bos.cvm.tamu.edu/bovarkdb.html	Bovine	Comparative maps; links to other databases
Cattle Genome Database, CSIRO, AU	http://spinal.tag.csiro.au/	Bovine	Comparative maps; genetic maps

and are generally found in noncoding regions. These repetitive regions are subject to additions and subtractions to the repetitive portion, which allows us to discriminate among animals carrying different alleles and to follow the inheritance of specific chromosomal regions. Microsatellite markers have provided the basis for individual and parentage identification in humans, dogs, cattle, and many other species. Approximately 2100 microsatellites have been mapped in cattle (http://locus.jouy.inra.fr/cgi-bin/lgbc/mapping/bovmap/Bovmap/main.pl). Figure 14.2

TABLE 14.2
View of Genomics from the Conceptual Evolution of Maps Leading to the Molecular Understanding of the Phenotype

Map Type	Description
Genetic map	Map that describes the relative location of genes along chromosomes by determining the frequency of recombination among genes within a reference population.
Physical maps	Physical maps have varying degrees of resolution but are based on physically marking the location of a gene.
Cytogenetic	Cytogenetic maps identify the physical location of genes on chromosomes in relation to the chromosome banding patterns or by direct *in situ* hybridization of markers to chromosomes.
Contig	Contig maps order DNA fragments spanning regions of chromosomes. These maps are used to isolate and characterize individual genes based on location in a chromosome.
Molecular maps	These are maps of the complete sequence of bases making up the DNA structure of chromosomes, roughly 3 billion base pairs (3000 Mb) in the human and cattle genomes. These maps provide the highest degree of resolution and provide all sequences upstream and downstream of the gene of interest. (Structural Genomics.)
Biochemical maps	These maps provide insight into the function of genes at the cellular level and contribute to efforts in functional genomics. Genome mapping at this level will benefit from comparative research in model organisms like humans, yeast, *Drosophila*, and mice. (Functional Genomics and Proteomics.)
Physiological maps	These maps will integrate knowledge to understand not only the function of genes but their interactions that determine the biological constitution of organisms and, ultimately, their phenotype.

661 ttatttcata atccactttc ttatctcctc cttcacaaaa ctgatgagaa attggtacaa
721 actctatcga caaaagatca cracttgatc ctcaatggca aaggcaaata tacattataa
781 atagcaaaac agctggcttg gaccatgttg ctggccactc atccagctga gagatttgaa
841 tgacatcata accctwgagm gggtaatgct agccagctgg tgttatttag aatacacaaa
901 aatgggggga aagaaaatgc actcatgtg**c acacacacac aaatacacac acacacacac**
961 acacaggttc aagttatgca gaaaaatatg aacaatggaa aaatcatttg cccctcagat

FIGURE 14.2 Partial sequence (exon 1) of bovine insulin-like growth factor 1(IGF1; GenBank accession number AF210383) gene showing a dinucleotide repeat microsatellite (ca repeat in bold) The sequence of primers used to amplify the microsatellite marker is underlined. This sequence, known as a sequence tag site (STS) is unique in the bovine genome.

shows an example of a microsatellite sequence within the gene that codes for insulin-like growth factor 1 (IGF1) located on bovine chromosome 5.

Coding genes, those sequences that will be transcribed into mRNA and then translated into protein, have been added to the different kinds of maps. Coding genes are dispersed throughout the chromosomes with noncoding DNA between them. In humans, coding genes have been identified approximately every 40 to 200 kb (Ishikawa et al., 1997; Glöckner et al., 1998). Some regions of chromosomes have many genes clustered together, while other regions are almost devoid of genes. Venter et al. (2001) found that 20% of the human genome has regions of more than 500,000 bases that contain no genes. The gene-rich regions will be used to compare bovine to human and other model organisms because these regions are likely to be highly conserved and have functional significance in all animals.

Coding genes have been mapped from sequence variation found within or near the gene by detecting single nucleotide changes or polymorphisms (SNPs). This type of variation has traditionally been identified by analyzing the DNA sequence with a restriction enzyme that recognizes the change and either cuts the DNA or leaves it intact. The polymorphism is then analyzed by separating the fragments and visualizing the lengths of DNA; where the enzyme recognizes the sequence, two shorter fragments are formed; where it does not, one large fragment of DNA remains. This has been the basis of restriction fragment length polymorphism analysis (RFLP), which was one of the first techniques used for typing DNA polymorphisms. Combined with the polymerase chain reaction (PCR), numerous bovine loci are routinely genotyped with techniques similar to those that were developed for the classification of milk protein genotypes (Medrano and Aguilar-Cordova, 1990).

SNP markers promise to be increasingly useful in the future to develop high-resolution maps. With the availability of sequencing of whole genomes, SNPs can now be identified dispersed across all chromosomes, presenting important advantages as markers for genome analysis. SNPs are widely distributed and provide more potential markers near a locus of interest. For example, in humans it has been estimated that an SNP can be found every 100 to 300 bases (http://www3.ncbi.nlm.nih.gov/SNP/). SNP maps are currently under development in a number of species; over 2.8 million SNPs have been placed in the human SNP database (http://www3.ncbi.nlm.nih.gov/SNP/). Some SNPs are located within the coding region of a gene affecting protein function which may be directly responsible for the variability of important traits between individuals. Other SNPs occur either "upstream" or "downstream" of the coding gene and may influence the regulation of gene expression. Others will occur in locations that do not interfere with the structure or production of a protein (Risch, 2000). Figure 14.3 depicts an SNP in bovine growth hormone located on bovine chromosome 19 that causes a change in the amino acid composition of growth hormone as a mature protein (Zhang et al., 1992). SNPs also have the advantage of being less likely to undergo a spontaneous mutation; thus, they are inherited with greater stability than microsatellites. These markers lend themselves to high-throughput genetic analysis at a lower cost than microsatellite markers.

Molecular markers, like microsatellites and SNPs, identify or mark unique segments of a chromosome, which allows the segment to be mapped, inheritance of a chromosome followed, and associations with production traits monitored. They also

```
1801 cacctcggac cgtgtctatg agaagctgaa ggacctggag gaaggcatcc tggccctgat
1861 gcgggtgggg atggcgttgt gggtcccttc catgtggggg ccatgcccgc cctctcctgg
1921 cttagccagg agaatgcacg tgggcttggg gagacagatc cctgctctct ccctctttct
1981 agcagtccag ccttgaccca ggggaaacct tttccccttt tgaaacctcc ttcctcgccc
2041 ttctccaagc ctgtagggga gggtggaaaa tggagcgggc aggagggagc tgctcctgag
                                               g
2101 ggcccttcgg cctctctgtc tctccctccc ttggcaggag ctggaagatg gcaccccccg
2161 ggctgggcag atcctcaagc agacctatga caaatttgac acaaacatgc gcagtgacga
2221 cgcgctgctc aagaactacg gtctgctctc ctgcttccgg aaggacctgc ataagacgga
```

FIGURE 14.3 Partial sequence of bovine growth hormone gene (GenBank accession number M57764.1). The positions of single nucleotide polymorphism (SNP), as well as the two possible alternative nucleotides, are shown in the box. The SNP changes the codon ctg corresponding to amino acid valine (Val) to gtg corresponding to amino acid leucine (Leu), therefore changing the amino acid sequence of the protein. The sequence of primers used to amplify the region covering the SNP is underlined. This sequence, known as a sequence tag site (STS) is unique in the bovine genome.

allow researchers to link the bovine map to model organisms. Linking the bovine map to that of a model organism like the mouse and human will become useful in assigning function to sequences and in studying particular functional properties of genes.

DNA COLLECTION

DNA is found in every nucleated cell in the body and has been collected from semen, muscle, fat, white blood cells found in blood and milk, skin, and epithelial cells collected from saliva. Methods of collection have become routine, requiring only minute amounts of the target tissue: a single drop of blood or several cells.

The traditional DNA collection systems of a tube of blood or a straw of semen are still commonplace, but other systems have been developed for ease of collection. There are systems that require a drop of blood blotted on a paper that is dried, covered, and stored at room temperature. There are ear tag systems that deposit a tissue sample in an enclosed container with the matching bar code information. Mouth swabs and hair follicles are another common collection method. These methods allow for rapid release of DNA from cells, allowing for minimal steps in screening of samples.

CURRENT USES OF GENETIC TECHNOLOGIES

The current uses of DNA-based technologies include animal identification or source verification, parentage verification or identification, diagnosis of genetic diseases and simply inherited traits, and marker-assisted selection (Table 14.3). Some have just become commercially available, and still others are in the research phase. Each of these will have economic justifications that will depend on which part of the system a producer contributes: seedstock, multiplier, or commercial, and whether he/she will retain ownership or contribute to a partnership program, like an alliance.

Animal identification and source verification are becoming increasingly important around the world. There are standard procedures that compare DNA, matching

TABLE 14.3
Summary of the Current Uses of Genetic Technologies Based on DNA Analysis in Cow–Calf Operations

Genetic Technology	Uses	Current Applications	Molecular Marker
Animal identification or source verification	Matching harvested products back to the source	Marketing tool for high-quality meats providing product assurance to customer Identification of original source of products Association of carcass data with progeny or grand-progeny of herd sires (aid to selection for carcass traits)	Panels of highly polymorphic microsatellites or SNPs
Parentage verification	Pedigree confirmation that can be performed on the animal at any age	Guarantee pedigree of embryo transfer calves and registered AI bulls Identification of sire in multiple-sire breeding operations	Panels of highly polymorphic microsatellites or SNPs
Diagnostic of simply inherited traits	Identification of carriers of recessive alleles	Production traits like polledness, coat color, or double-muscling Carriers of genetic diseases producing malformations or reproductive problems like white heifer disease	Microsatellites or SNPs
Marker-assisted selection (MAS)	Association of marker alleles with complex polygenic traits within families and across families (future)	Aid to selection for production traits, allowing increased accuracy for the identification of superior animals. Selection for tenderness and marbling (Ex: Carcass Merit Project), metabolic efficiency, disease resistance, and others in the future Early selection to reduce generation interval in selection programs	Microsatellites or SNPs (future use when high-density maps are developed)

specific marker genotypes from birth with those taken later in life or matching harvested product back to the original source. All marker types can be effectively used to match DNA samples and a probability that the two samples are identical can be assigned. DNA remains constant in the tissues of animals from conception through consumption. Thus, it becomes the ultimate source of animal identification because it cannot be altered.

Parentage verification and identification is a common use of DNA technology. Blood-type markers have been traditionally used for parentage verification and are reliable for most parentage disputes. However, with only 12 blood-group systems available to determine the genetic profile of an individual animal, this method is limited in parentage disputes involving closely related individuals or in large mating groups. DNA markers provide a virtually unlimited number of genetic determinants that can be used to identify specific chromosomal segments in progeny that link to its parents. The International Society of Animal Genetics (ISAG) has identified a panel of markers recommended for this purpose and provides a laboratory certification program. Details concerning the guidelines for certification can be found at: http://www.wisc.edu/animalsci/isag/guidelines.html.

Parentage verification provides assurance of the validity of the assumed parents of an animal. This procedure requires DNA samples from the sire and dam in order to determine the probability that a calf inherited the specific genotype from the putative parents. If the calf has a genotype consistent with the parents' genotypes, the only error possible is that the calf, just through chance, has inherited a genotype compatible with both parents—usually a small probability given several markers.

Parentage verification has been used to guarantee the pedigree of embryo transfer calves and of registered bulls used in AI programs. This technology has been valuable to ensure the integrity of the program and provide assurances to potential buyers. It also allows breed associations to guarantee the integrity of the pedigree information.

Parentage identification analysis assigns parents based on a panel of potential parents. This procedure relies on having DNA samples from all potential parents. Often, dams are matched to progeny at birth, but the sire is unknown. To determine the sire of a calf, all potential sire genotypes are compared to the genotypes of the progeny. Bulls are "excluded" as the potential sire if they have a genotype incompatible with the genotype of the progeny. At a single genetic marker, parents can have only two copies of the gene: one from their sire and one from their dam. There can be many different forms of the gene at each marker, called alleles, but each parent and each progeny can only have two copies. One of those two copies must be passed on to their progeny. Bulls are excluded until only one bull remains that has a genotype consistent with the calf's genotype. Thus, in order to successfully match calves with their sire, DNA samples from all bulls that possibly could have mated cows must be included in the analysis.

Parentage assignment can also be an important genetic evaluation tool when reliable information concerning pedigree structure is not available. For example, commercial herds that rely on multiple-sire breeding pastures are unable to determine the value of the progeny produced from a given sire unless they have used a phenotypic marker (like coat color or Brahman influence) or DNA typing. There are a limited number of phenotypic markers available; thus, the evaluation from visual markers compares breed performance instead of the genetic potential of individual sires. DNA typing can be used as a progeny testing tool by assigning calves to their individual sires based on inheritance of markers. For traits easily measured in the bull (like weaning and yearling weights and growth rate), this additional information usually adds little to the estimate of genetic merit of a sire (DeNise, 1999). However, for those traits that cannot be measured directly in the bull (for example, carcass or milk production traits) parentage verification may provide additional information to improve the genetic potential of future generations for traits that are economically important. The new models of marketing cattle through strategic alliances and retained ownership provide commercial cattle producers a reason to be interested in the genetic merit of bulls for carcass traits. This technology will allow producers with multiple-sire breeding pastures to identify bulls not suited to their marketing or management program.

Parentage assignment has not been widely adopted in commercial populations because the benefits of DNA testing all calves has not been justified given the current costs of testing. New strategies of genotyping animals with extreme phenotypes for specific traits in the breeding objective provides a way to use the phenotypic information in genetic evaluation of bulls at a reasonable cost (DeNise, 1999).

In the future, there may be opportunities for reducing the cost of DNA typing. We have already seen a reduction in the costs of testing because of improvements in the technology, but it remains to be seen how low the costs can be with current procedures. SNP markers may be able to reduce the cost of individual testing by incorporating completely new technologies.

MARKER-ASSISTED SELECTION

Genetic markers can be used to identify inheritance of specific genes of animals from their parents. If a marker has been tested and validated, it can serve as a road sign for a specific gene influencing a complex trait called a quantitative trait locus or QTL. Marker-assisted selection is used within a family that has two different copies or alleles of a gene influencing a trait, one preferred over the other. The genetic marker can be used to determine which progeny have inherited the favorable form of the QTL.

Marker-assisted selection requires a three-phase testing program in order to utilize the technology (Davis and DeNise, 1998). First, the QTL location and size must be identified in a research population using widely spaced markers in a preliminary screening procedure. Then the location must be refined by identifying markers on each side of the QTL, called flanking markers. These flanking markers must be in close proximity so that the accuracy of predicting inheritance of the favorable gene is high. In the initial screening, the size of the effect of substituting one gene for another is estimated. If the QTL accounts for little of the variation seen in the trait, it will not be cost effective to use the test in a selection program. Traditional selection will be much more effective. In the second phase, the QTLs must be verified or validated in commercially important populations. Finally, if the location and effects of the QTL are validated, the technology can be commercialized for distribution to the population at large.

If the marker identifies the exact change in sequence that causes the unique phenotype or if the marker is tightly linked to the QTL, we have a diagnostic tool that can be used across families. Diagnostic tests for recessive, single-gene effects have been developed for a number of traits: genetic diseases such as bovine leukocyte adhesion deficiency (BLAD; Shuster et al., 1992), double-muscling (myostatin; McPherron and Lee, 1997; Smith et al., 1997), and simply inherited phenotypic traits such as carriers of the red coat color gene (Klungland et al., 1995). These tests were often developed by comparing phenotypic characteristics of cattle with human or mouse model organisms. Once the gene has been identified in another organism, it can usually be sequenced in the bovine and mutants uncovered. Many of these tests have been patented by the scientists or companies that discovered the mutation; thus, to use the test, a royalty is paid to the owner of the patent. Diagnostic tests have been invaluable in identifying carrier animals that do not express the gene and in allowing breeders to either increase the use of those animals or eliminate them from the breeding population. Molecular breeding (evaluating the genotype by "reading" the DNA) is much more economical than relying on an expensive and lengthy progeny testing program to uncover carriers of the mutant gene.

Marker-assisted selection can be utilized in a limited way today and will be used with greater efficiency in the future. Research has identified regions of chromosomes that influence traits of economic importance. A study from the USDA–ARS U.S. Meat Animal Research Center, one of the few published studies, has described significant effects for genes segregating on chromosome 2 for retail product yield, and at the suggestive level for the components of retail product yield: rib meat, marbling, and fat thickness (Stone et al., 1999). Chromosome 5 had significant effects for bone weight and dressing percentage, and chromosome 13 had suggestive significance for retail product yield, rib meat, rib fat, and dressing percentage (Stone et al., 1999). Chromosomes 1 and 7 had suggestively significant effects on birth weight, and chromosome 1 had suggestively significant effects for weaning and yearling weight in addition to birth weight (Stone et al., 1999). Chromosomes 11, 14, 18, and 26 had suggestive levels of significance for several consumption traits (Stone et al., 1999). Another gene influencing tenderness was identified on chromosome 15, accounting for 26% of the variation in tenderness (Keele et al., 1999).

The goal of another project at the USDA Meat Animal Research Center is to identify the genes that influence ovulation rate and twinning in a reference population of cattle selected for increased incidence of twinning (Kappes et al., 1999). They have identified regions on chromosomes 5, 9, 10, and 22 that influence ovulation rate, and regions on chromosomes 4 and 28 that appear to influence twinning rate, but not ovulation rate. In an earlier study on the population, Blattman et al. (1996) found a region on chromosome 7 that influenced ovulation rate.

The Carcass Merit Project sponsored by the National Cattleman's Beef Association and partnering breed associations is evaluating markers identified in a research population developed at Texas A&M University (Tanksley et al., 1999). Sixteen breeds are enrolled in the project to develop EPDs for carcass quality traits like marbling, tenderness, and ribeye area. Ten bulls from each breed will be used to validate whether the markers found in the research population can be used in commercial populations to identify animals segregating genes that influence tenderness and marbling. The chromosomal locations of at least five genes influencing beef tenderness and another four genes influencing marbling are under validation in the program (http://www.beef.org/library/publications/research/genomexe.htm). During the evaluation stage, the amount of variation that can be accounted for by specific genes within each breed will be estimated and programs developed to capitalize on the information. The critical components of this strategy are that it will only be successful in families that have been previously tested, bulls must be heterozygous for the QTL to identify them, and they must be heterozygous for the markers. It is hoped that the information from this project will help generate a more uniform and higher-quality product for consumers.

SNP maps provide an opportunity to create a unique tool for identifying QTLs without relying on within-family studies. Because SNPs are found in greater frequency than microsatellite markers, a map of SNP markers could be created of such a density that QTLs could be identified across a commercial population without family structure. The QTLs could be identified directly in the population of interest, and all QTLs influencing a trait could be evaluated in a single analysis. These types of studies are called association studies and have been used to identify disease-causing variations in humans (Horikawa et al., 2000; Jorde, 2000).

Markers may also be used in the future to manage animals with genetic potentials to meet specific market requirements. For example, if the molecular genotype predicts that an animal is likely to have marbling scores to reach choice and prime quality grades, producers could manage those animals for optimum marbling scores and market them on a quality grid pricing program. Likewise, if the animal has a genotype for lean meat production with a low propensity to marble, the animals could be aggressively implanted, fed fewer days, and marketed on a red-meat grid pricing system. Breeds have been the primary determinant in the past for predicting performance in the feedlot and for the quality and yield grades to expect. However, while breeds provide some information, they are not good predictors of final outcomes. Genotyping animals entering the feedlot can reduce the risk of feeding and marketing animals by predicting their genetic potential for specific traits.

Producers will have a plethora of markers available to them in the future. It remains to be seen how useful these will be in genetic improvement. Certainly markers will be able to improve the accuracy of selection of individual animals, but the increase in accuracy may not be worth the cost of the test in relation to the value of expected improvement. Certainly for traits of high heritability and easily measured on potential parents, markers will probably not provide any economic advantage over traditional estimates of genetic merit. Unless many of the genes of substantial size that influence a quantitative trait have been mapped prior to selection, genetic improvement based only on markers cannot be assured.

FUTURE TECHNOLOGIES

Future technologies will rely on integration of resources from every species to develop hypotheses and test new ideas. Not only will the genetic code be important, but new technologies will delve more deeply into the kinds of genes made into proteins at different physiological stages and in specific tissues. Not all genes are transcribed in all tissues; thus, interactions among genes, indicator genes, and protein products will be important pieces of information in the effort to solve important management and genetic problems of the future.

This new age of genomics will require a multidisciplinary approach to mine the genetic data and unravel the complexities and interactions among gene products in the animal. In the next phase of genomic research, genes mined from sequence information will be used to understand the function of gene products at the cellular, organ, and whole animal level. The power of this technology lies in developing an accessible database of information that connects and summarizes all genomic data, from the DNA sequences to gene function, from homology with model organisms to actual production characteristics.

BIOINFORMATICS AND COMPARATIVE GENOMICS

A feature of this new era is the integration of computer science into genetics. This is not foreign to animal breeders who have used computers for decades to analyze performance records and predict genetic worth. In the new age of genetic evaluation, the procedures will become more complex as we integrate genotype information into the standard genetic evaluation programs. At the gene level, large amounts of sequence data are produced daily that must be analyzed, annotated, and stored in databases that facilitate access by the scientific community. Software is being written to sort out the important information contained in the data and to incorporate information from different sources.

The sequence information generated from model organisms like the nematode *Caenorhabditis elegans* (sequenced in 1998), the fruit fly *Drosophila melanogaster* (sequenced in 2000), and human *Homo sapiens* (sequenced in 2000), and the mouse *Mus musculus* (sequenced in 2001) provides "piles of information, but only flakes of knowledge" (Eisenberg et al., 2000). If you wonder what a cattle breeder can learn from organisms like these, you may be surprised that other organisms are providing insight into the functionality of specific genes and providing new genes to study in cattle. While producers are not likely to explore genomes of these exotic organisms themselves, animal science researchers will be using a technique called comparative genomics to explore new genes and sequences that have been obtained from other species.

There are examples that demonstrate the power of utilizing comparative genetic analysis of traits of economic importance using model organisms. The myostatin gene that causes the double-muscling phenotype found in a number of breeds was first identified in mice using a technique called a knockout, where the function of the gene is eliminated (McPherron and Lee, 1997; Lee and McPherron, 1999). Mice lacking functional myostatin genes have a generalized increase in muscle mass, suggesting that the gene could be a negative regulator of muscle growth. The gene is highly conserved among species and the cattle double-muscling gene mapped to a region that identified myostatin as a candidate for the gene effect from comparing the genetic maps of other species (Smith et al., 1997). The search for mutations in other species demonstrated that the double-muscled breeds of cattle have mutations on the myostatin coding sequence (McPherron and Lee, 1997). The phenotype of myostatin knockout mice is very similar but not identical to the one seen in cattle, implying that even when the same gene is mutated, species-specific genetic factors modulate the influence of the mutation on the phenotype. Therefore, the last step in the process of gene identification will always be the characterization of the gene in the species of interest. However, substantial knowledge about the gene can be gained in the parallel analysis of model organisms.

MICROARRAY TECHNOLOGY AND EXPRESSION ANALYSIS

The next-generation technology, microarrays, are likely to increase the power of genomics studies, reduce the cost of delivering information, and revolutionize selective breeding. A microarray creates thousands of interrogation sites (10,000 gene arrays are not uncommon) on a single, small, solid support, either a filter or a glass slide, that can be used to evaluate entire populations of expressed genes from a single tissue at one time or genotype for multiple loci. The solid-based support systems have been called chips, and scientists foresee the day when the entire genome of an organism may be placed on a single chip to be used in a variety of experiments (Lander, 1999).

The genotyping arrays have the potential of reducing the costs of parentage verification and identification. This type of array, called an oligonucleotide array, can determine the differences between SNP markers for thousands of locations at a single time, on a single chip (Lipshutz et al., 1999). For establishing this technology, new strategies will have to be developed that will permit the rapid identification of a large number of SNPs in the genome. This system is likely to be the method of choice in the future for animal identification procedures.

Microarray systems can also be used to screen entire populations of expressed genes from target tissues in a procedure described as functional genomics. This analysis has been used to explore the complex interactions among the genes for drug discovery (Debouck and Goodfellow, 1999), pathogenesis of disease (Brown and Botstein, 1999), and tumorigenesis (DeRisi et al., 1996; Cole et al., 1999). Producers can expect new products on the market that diagnose disease before symptoms are exhibited, new products to enhance performance based on cellular studies, and new products for predicting which animals should be managed to reach specific market endpoints.

CURRENT STATUS OF COMMERCIAL GENOMIC TECHNOLOGIES

There are a number of commercial companies offering a variety of tests; and the number and types of tests will continue to grow in the future. Microsatellite and/or SNP marker analysis are available from several companies for source verification, parentage verification, and parentage identification.

There are benefits beyond simply knowing the sire of a calf in commercial production systems. Product and consumption traits have become an important economic consideration for cow–calf producers in recent years with the integration and partnerships formed within the industry. Melton (1995) estimated that the emphasis placed on a cow–calf producer switching from a calf marketing system to a retained ownership system should result in a change in emphasis on reproduction from 47% to

31%, production from 24% to 29%, and consumption from 30% to 40%. For the first time, seedstock producers can access carcass information on progeny or grand-progeny of their herd sires in commercial herds, outside planned progeny testing programs. If their bull or semen buyers are paid on a grid system, the carcass data are collected and could potentially be used in genetic evaluation. The difficulty has been to identify the sires of calves going through the system, especially in multiple-sire breeding programs. DNA analysis can be used in multiple-sire systems to link sires with the phenotypes of their progeny, thus, supplying data for genetic evaluation programs. The genetic evaluation program could be used by both the commercial herd and the bull supplier to make genetic improvement in the economically important traits. The major limitation on implementing these procedures has been the cost of DNA testing every animal. Strategies for genotyping only the extreme phenotypes have been offered as a way to reduce the cost of testing for genetic evaluation purposes (DeNise, 1999). This procedure requires that all potential sires have been identified and DNA samples are available.

Marker-assisted selection is becoming a reality with the validation of markers identified in research projects. The Carcass Merit Project is in the validation stage for markers that have been shown to influence tenderness and marbling, and the licensing of the commercial test has already been completed. There have been popular press reports of markers for marbling in specific breeds that have either begun marketing the test in the United States, or hope to market it soon. More markers will be available in the near future to cow–calf producers to utilize in marker-assisted selection programs. The decision to incorporate markers into selection programs will depend on the amount of variation that can be accounted for by the markers, the estimated frequency of the QTLs in the population, and the economic benefits of the technology. Some markers will have little influence on the characteristic, either because the effect itself is small or the frequency of the favorable QTL is already fairly high in the population; thus, they will probably not be utilized because the benefit of the technology will not be greater than the costs of implementation. However, as we learn more about the genes influencing economically important traits, we will develop strategies for capitalizing on the information to improve the value of the product.

Beyond marker-assisted selection programs that can only be used within families, diagnostic tests will be developed that can be used across families to predict important characteristics like carcass value, metabolic efficiency, disease resistance, and others. Tests are currently available to identify carriers of double-muscling and some genetic diseases, and there will be more available in the future. As more is learned about QTLs segregating within families, the genes actually causing the effect will be uncovered, tests developed, validated, and offered commercially. SNPs provide the opportunity to uncover genes very precisely without actually identifying the specific mutation. Because SNPs are tightly spaced, diagnostic tests based on SNPs may be available in the future without the exact mutation that causes the effect being known. A number of research programs in humans are trying to uncover the cause of complex genetic diseases like type II diabetes using SNP markers.

Breed associations are becoming more involved in the molecular technology as they switch from parentage verification with blood typing to DNA typing.

Samples collected for parentage verification can be stored for additional tests in the future or can be used immediately for further DNA analysis. Breed associations will be offering new technologies to improve estimates of genetic merit by incorporating markers into estimates of EPDs, probably the first of which will be tenderness.

These technologies promise to yield important management and genetic tools in the future. Information mined from the bovine genome will be used to improve profitability in cow–calf operations by identifying genetically superior bulls and characterizing cows as to their optimum performance in particular environmental conditions. Diagnosis of metabolic disease, bacterial and viral infections, and other types of ailments that can influence not only cow health and production but also food safety will be possible. Allied industries will be able to use the data to assist in rational development of performance-enhancing products and interventions. Producers will have to evaluate the cost vs. benefits of the technology and make choices on where their dollars are best spent for their specific operation.

REFERENCES

Blattman, A.N., B.W. Kirkpatrick, and K.E. Gregory. 1996. A search for quantitative trait loci for ovulation rate in cattle. *Anim. Genet.* 27:157.

Boguski, M.S. and G.D. Schuler 1995. ESTablishing a human transcript map. *Nature Genet.* 10:369.

Brown, P.O. and D. Botstein. 1999. Exploring the new world of the genome with DNA microarrays. *Nature Genet.* Suppl. 21:33.

Cole, K.A., D.B. Drizman, and M.R. Emmert-Buck. 1999. The genetics of cancer—a 3D model. *Nature Genet. Suppl.* 21:38.

Davis, G.P. and S.K. DeNise, 1998. The impact of genetic markers on selection. *J. Anim. Sci.* 76:2331.

Debouck, C. and P.N. Goodfellow. 1999. DNA microarrays in drug discovery and development. *Nature Genet. Suppl.* 21:48.

DeNise, S.K. 1999. Using parentage analysis in commercial beef operations. *Proc. Beef Improv. Fed. Thirty-first Ann. Res. Symp. Ann. Mt. June,* Roanoke, VA. pp. 183.

DeRisi J., L. Penland, P.O. Brown, M.L. Bitner, P.S. Meltzer, M. Ray, Y. Chen, Y.A. Su, and J. M. Trent. 1996. Use of a cDNA microarray to analyse gene expression patterns in human cancer. *Nature Genet.* 14:457.

Eisenberg, D., E.M. Marcotte, I. Xenarios, and T.O. Yeates. 2000. Protein function in the post-genomic era. *Nature* 405:823.

Glöckner G, S. Scherer, R. Schattevoy, A. Boright, J. Weber, L.C. Tsui, and A. Rosenthal. 1998. Large-scale sequencing of two regions in human chromosome 7q22: Analysis of 650 kb of genomic sequence around the EPO and CUTL1 loci reveals17 genes. *Genome Res.* 8:1060.

Hieter, P. and M. Boguski. 1997. Functional genomics: it's all how you read it. *Science* 278:601.

Horikawa, Y., et al. 2000. Genetic variation in the gene encoding calpian-10 is associated with type 2 diabetes mellitus. *Nature Genet.* 26:163.

International Human Genome Sequencing Consortium. 2001. Initial sequencing and analysis of the human genome. *Nature* 409:860.

Ishikawa, S., M. Kai, M. Tamari, Y. Takei, K. Takeuchi, H. Bandou, Y. Yamane, M. Ogawa, and Y. Nakamura. 1997. Sequence analysis of a 685-kb genomic region on chromosome 3p22-p21.3 that is homozygously deleted in a lung carcinoma cell line. *DNA Res.* 4:35.

Jorde, L.B. 2000. Linkage disequilibrium and the search for complex disease genes. *Genome Research* 10:1435.

Kappes, S.M., G.L. Bennett, and R.M. Thallman. 1999. Identification of genes influencing reproduction in cattle. *Proc. Beef Improv. Fed. Thirty-first Ann. Res. Symp. Ann. Mt.* June, Roanoke, VA. pp. 191.

Keele, J.W., S.D. Shackelford, S.M. Kappes, M. Koohmaraie, and R.T. Stone. 1999. A region on bovine chromosome 15 influences beef longissimus tenderness in steers. *J. Anim. Sci.* 77:1364.

Klungland, H., D.I. Vage, L. Gomez-Raya, S. Adalsteinsson, and S. Lien. 1995. The role of melanocyte-stimulating hormone (MSH) receptor in bovine coat color determination. *Mammal. Genome* 6:636.

Lander, E.S. 1999. Array of hope. *Nature Genet. Suppl.* 21:3.

Lee, S.J. and A.C. McPherron. 1999. Myostatin and the control of skeletal muscle mass. *Curr. Opin. Genet. Dev.* 9:604.

Lipshutz, R.J., S.P.A. Fodor, T.R. Gingeras, and D.J. Lockhart. 1999. High density synthetic oligonucleotide arrays. *Nature Genet. Suppl.* 21:20.

McPherron, A.C., A.M. Lawler, and S.J. Lee. 1997. Regulation of skeletal muscle mass in mice by a new TGF-beta superfamily member. *Nature* 387:83.

McPherron, A.C. and S.J. Lee. 1997. Double muscling in cattle due to mutations in the myostatin gene. *Proc. Natl. Acad. Sci. USA* 94:12457.

Medrano, J.F. and E. Aguilar-Cordova.1990. Genotyping of bovine kappa-casein loci following DNA-sequence amplification. *Biotechnology* 8:144.

Melton, B. 1995. Economic differences between commercial and seedstock genetic decisions. *Proc. Composite Cattle Breeders Inter. Assoc.* North Platte, NE, August.

Rebeiz M. and H.A. Lewin. 2000. Compass of 47,787 cattle ESTs. *Anim. Biotechnol.* 11:75.

Risch, N.J. 2000. Searching for genetic determinants in the new millennium. *Nature* 405:847.

Schuler, G. D. 1997. Pieces of the puzzle: expressed sequence tags and the catalog of human genes. *J. Mol. Med.* 75:694.

Shuster, D.E., M.E. Kehrli, M.R. Ackermann, and R.O. Golbert. 1992. Identification and prevalence of a genetic defect that causes leukocyte adhesion deficiency in Holstein cattle. *Proc. Natl. Acad. Sci. USA* 89:9225.

Smith, T.P., N.L. Lopez-Corrales, S.M. Kappes, and T.S. Sonstegard. 1997. Myostatin maps to the interval containing the bovine mh locus. *Mamm. Genome* 8:742.

Stone, R.T., J.W. Keele, S.D. Shackelford, S.M. Kappes, and M. Koohmaraie. 1999. A primary screen of the bovine genome for quantitative trait loci affecting carcass and growth traits. *J. Anim. Sci.* 77:1379.

Tanksley, S.M., S.L.F. Davis, S.K. Davis, and J.F. Taylor. 1999. Marker-assisted selection for meat quality in outbred cattle populations. *Proc. Beef Improv. Fed. Thirty-first Ann. Res. Symp. Ann. Mt.* June, Roanoke, VA. pp. 250.

Venter, J.C., M.D. Adams, E.W. Myers, P.W. Li, R.J. Mural, G.G. Sutton, H.O. Smith, M. Yandell, et al. 2001. The sequence of the human genome. *Science* 291:1304.

Zhang, H.M., D.R. Brown, S.K. DeNise, and R.L. Ax. 1992. Nucleotide sequence determination of a bovine somatotropin allele. *Anim. Genet.* 23:578.

15 Somatic Cell Cloning in the Beef Industry

Audy Spell and James M. Robl

CONTENTS

Technically, cloning is the production of multiple genetically identical animals. Surprisingly, genetically identical animals have been produced for over 30 years, using a technique of dividing embryos into two or more portions to produce multiple embryos. This technique was termed "splitting" and, consequently, did not generate as much public interest, or concern, as the current technique of nuclear transplantation which, for better or worse, is commonly referred to as "cloning." Although splitting has been done commercially for many years, its application is limited by the fact that only embryos can be divided and only two copies can be made. With each division of the embryo, the resulting portions become smaller and less viable. Nuclear transplantation is an entirely different approach and has the potential of producing an unlimited number of genetically identical animals (Wilmut et al., 1997; Cibelli et al., 1998; Stice et al., 1998; Chan, 1999; Kato et al., 1999; Zakhartchenko et al., 1999; Polejaeva and Campbell, 2000).

The theoretical basis for using nuclear transplantation for cloning is that all nuclei from an organism, with only a few exceptions, contain the exact same set of genes (Wilmut et al., 1997; Stice et al., 1998; Wakayama et al., 1998; Kato et al., 1999, 1998; Wakayama and Yanagimachi, 1999; Zakhartchenko et al., 1999). A skin cell contains the same set of genes as a liver cell, a kidney cell, or any other cell in the body. Furthermore, small cell samples can be taken from an individual and greatly expanded in culture. Therefore, many genetic copies of an individual are technically available for making clones.

0-8493-1117-9/02/$0.00+$1.50

In each of these cells, however, a specific set of genes is used, which results in each cell type having a unique function. We, therefore, need a method of turning off the genes that these cells are using in their original form and turning on the genes appropriate for each cell type in a complete, fully formed individual (Stice et al., 1998). The embryo does this naturally, so the objective is to turn the cell into an embryo. The simple-minded approach that has been used is the removal of the genetic material from an egg and its replacement with the genetic material from one of the millions of somatic, or body cells, from the individual to be cloned. And, believe it or not, this actually works.

Techniques for the genetic modification of animals have also been available since the early 1980s (Chan, 1999; Piedrahita et al., 1999). As with cloning, great interest was generated for application to farm animals, especially following a report of genetically modified mice that grew to 40% larger than nontransgenic littermates. The method of making genetic modifications in animals at this time was to microinject a few hundred to a few thousand copies of a gene into the pronucleus of a newly fertilized embryo (Chan, 1999; Piedrahita et al., 1999). As with cloning, this method was used to produce a few transgenic cattle, sheep, goats, and pigs. However, the method was limited, as for cloning, by the availability and expense of embryos and recipients. The nuclear transplantation method using cultured cells for nuclear donors has provided a much simpler and more efficient method for genetically modifying agricultural animals (Cibelli et al., 1998; Stice et al., 1998; Chan, 1999; Piedrahita et al., 1999; Zakhartchenko et al., 1999; Polejaeva and Campbell, 2000). Also, for the first time, it is now possible to insert a length of DNA into a predetermined site in the genome of agricultural animals (Cibelli et al., 1998; Chan, 1999; Stice et al., 1998; Piedrahita et al., 1999; Zakhartchenko et al., 1999; Polejaeva and Campbell, 2000).

HISTORY OF MAMMALIAN CLONING RESEARCH

The method of nuclear transplantation was first developed and used successfully in amphibians in the 1960s (Wilson, et al., 1995). Although several attempts were made to develop an efficient procedure for use in mammals over the years, it wasn't until 1983 that a technique was developed in mice that led to the first work in agricultural species (Wilson, et al., 1995). This method is elegant and highly efficient.

The method first involves the removal of the genetic material from a recipient cell, using a unique noninvasive approach. Although a sharpened micropipet is used for the procedure, the cell membrane is so resilient that it is not penetrated (Robl et al., 1987). The pipet can be moved adjacent to the genetic material and, with aspiration, the genetic material is sucked into a membrane-bound pocket, which is then pinched off from the cell. The membranes seal, and the cell is perfectly intact.

Completion of the nuclear transplant process required the development of a method for inserting a nucleus into the enucleated recipient cytoplast. To avoid disrupting the recipient cell membrane, methods of cell fusion were attempted and proved to be successful (Robl et al., 1987). The resulting method for transplanting nuclei from one cell to another in mice was efficient and provided hope that similar methods could be practical for agricultural species.

Work began immediately to develop nuclear transplantation procedures for cows, pigs, and sheep. This work was particularly difficult because efficient *in vitro* culture methods for oocytes and embryos had not yet been developed for these species. Ovulated oocytes and embryos, recovered from superovulated animals sent to slaughter, were used. Resulting nuclear transplant embryos were transferred to the oviducts of sheep for several days before being recovered and transferred into recipient cows as blastocyst stage embryos (Robl et al., 1987).

Furthermore, methods developed for the mouse were not directly transferable to cows, sheep, and pigs. The Sendai virus fusion system did not work at all. The cytoplasm of embryos in these species is much more opaque than the mouse, making enucleation techniques more difficult, and the size of the cells is different.

However, one of the greatest challenges in cloning embryos from agricultural species was not related to these technical difficulties. Interestingly, not long after the highly efficient procedure for nuclear transplantation in mice had been published, a second paper was published by the same group of investigators reporting that the procedure would not work for cloning and, in fact, that "cloning by simple nuclear transplantation was biologically impossible." The mouse proved to be a very poor model for cloning, and real success with this species was not realized until recently. Fortunately, work persisted on cloning agricultural species, and the first success was reported in the sheep in 1986, in the cow in 1987, and in the pig in 1989 (Robl et al., 1987).

Though successful, all cloning work at this time was based on a fallacy. Prior work in amphibians, and work that still stands today, indicated that cloning success decreased as cell differentiation increased. Differentiation is the specialization that occurs in cells as the various organs and tissues form during embryo development. The thought was that differentiation was the result of a specific set of genes being turned on and the coincident inactivation of many other genes not needed in that specific cell type. It was thought that the inactivation of genes was an irreversible event; therefore, differentiation resulted in a cell having a specific function that could not be changed.

Consequently, cloning was done in agricultural animals, until recently, using donor cells from early embryos (Robl et al., 1987). These cells were thought to be less differentiated than cells from either fetuses or adult animals. One of the disadvantages of using early embryos is that the number of nuclear donor cells is limited, so the number of clones that can be produced is limited, and efficiency is of utmost importance. For example, if a 32-cell embryo is used and the efficiency is 25% blastocyst production and 25% of these embryos survive to term, the average number of clones produced is two; the same as with embryo splitting.

During the decade following the first successes with cloning, tremendous effort was placed on improving the efficiency of the embryo cell cloning procedure. One aspect of this effort was to develop and improve *in vitro* oocyte maturation and embryo culture systems in the cow and pig. Success was first realized in the cow, which greatly facilitated further research by making cow gametes and embryos readily available. A second aspect was to develop a better understanding of early development in agricultural species. Prior to this time, nearly all work on mammalian development had been done in the mouse, and very little was known about the normal process of embryo development in cows and pigs. This information was necessary for designing cloning manipulations that would simulate normal development as closely as possible. The third aspect of improving efficiency was developing novel approaches for manipulating embryos and refining these new methodologies. Ultimately, the embryo cell cloning procedure was proven not to be practical, and interest in nuclear transplantation waned.

In 1997, the surprising and revolutionary discovery was made that differentiated cells could support development to term following fusion with an enucleated egg (Wilmut et al., 1997). This first work was done in sheep and was verified the following year in the cow. In the 3 years since this discovery was made, interest in cloning research has exploded. To date, somatic cell cloning has been successful in sheep, cow, goat, mouse, and pig. Hundreds of cloned cattle and mice have been produced. Still, work remains on understanding the process and improving efficiency.

HISTORY OF GENETIC MODIFICATION RESEARCH IN CATTLE

Although cloning and genetic modification are two separate and different technologies, they have developed in parallel because they both involve manipulation of early embryos. Until recently, the only method available for genetically modifying cattle was the injection of one of the pronuclei in newly fertilized embryos. This method was quite effective in mice and has been used to make thousands of lines of transgenic mice. The efficiency in mice is about 1 to 3%, using embryos recovered directly from the mouse and transferred back immediately after injection. This level of efficiency is satisfactory in the mouse where embryos are inexpensive and easy to obtain. In the cow, early attempts at producing transgenic animals resulted in at least a tenfold lower efficiency, and the procedure was only used by a few groups, and most efforts were aimed at using transgenic cows for the production of human proteins in milk. Various attempts have been made at developing alternative methods of making transgenic cattle. The use of retrovirus DNA to carry the transgene into the egg has been successful. Many attempts have been made at coating sperm with DNA, with the hope of the sperm carrying the DNA into the egg at fertilization. None of these attempts was successful.

The first calves produced in 1998 using the somatic cell cloning procedure were genetically modified. This procedure is now being used by several companies for the production of human therapeutic proteins in either milk or blood of cows.

NUCLEAR TRANSPLANTATION METHOD FOR CLONING CATTLE

The nuclear transplantation method for cloning cattle begins with recovery and preparation of the donor cells from the animal to be cloned. Tissue is recovered as a small disc of skin from the back of the ear. The ear is shaved and washed with a disinfectant solution. A trochear is used to cut the disc of skin, which is removed with a forceps, dipped in 70% ethanol, and placed in a tube containing a sterile saline solution. Good samples have been recovered from animals of any age and, in some cases, animals that have been dead for several hours. The sample is then shipped to the laboratory by overnight express mail.

At the laboratory, the skin sample is washed, cut into small pieces, and digested with enzymes to free the fibroblast cells from the tissue. Interestingly, this process is not necessary, and fibroblasts will actually creep out of the edges of the tissue, attach to the bottom of the culture dish, and begin growing without any processing of the tissue at all. Generally, the cells are grown for several weeks, removed from the bottom of the dish with an enzyme treatment, and then cryopreserved. Several days prior to nuclear transplantation, a vial of the cells is thawed and grown until the cells cover the bottom of the dish (a confluent layer). The cells are then passed into multiple dishes to be used for several days of nuclear transplantation.

Oocytes are recovered from the ovaries of slaughtered animals. Ovaries are collected at the slaughterhouse and transported to the laboratory at room temperature. The oocytes are collected by aspirating antral follicles, using a needle attached to a vacuum pump. The oocytes, in follicular fluid, are then sorted under a dissecting microscope, and only oocytes with several layers of intact cumulus cells (the cells surrounding the egg) and a uniform cytoplasm are retained for use the following day as nuclear recipients.

Oocytes collected directly from ovarian follicles are not ready to be used in nuclear transplantation. They must undergo a process called maturation, which takes about 18 hours. After maturation, the cumulus cells are removed from around the oocytes by shaking the cells vigorously in an enzyme solution. The mature oocyte is arrested, with condensed chromosomes, in the second metaphase of meiosis. It has extruded half of its DNA in a small cell called a polar body. The chromosomes are not readily visible in the oocyte and must be labeled with a fluorescent dye for removal.

Chromosome removal is done under a microscope at 400 × magnification because the egg is only about 0.1 mm in diameter, about the size of the smallest speck of dust that can be seen. The microscope is set up with micromanipulators. The micromanipulators translate the very coarse movement of the hand into a very fine movement of a set of tiny glass micropipets. The micropipets are used to hold the egg and to remove the chromosomes. Chromosome removal involves fixing the egg in place with the holding pipet and inserting the enucleation pipet through the zona pellucida, the soft shell surrounding the egg. The chromosomes are located by a flash of UV light, which makes the fluorescent dye visible. The enucleation pipet is moved adjacent to the chromosomes, and, with aspiration, the chromosomes are removed in a membrane-enclosed bleb.

Donor cells are simply removed from the dishes using an enzyme treatment. They are then picked up individually and placed between the zona pellucida and the egg. These couplets are then fused together using a high-voltage electrical shock. The electrical pulse creates a charge across the cell membranes, pulling the nuclear donor cell tightly against the recipient cytoplast. At some point, the charge becomes sufficiently high that it ruptures the membranes, creating many tiny holes between the two cells. When the holes reseal, the two cells fuse together.

The next step in the process involves stimulating the nuclear transfer embryo to divide. Normally, when the egg is fertilized, the sperm imparts a signal that initiates cell division. Because we are omitting the sperm in nuclear transfer, we have to use an artificial stimulus to initiate cell division. There are many approaches to doing this. The procedure we use consists of treating the nuclear transfer embryos with a chemical that elevates intracellular calcium following by a chemical that inhibits protein synthesis. Egg activation is a complex process and is currently an important area of further research.

The nuclear transplant embryos are then placed in culture for 7 to 9 days. During this time, cell division takes place, and some differentiation occurs so that the embryos have about 80 to 200 cells, and cells that will form the placenta and fetus are clearly distinguishable. Embryos are then transplanted into recipients for development to term.

GENETIC MODIFICATION METHOD IN CATTLE

Genetic modification involves a couple of additional steps in the nuclear transfer procedure. After the donor fibroblast cells are grown in culture for a few weeks, the cells are removed from the dish, using an enzyme treatment. They are mixed with the DNA that is to be inserted into the cell and, using an electrical pulse to create holes in the membrane, the DNA is sucked into the cells, and, in a few rare cases, the DNA will incorporate into the cells' DNA.

The gene construct that is put into the cells usually has a few components that have important functions. The DNA must obviously contain a gene sequence. It usually contains a sequence that controls when and where the gene is to be expressed. It also contains a selection gene with its own control region. The selection gene is important because it provides a way of selecting only those cells that have incorporated the DNA into a place where the genes can function properly.

After transfection (insertion of the DNA into the cells), the cells are grown in a culture media that contains a chemical which kills all cells that are not expressing the selection gene (Cibelli et al., 1998). Most cells will die in a few days, but a few small colonies of cells will begin to appear and eventually will fill the dish. These are the transgenic cells. Unfortunately, with this approach there is no way of predicting where the gene is integrated and if it might cause some harmful mutation. Also, there is no way of determining if the gene will function properly in the animal (Chan, 1999; Piedrahita et al., 1999). Although these factors are not usually a problem, we cannot predict how the gene will function without producing the animal.

RESULTS OF SOMATIC CELL CLONING IN CATTLE

To date, there are about 150 calves that have been produced using somatic cell nuclear transplantation. About 30 different laboratories, with most being in Japan, have produced these calves. The overall development rate from nuclear transfer to healthy calf at term is about 1 to 3%.

A more detailed analysis provides information on when losses occur. The results can vary greatly depending on many factors—some known and many unknown. In general, development of cloned embryos in culture to the blastocyst stage at day 7 to 9 is about 15 to 20% but can vary from 5 to 50%. Pregnancy rates, transferring either one or two embryos per recipient, are generally high (40 to 60%) at day 35, but substantial losses occur by day 60 to 90, resulting in pregnancy rates of about 25%. Although some losses occur during the middle trimester of gestation, they are generally low. In the last trimester, substantial losses occur, resulting in about 10 to 20% of the calves surviving to term (Hill et al., 1999). Unfortunately, perinatal and postnatal losses can be as high as 50%, but with intensive care can be reduced to about 10 to 20%.

Prenatal, perinatal, and postnatal losses have been documented in some detail. Prenatal losses generally occur as a result of fluid accumulation in the placenta (hydrops). The fluid accumulation becomes sufficiently high that the cow aborts the calf. On gross examination, the calf may look normal. Most calves have been taken by Caesarian section. Hydrops is noted in many of these pregnancies (Hill et al., 1999). In addition, the calves have an extraordinarily large umbilicus (Hill et al., 1999). They also typically have breathing difficulties and benefit from being placed on oxygen for a few hours (Hill et al., 1999). They also may have a variety of other problems, such as fluid accumulation in various organs. By the first few days after birth, most calves appear normal but may exhibit digestive problems. By 60 days of age the calves have generally outgrown any defects they might have had at birth.

The defects that are seen are remarkably consistent from lab to lab and appear to stem from gestational problems as opposed to genetic defects. Analysis of early conceptuses indicates that the placenta is likely the cause of most of the problems that have been observed. For unknown reasons, placental attachment to the uterine lining does not occur properly in cloned embryos. This lack of adequate attachment may result in the fluid accumulation and large umbilical cords. The gestational defect is sometimes observed in calves derived from *in vitro*-produced embryos, which indicates that the defect may be the result of poor culture systems. However, it is unlikely that *in vitro* culture is the only cause of the problem (Cibelli et al., 1998).

RESULTS OF CLONING FOR THE PRODUCTION OF TRANSGENIC CATTLE

Production of transgenics by nuclear transplantation has the great advantage over injecting newly fertilized embryos in that all offspring produced are transgenic. This greatly increases the efficiency of producing genetically modified cattle. Approximately tenfold fewer recipient animals are needed to make a transgenic calf with cloning compared to the injection procedure. This makes the process much more affordable.

In terms of efficiency, cloning from transgenic cells is no different than cloning from nontransgenic cells with one exception. To genetically modify a cell line, the cells must be cultured *in vitro* for an extended period of time. Any cells from the body have a finite life span in culture. Fibroblasts cells from cows can divide about 25 to 35 times in culture, and then they will stop. Interestingly, the cells do not die; they just stop dividing. This added time in culture might reduce the developmental potential of the cells, therefore reducing pregnancy rates. At this time, however, there is no conclusive evidence to verify that cells cultured to the end of their life span are any less viable than freshly recovered cells (Cibelli et al., 1998).

CLONING IN THE CATTLE INDUSTRY

Somatic cell cloning has great promise, but the limitations of low pregnancy rates and calf survival restrict its current use. At this time, somatic cell cloning is still in the research phase. Considerable work needs to be done to improve survival rates and evaluate variations in results before cloning can be commercialized. The second phase will be small-scale commercialization of the technology to multiply animals of high value. As efficiency and the quality of embryos improve, cryopreservation will become feasible, and large numbers of embryos will be sold in straws just as semen is today.

The ability to clone adult animals from easily cultured skin cells presents several opportunities to the beef industry. One opportunity is to preserve valued genotypes. To maintain the genetics of any animal in the herd indefinitely, a skin sample can be taken, processed, and the cells frozen for future use. Unlike semen, the cells contain a full copy of the genetic material rather than just half the chromosomes. Unlike embryos, the phenotype of the animal is known.

Cloning may also present a unique opportunity for marketing genetics. Evaluation of an animal by performance and progeny testing is costly and takes several years. Cloning may be useful in increasing the return from fully tested animals. These animals would be of great value to producers because of the predictability of the outcome.

Cloning could facilitate integration of segments of the beef industry for the production of specialty brand name products. Highly uniform brand name products would be helpful in increasing both value and consumption of beef products. Examples of possible products are high-lean, low-fat beef from double-muscled animals or high-quality beef from Japanese Weygu animals. Increased uniformity of beef products will improve consumer appeal and reduce the cost of production.

TRANSGENICS IN THE BEEF INDUSTRY

The production of genetically modified organisms (GMOs) for food consumption has been criticized extensively in the popular press. At this time, there are no transgenic animal products on the market for food consumption, and it will likely be several years before any can possibly be evaluated for commercial agricultural use. All of the recent concern has been about genetically modified crops. The concerns about transgenic crops relate to the insertion of herbicide and pesticide genes. Some individuals have expressed concern about these genes moving into weed species, which would then be difficult to control or that natural pesticides may kill nontarget insects. Some have also expressed concern about potential negative health benefits from consuming the products of these genes. None of these concerns is applicable to transgenic cattle.

There are several genetic modifications that have been considered beneficial for beef production. In the beef industry, bull calves are significantly more valuable than heifer calves. A modification that resulted in bulls siring only male calves could have an important economic impact in the industry. Inherent disease resistance could be important in some conditions, such as shipping, for the beef industry. In tropical regions, ticks are a severe problem and inherent resistance to ticks would be beneficial. Other traits such as growth rate, feed efficiency, yield, and quality are genetically complex and are likely to be improved more efficiently by standard selection approaches. With full consumer and producer acceptance, genetic modifications could have a beneficial, but limited, impact on beef production with no harmful side effects.

REFERENCES

Chan, A.W.S. 1999. Transgenic animals: current and alternative strategies. *Cloning* 1:25.

Cibelli, J.B., S.L. Stice, P.J. Golueke, J.J. Kane, J. Jerry, C. Blackwell, F.A. Ponce de Leon, and J.M. Robl. 1998. Cloned transgenic calves produced from nonquiescent fetal fibroblast. *Science* 280:1256.

Hill, J.R., A.J. Roussel, J.B. Cibelli, J.F. Edwards, N.L. Hooper, M.W. Miller, J.A. Thompson, C.R. Looney, M.E. Westhusin, J.M. Robl, and S.L. Stice. 1999. Clinical and pathological features of cloned transgenic calves and fetuses (13 case studies). *Theriogenology* 51:1451.

Kato, Y., A. Yabuuchi, M. Nami, J.Y. Kato, and Y. Tsunoda. 1999. Development potential of mouse follicular epithelial cells and cumulus cells after nuclear transfer. *Biol. Reprod.* 61:1110.

Kato, Y., T. Tani, Y. Sotomaru, K. Kurokawa, J.Y. Kato, H. Doguchi, H. Yasue, and Y. Tsunoda. 1998. Eight calves cloned from somatic cells of a single adult. *Science* 282:2095.

Piedrahita, J.A., P. Dunne, C.K. Lee, K. Moore, E. Rucker, and J.C. Vazquez. 1999. Review: use of embryonic and somatic cells for production of transgenic domestic animals. *Cloning* 1:73.

Polejaeva, I.A. and K.H.S. Campbell. 2000. New advances in somatic cell nuclear transfer: application in transgenesis. *Theriogenology* 53:117.

Robl, J., R. Prather, F. Barnes, and W. Eyestone. 1987. Nuclear transplantation in bovine embryos. *J. Anim. Sci.* 64:642.

Stice, S.L., J.M. Robl, F.A. Ponce de Leon, J. Jerry, P.G. Golueke, J.B. Cibelli, and J.J. Kane. 1998. Cloning: new breakthroughs leading to commercial opportunities. *Theriogenology* 49:129.

Wakayama, T., A.C.F. Perry, M. Zuccontti, K.R. Johnson, and R.I. Yanagimachi. 1998. Full-term development of mice from enucleated oocytes injected with cumulus cell nuclei. *Nature* 394:369.

Wakayama, T. and R. Yanagimachi. 1999. Cloning of male mice from adult tail-tip cells. *Nature Genet.* 22:127.

Wilmut, I., A.E. Schnieke, J. McWhir, A.J. Kind, and K.H.S. Campbell. 1997. Viable offspring derived from fetal and adult mammalian cells. *Nature* 385:810.

Wilson, J.M., J.D. Williams, K.R. Bodioli, C.R. Looney, M.E. Westhusin, and D.F. McCalla. 1995. Comparison of birth weight and growth characteristics of bovine calves produced by nuclear transfer (cloning), embryo transfer, and natural mating. *Anim. Reprod. Sci.* 38:73.

Zakhartchenko, V., G. Durcova-Hills, M. Stojkovic, W. Schernthaner, K. Prelle, R. Steinborn, M. Muller, G. Brem, and E. Wolf. 1999. Effects of serum starvation and re-cloning on the efficiency of nuclear transfer using bovine fetal fibroblast. *J. Reprod. Fertil.* 115:325.

Zakhartchenko, V., R. Alberio, M. Stojkovic, K. Prelle, W. Schernthaner, P. Stojkovic, H. Wenigerkind, R. Wanke, M. Duchler, R. Steinborn, M. Mueller, G. Brem, and E. Wolf. 1999. Adult cloning in cattle: potential of nuclei from a permanent cell line and from primary cultures. *Mol. Reprod. Develop.* 54:264.

16 Alternative Methods to Micromanipulation for Producing Transgenic Cattle

M. Shemesh, M. Gurevich, E. Harel-Markowitz, L. Benvenisti, L. S. Shore, and Y. Stram

CONTENTS

Transgenic technology holds great promise as a way to introduce exogenous DNA into animals of commercial importance in order to modify their growth characteristics, to modify the production of desired animal products, or to convert the animals into factories for the production of specific proteins or other substances of pharmaceutical interest (Moore, 2001).

Early transgenic mice were produced by microinjection of exogenous DNA into the pronuclei of early embryos (Palmiter et al., 1983) or by transfection using viral vectors (Egletis et al., 1985). However, these techniques have proved to be of limited use in producing transgenic lines of livestock, as their efficiency in larger animals is extremely low (Moore, 2001). This fact, coupled with the high cost of research with domesticated animals, has made the production of transgenic animals of commercial interest extremely expensive. The poor viability of embryos after DNA microinjection and low integration rate of exogenous genes so introduced are considered to be mainly responsible for the low success rates in the production of

0-8493-1117-9/02/$0.00+$1.50

transgenic farm animals. A description of the various methods for producing transgenic animals by micromanipulation can be found elsewhere in this book (Moore, 2002; Spell and Robl, 2002). However, all of these methods using micromanipulation require extensive practice, and the technique itself is quite costly.

The present chapter describes three alternative methods to micromanipulation in cattle: 1) microprojectile bombardment; 2) electroporation of sperm; and 3) transfection of sperm using lipofection and restriction enzyme-mediated insertion. The efficacy of these techniques was evaluated with a unique reporting system, green fluorescent protein (GFP) from the jellyfish *Aequorea victoria*. Green fluorescent protein has been used as a fluorescent marker for gene expression in a variety of organisms ranging from bacteria to higher plants and animals, including bovine (Chan et al., 1997). The cloning of the gene for GFP (Inoue and Tsuji, 1994) and its subsequent expression in heterologous systems have established GFP as a valuable genetic reporter system.

ALTERNATIVE METHODS TO MICROMANIPULATION

Microprojectile Bombardment

Zelenin et al. (1993) were the first to describe a device for direct insertion of a gene into mammalian blastocysts without any manipulation. Mouse cells of developing embryos at the 2- to 4-cell, morula, and blastocyst stage were bombarded by high-velocity tungsten microprojectiles. About 70% of the developing embryos survived the bombardment. Penetration of the tungsten microparticles into the embryo cell nuclei was found at all stages investigated, and tungsten particle localization on mitotic chromosomes was demonstrated. The most important results were obtained in experiments with blastocysts. In three cases of blastocyst bombardment, the presence of transferred plasmid DNA (pSV-neo) was demonstrated. Transfected cells were shown to be located in the fetal membrane as well as in the embryo. The bombardment of mouse culture cells resulted in their transfection (G418-resistant clones). This approach had the advantage of immediate DNA incorporation into the nucleus of the cells, and also that it could be used for cells that are difficult or impossible to transfect by other methods.

We adapted this technique for murine and bovine cells using high-velocity tungsten microprojectiles coated with plasmid GFP DNA. The particle gun was the Biolistic PDS-1000/He particle delivery system (Biorad, Hercules, CA). Target cells were placed approximately 10 to 20 cm from the end of the device. Calcium-phosphate precipitation was used to coat the tungsten particles with plasmid GFP DNA. Ten milligrams of tungsten particles coated with 1 μg plasmid supercoiled DNA were used for one shot. Immediately after the shot, plates were placed in CO_2 (5%) incubator for 15 minutes to restore embryonic membrane lesions, and growth medium was added.

We initially bombarded murine one-cell embryos (fertilized oocyte before first division); two-cell, eight-cell, morula, or blastocysts and determined the gene expression (GFP) by reverse transcriptase polymerase chain reaction (RT–PCR) in the developed blastomeres and emission of specific fluorescent green light. A culture of murine embryonic cells resulting from the bombardment of a blastocyst is shown in Figure 16.1. The bombardment delivered DNA successfully in both one-cell and

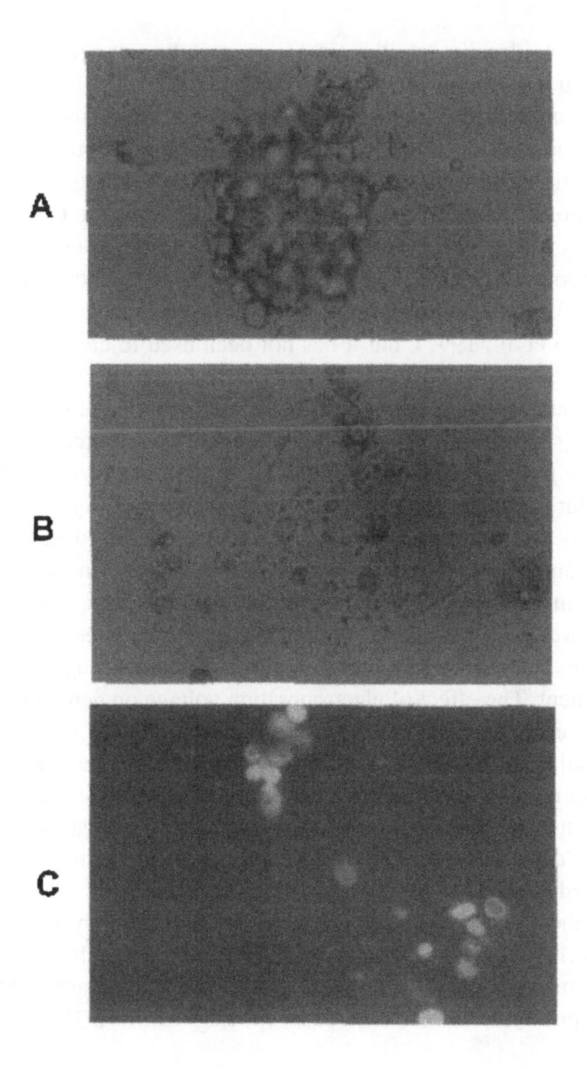

FIGURE 16.1 (A) Culture of murine blastocyst on feeder cells for 96 hours. The cells were transfected by enhanced green fluorescent protein (pGFP) gene using microprojectile bombardment. (B) Inner cell mass was removed from the trophoblast cells shown in panel A and dissociated. The cell clumps were then cultured on feeder cells for 48 hours. (C) Expression of GFP following illumination of cells shown in panel B with 450 to 490 nm light.

blastocysts. In contrast, when the stages from two cells to morula were used, bombardment damaged the embryonic blastomeres and reduced their viability. Similar results were found for bovine embryos produced by *in vitro* fertilization (IVF).

Electroporation of Sperm

The system of sperm-mediated gene transfer has triggered considerable controversy about its efficacy in promoting transgenesis (Brinster et al., 1989; Lavitrano et al., 1989; Maione et al., 1997). However, Perry et al. (1999) and Chan et al. (2000) have shown that intracytoplasmic injection of sperm encoding exogenous DNA can be used to produce transgenic mice and monkey embryos.

An attractive method to introduce new genetic information into bovine embryos would be electroporation of sperm cells. Electroporation has previously been used for obtaining stable transformants in eukaryotic cell lines (Knutson and Yee, 1987). Electroporated bovine sperm cells have been used for IVF for DNA transfer into oocytes (Gagn et al., 1991). However, the authors did not demonstrate that the gene was incorporated into the chromosome before fertilization. Electroporated sperm have been used to introduce exogenous DNA fragments in fish (Buono and Linser, 1992; Tsai et al., 1997), but it has not been used to create transgenic farm animals.

In order to evaluate the possibility of using electroporation of bull sperm, we first determined if the sperm captures plasmid green fluorescent protein (pGFP) and then determined if the gene is expressed in the early embryo following IVF with transfected sperm (Gurevich et al., 1993). To determine the percentage of pGFP retained by electroporated spermatozoa, 10^7 sperm cells were incubated in the presence of ^{32}P radiolabeled nucleotides and an unlabeled (1 μg/0.4 ml) plasmid containing green fluorescent protein (pGFP) electroporated over the voltage range from 0 to 100 volts. After extensive washings of the electroporated sperm cells, free plasmids were not present in the supernatant, but radiolabeled plasmids were strongly captured by the sperm cell sediment. The effect of electroporation voltage on gene insertion into the electroporated sperm cell and the effect on motility are shown in Figures 16.2 and 16.3. The optimal voltage was 50 volts for 40 mseconds for gene insertion without an adverse effect on motility over the range tested. This dosage was therefore used for the subsequent experiments for gene insertion. The levels of captured radiolabeled plasmid GFP in the electroporated sperm cells were significantly higher than the nonelectroporated controls ($P < 0.01$).

To evaluate the pGFP insertion by electroporation into sperm cells, we compared electroporated cells with nonelectroporated cells by exposing them to DNase I and performing polymerase chain reaction (PCR) to amplify the plasmid DNA, using a pair of specific primers. The 911 base pair (bp) fragment of pGFP was detected in

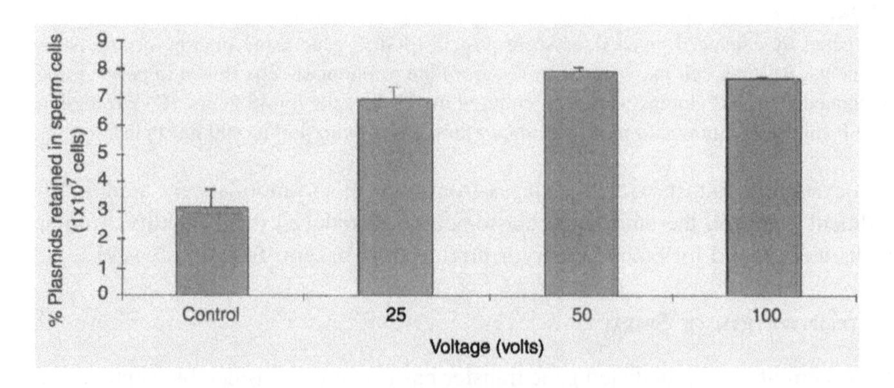

FIGURE 16.2 Effect of voltage range (0, 25, 50, and 100 volts) on percentage of radiolabeled GFP captured by the sperm cells, following electroporation.

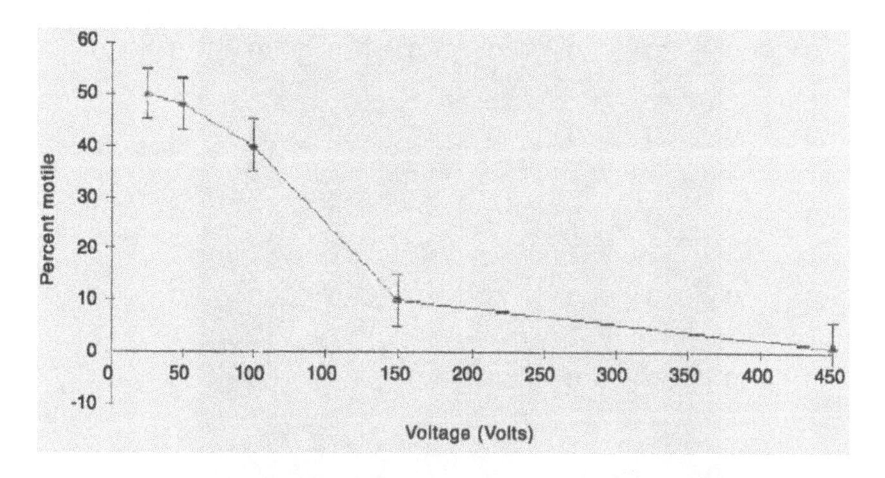

FIGURE 16.3 Effect of electroporation voltage strength on subsequent sperm motility as determined by video-enhanced images.

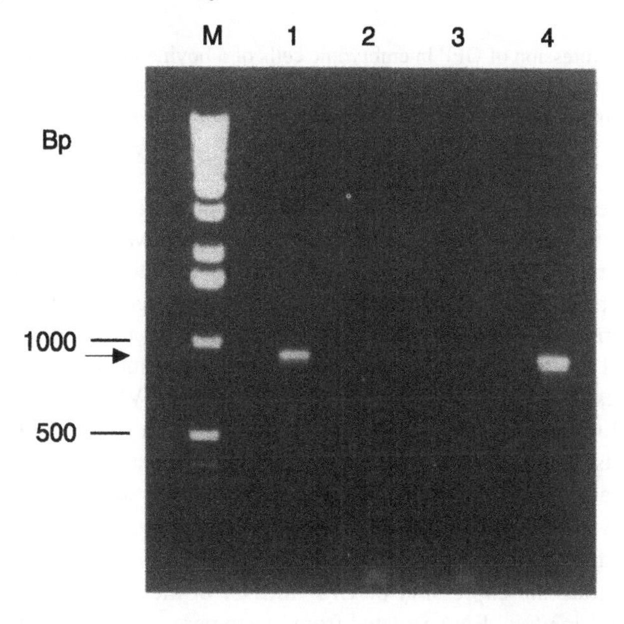

FIGURE 16.4 PCR of DNA from DNase-treated electroporated and nonelectroporated sperm cells. Lane 1 represents DNA from pGFP electroporated and DNase-treated sperm. Lane 2 represents DNA from pGFP nonelectroporated DNase-treated sperm cells. Lane 3 represents negative control, total DNA from untreated sperm cells. Lane 4 represents positive control of pGFP fragment of 911 bp.

electroporated sperm cell DNA but not in nonelectroporated controls (Figure 16.4). The lack of PCR product in the nonelectroporated cells following DNase treatment suggests radiolabeled pGFP binding seen in the nonelectroporated cells was the result of surface membrane interactions rather than internalized plasmid. This is opposed to captured pGFP in the electroporated cells that could not be digested by DNase.

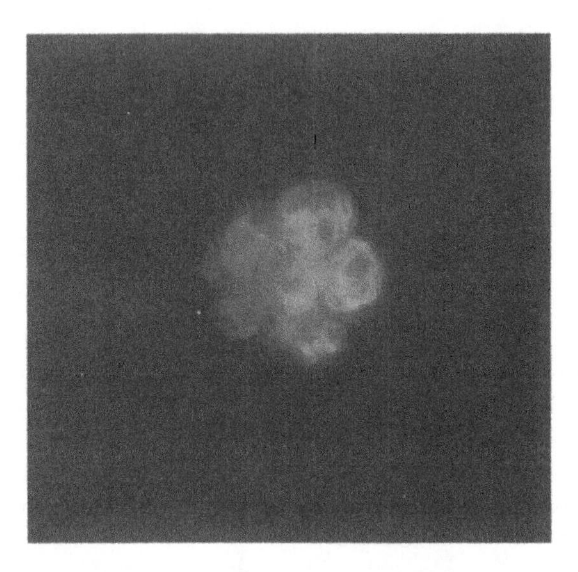

FIGURE 16.5 Expression of GFP in embryonic cells of a bovine morula. This early embryo resulted from an oocyte fertilized *in vitro* with bovine sperm transfected by electroporation. As determined by light microscopy, the embryo absorbed blue light and emitted green light at 488 nm.

The use of electroporated sperm cells in IVF resulted in a decrease in the number of cleaved oocytes (69 vs. 42%). Green fluorescent protein was fully expressed in all of the blastomeres in 7 to 12% of the resultant morula, as determined using a fluorescence microscope (Figure 16.5). Green fluorescent protein was also partially expressed in a mosaic form in other morula. Using fluorescence-activated cell sorter analysis for mechanized analyses of GFP expression in the bovine embryo, it was found that GFP was expressed in 30% of the embryos resulting from IVF by pGFP transfected sperm cells. However, 5.6% of the 2000 embryos resulting from IVF by nontransfected sperm also expressed some green light activity compatible with GFP, i.e., the false positive rate was 5.6%. This can be compared with a rate of 22% obtained for bovine blastocysts resulting from IVF using sperm which spontaneously took up a pSV_2CAT construct obtained by Sperandio et al. (1996).

To demonstrate the expression of GFP–mRNA in the early embryo, RT–PCR was performed using total RNA isolated from blastomeres resulting from GFP transfected and nontransfected sperm cells. An identical amount of RNA (1 μg) was used in all RT–PCR reactions, and PCR products of pGFP were compared in both embryonic tissues. Results are illustrated in Figure 16.6. The 313 bp target fragment specific to GFP gene was found only in the RT–PCR reaction product of total RNA from blastomeres that originated from IVF of pGFP transfected sperm cells.

Although it was demonstrated that transfected sperm could be utilized successfully in IVF, there was a significant decline in the number of cleaved oocytes in the presence of electroporated sperm cells. This decrease in fertilization rate can be explained by a premature acrosomal reaction or by possible membrane changes associated with signal transduction of the sperm cells.

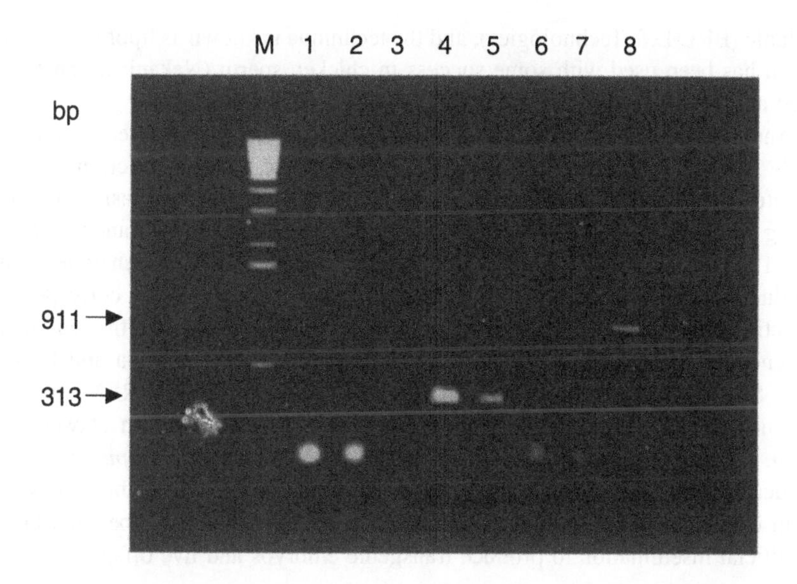

FIGURE 16.6 RT–PCR products to evaluate GFP–mRNA expression in the bovine morula. The left-most lane (M) is a DNA ladder; lane 1 represents negative control in the absence of RNA; lane 2 represents product from blastomeres from nontransfected sperm cell; lanes 3 and 4, products from blastomeres originating from pGFP-transfected sperm cells in which the GFP fragment of 313 bp is evident; lane 5 is the PCR product of pGFP as a positive control for the 313 bp fragment; and lanes 6 and 7, fragment of 911 bp PCR product, resulted from specific primer to the plasmid (same tissues as lanes 3 and 4). Lane 8 is a positive control for 911 bp product of the plasmid. The 313 bp target fragment specific to GFP and 911 bp target fragment specific to pGFP (plasmid) are indicated by arrows.

We concluded from the electroporation study that 1) bovine sperm can capture foreign DNA before fertilization, 2) that the transfected bovine sperm are capable of fertilization, and 3) that the resultant embryos express the transfected gene (GFP) and its mRNA. Therefore, electroporated sperm could provide transgenic bovine embryos by a simple inexpensive and less traumatic process than microinjection.

Sperm-Mediated Transfer Using Lipofection and Restriction Enzyme-Mediated Transfer

In addition to electroporation, other techniques for introducing foreign DNA into the sperm are spontaneous binding or lipofection. These techniques have been used to produce genetically transformed mice (Lavitrano et al., 1989; Bachiller et al., 1991), chickens (Nakanishi and Iritani, 1993), pigs (Gondolfi et al., 1996; Sperandio et al., 1996), and cows (Schellander et al., 1995). In general, the technique was not very efficient, e.g., in the cattle experiments, only one of 64 calves was transgenic (Schellander et al., 1995).

Felgner and co-workers have reported innovative methods and compositions for making liposomes that greatly improved their efficiency for transporting exogenous DNA into cells (Felgner et al., 1987). These cationic liposomes are now commercially available (BRL-Life Technologies), and the technique is known as lipofection. Lipofection has been used with some success in chicken sperm (Nakanishi and Iritani, 1993) and murine eggs and embryos (Carballada et al., 2000).

Another method developed for integrating exogenous DNA into yeasts and slime molds is restriction enzyme-mediated integration (REMI). Restriction enzyme-mediated integration utilizes a linear DNA that is derived from a plasmid DNA by cutting that plasmid with a restriction enzyme that generates single-stranded cohesive ends. The linear, cohesive-ended DNA, together with the restriction enzyme, is then introduced into the target cells by lipofection or electroporation. The corresponding restriction enzyme is then thought to cut the genomic DNA at sites that enable the exogenous DNA to integrate via its matching cohesive ends (Kuspa and Loomis, 1996; Schiestl and Petes, 1991). Kroll and Amaya (1996) used REMI to insert exogenous DNA into nuclei isolated from *Xenopus laevis* (African clawed toad) sperm. The isolated nuclei were then manually injected into *Xenopus* oocytes to produce transgenic embryos. The present study was conducted to produce transgenic sperm by REMI and to demonstrate that the transgenic sperm can be used in IVF or artificial insemination to produce transgenic embryos and live offspring.

Production of Transgenic Sperm Using Lipofection and Restriction Enzyme-Mediated Integration (REMI)

Bovine sperm cells were lipofected with *Not* I linearized pGFP and its corresponding restriction enzyme using lipid mediators. Retention of plasmid DNA in sperm was determined by using radiolabeled GFP and expression of GFP in morulas as described above for electroporation with similar results.

To demonstrate the effect of REMI on GFP integration into the sperm genomic DNA, sperm DNA was extracted and digested by *Eco* R I. Digested DNA was separated by electroporation, and sites in the digested DNA lanes (1000 to 5000 bp) were excised by needle pricking. The pricked DNA sites were then used in PCR for each site using specific GFP primers. The GFP sequence was found exclusively in DNA from lipofected sperm cells with a size of 1600 to 4000 bp but not in those with a smaller (1000 bp) or larger (5000 bp) size or in the controls (Figure 16.7). This observation was confirmed by Southern blot analysis (Figure 16.8). DNA was digested with *Hin*d III, and two different fragments about 4.1 kb and 4.9 kb were illuminated with our specific probe of 313 bp for GFP in addition to the 0.7 kb band of free GFP. This 0.7 kb band was free GFP from digested unintegrated cytoplasmic pGFP as evidenced by the control digested plasmid. These experiments indicate that GFP is integrated into the sperm genomic DNA at the *Not* I sensitive site. Presumably the two bands resulting from the *Hin*d III digestion represent two separate sites of digestion on the same genomic sequence.

Production of Transgenic Calves

Bovine sperm cells were lipofected with *Not* I-linearized pGFP and its corresponding restriction enzyme as described in Shemesh et al. (2000). The sperm was then used for artificial insemination in six cows. Three of the six cows became pregnant (50%),

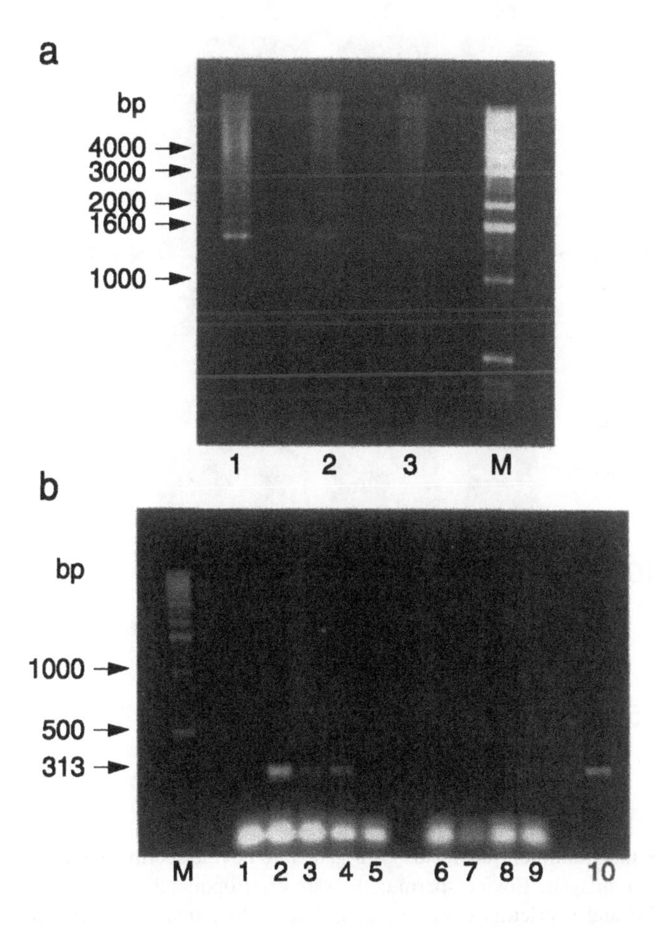

FIGURE 16.7 Integration of transfected GFP gene into the bovine sperm genome, using lipofection and restriction enzyme-mediated integration (REMI). Bovine sperm cells were transfected with *Not* I restriction enzyme and *Not* I linearized pGFP. Twenty-four hours after transfection, DNA was extracted and digested by *Eco*R I. Following separation of the digested DNA by electrophoresis on 1% agarose, designated sites in the digested DNA lanes were needle pricked. The pricked DNA sites were then used as templates in PCR, using specific GFP primers #1 and #2. Panel a: electrophoresis of transfected sperm cell DNA digested by *Eco*R I. Lane 1—DNA from sperm cells transfected with the corresponding restriction enzyme; lane 2—negative control; digested DNA from nontransfected sperm cells mixed with pGFP after DNA extraction; lane 3—DNA from control transfected sperm cells (incubated with pGFP in the absence of *Not* I restriction enzyme); lane M—1 Kbp ladder as a size marker. Panel b: PCR analysis of needle-pricked samples recovered from Figure 16.7a. Lane 1—DNA sample from Figure 16.7a at size of 1000 bp; lane 2—the same as lane 1 at size of 1800 bp; lane 3—same as lanes 1 and 2 at size of 3000 bp; lane 4—same as lanes 1 to 3 at 4000 bp; lane 5—same as lanes 1 to 4 at size of 5000 bp. Lanes 6 through 9—DNA samples from Figure 16.7a, lane 2 using the same size range as those for lanes 1 to 4. Lane 10—PCR product of pGFP as a positive control for the 313 bp fragment. Lane M—1 Kbp ladder as a size marker.

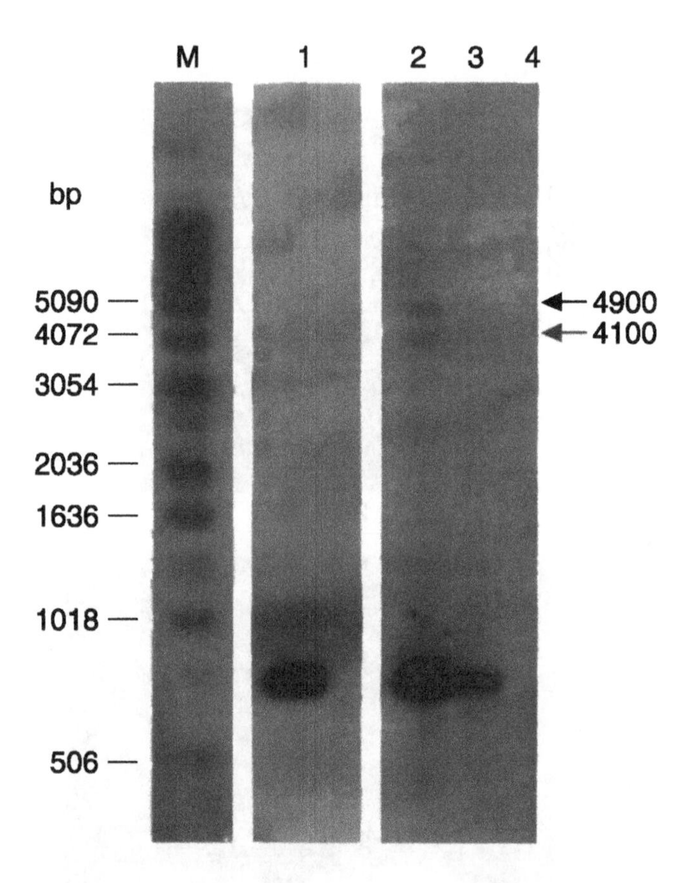

FIGURE 16.8 Integration of transfected GFP gene into bovine sperm genome as determined by Southern blot analysis. Bovine sperm cells were electroporated in the presence of *Not* I-linearized pGFP and restriction enzyme *Not* I. Twenty-four hours after transfection, DNA samples were digested by *Hin*d III, resolved on 1% agarose, and transferred onto Nytran-plus membrane filters. The filter was hybridized with the specific PCR product, 313 bp fragment of GFP gene, which was labeled with random primer biotin labeling of DNA Kit (BioLabs). M—Marker DNA (1 Kb ladder, BRL). Lane 1—free GFP obtained by double digesting of pGFP with *Not* I and *Hin*d III. Lane 2—total DNA extracted from sperm transfected with restriction enzyme-mediated integration (REMI). Lane 3—total DNA extracted from sperm transfected without restriction enzyme. Lane 4—total DNA extracted from nontransfected sperm digested with *Hin*d III endonuclease.

one of which aborted. At 2 months of age, blood samples were withdrawn from the two live offspring and used for fluorescence-activated cell sorter (FACS) analysis, PCR, RT–PCR, and Southern blot analysis.

Flow cytometry was used to evaluate GFP expression as determined by green fluorescence emission by bovine lymphocytes. Sixty percent of the cells demonstrated specific emission of green light in both of the calves tested (Figure 16.9). Both calves were positive for expression of a specific fragment of RT–PCR and PCR product in the lymphocytes (Figure 16.10).

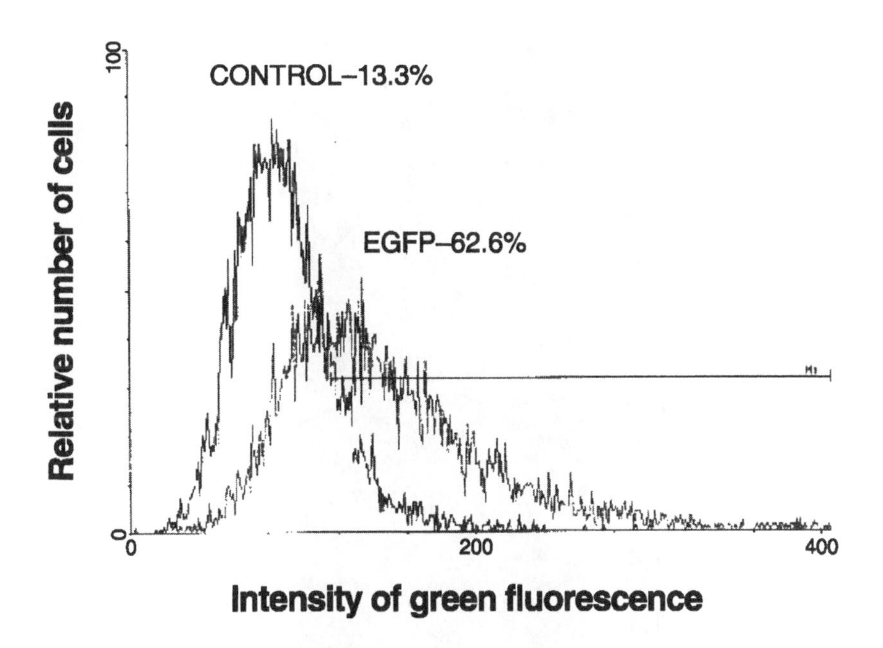

FIGURE 16.9 Fluorescence-activated cell sorting (FACS) of lymphocytes from transgenic cows. Flow cytometry was used to evaluate GFP expression as determined by green fluorescence emission in bovine lymphocytes. Lymphocytes taken from 2-month-old calves born to cows artificially inseminated with *Not* I-linearized pGFP and restriction enzyme *Not* I-lipofected sperm cells.

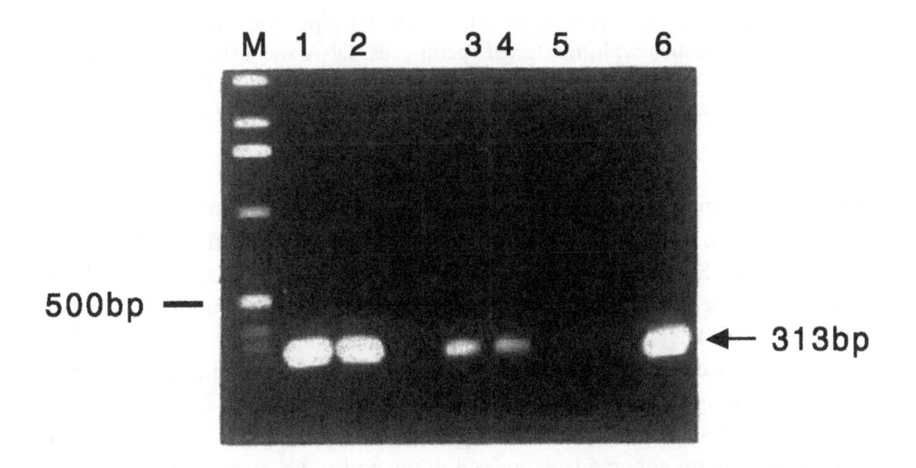

FIGURE 16.10 PCR and RT–PCR products extracted from lymphocytes of GFP transgenic calves. Left-most lane (M) is DNA ladder. Lanes 1 and 2—DNA extracted from bovine lymphocytes used in PCR reaction. Lanes 3 and 4—DNA extracted from the same lymphocytes used in RT–PCR reaction. Lane 5—negative control (lymphocytes from nontransgenic calves). Lane 6—positive control (GFP plasmid).

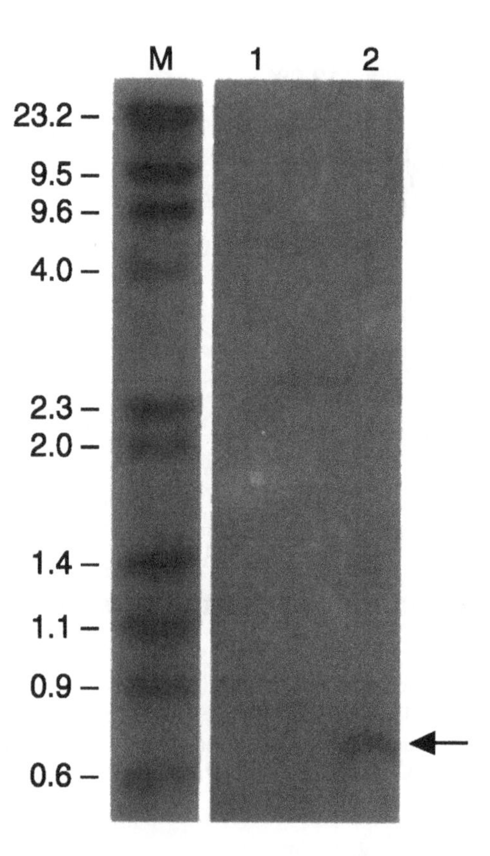

FIGURE 16.11 Southern blot analysis for total DNA of lymphocytes from calves resulted from artificial insemination with transfected sperm cells. DNA was extracted and digested with *Hin*d III + *Not* I. Lane M—DNA ladder. Lanes 1 and 2—DNA from PCR and RT–PCR positive calves, respectively. Arrow indicates positive fragment of about 0.7 kbp.

Southern blot analysis was used to determine GFP integration into the total DNA of the lymphocytes; the DNA was extracted and digested with *Hin*d III + *Not* I. The probe used for the Southern blot analysis was a biotinylated 313 bp GFP amplified fragment. A fragment of 700 bp was illuminated by our probe in one of the two calves (Figure 16.11).

In summary, we found that in using REMI in conjunction with lipofection of sperm cells: 1) foreign DNA was integrated into the sperm genome; 2) the transfected sperm can be used successfully in IVF to produce bovine embryos; 3) the gene integrated into the sperm genome is expressed in the embryo as demonstrated by both transcription and translation; and 4) the transfected sperm can be used for artificial insemination to produce transgenic calves.

The expression of GFP in the sperm cells could have resulted from episomal attachment or true integration into the genome. Furthermore, some of the embryos may have resulted from integration of the REMI construct directly into the oocyte genome. This is possible, as it was not determined what percentage of the sperm used in artificial insemination was actually transgenic. However, the data strongly suggest that GFP was integrated into a unique site in the sperm cell genome before fertilization.

We found that the REMI-lipofected sperm could be used in both IVF and artificial insemination. Following IVF, morulas having full or mosaic expression were produced (30%). However, since both calves produced by artificial insemination were positive for GFP (as evidenced by PCR, RT–PCR, and FACS), it would appear that artificial insemination is superior to IVF for producing transgenic calves. Although only two GFP calves were produced, an additional two transgenic calves were produced containing a different construct using the same method, so overall four out of four calves were positive. When the same method was applied to chickens, 89% (17/19) of the first-generation chicks were positive for GFP, and 83% of the second generation were also positive.

SUMMARY

Three methods for introducing genes into cattle were evaluated. It was found that microprojectile bombardment was suitable for bovine blastocysts and one-cell embryos. Electroporation of bovine sperm followed by artificial insemination was found to be effective in producing early embryos expressing GFP. However, at the present time, live births of calves whose cells express a foreign DNA were demonstrated only using lipofection and restriction enzyme-mediated insertion. The use of REMI and lipofection of sperm for gene transfer is a highly efficient (80 to 95%) way of producing transgenic animals. Since the eggs are fertilized by artificial insemination, the trauma to the oocyte is eliminated and all live calves born were of normal size and have survived for at least 2 years with no noticeable pathology.

REFERENCES

Bachiller, D., K. Schellander, J. Peli, and U. Rüther. 1991. Liposome-mediated DNA uptake by sperm cells. *Mol. Reprod. Dev.* 30:194.

Brinster, R.L., E.P. Sandgren, R.R. Behringer, and R.D. Palmiter. 1989. No simple solution for making transgenic mice. *Cell* 59:239.

Buono, R.J. and P.J. Linser. 1992. Transient expression of RSVCAT in transgenic zebrafish made by electroporation. *Mol. Mar. Biol. Biotechnol.* 1:271.

Carballada, R., T. Degefa, and P. Esponda. 2000. Transfection of mouse eggs and embryos using DNA combined to cationic liposomes. *Mol. Reprod. Dev.* 56:360.

Chan, A.W.S., G. Kukolj, A.M. Skalka, and R.D. Bremel. 1997. Expression of green fluorescence protein in mammalian embryos: a novel reporter gene for the study of transgenesis and embryo development. *Theriogenology* 47:222.

Chan, A.W.S., C.M. Luetjens, T. Dominko, J. Ramalho-Santos, C.R. Simerly, L. Hewitson, and G. Schatten. 2000. Foreign DNA transmission by ICSI: injection of spermatozoa bound with exogenous DNA results in embryonic GFP expression and live rhesus monkey births. *Mol. Hum. Reprod.* 6:26.

Egletis, M.A., P. Kantoff, E. Gilboa, and W.F. Anderson. 1985. Gene expression in mice after high efficiency retroviral mediated gene transfer. *Science* 230:1395.

Felgner, P.L., T.R. Gadek, M. Holm, R. Roman, H.W. Chan, M. Wenz, J.P. Northrop, G.M. Ringold, and M. Danielsen. 1987. Lipofection: a highly efficient, lipid-mediated DNA-transfection procedure. *Proc. Natl. Acad. Sci. USA* 21:7413.

Gagné, M.B., F. Pothier, and M.A. Sirard. 1991. Electroporation of bovine spermatozoa to carry foreign DNA in oocytes. *Mol. Reprod. Dev.* 29:6.

Gandolfi, F., M. Terqui, S. Modina, T.A. Brevini, P. Ajmone-Marsan, F. Foulon-Gouz, and M. Fourot. 1996. Failure to produce transgenic offspring by intra-tubal insemination of gilts with DNA-treated sperm. *Reprod. Fertil. Dev.* 8:1055.

Gurevich, M., M. Shemesh, S. Marcus, E. Harel Markowitz, and L.S. Shore. (1993). *In vitro* fertilization of oocytes obtained from Israeli Holstein cows. *Isr. J. Vet. Med.* 48:84.

Inoue, S. and F.I. Tsuji. 1994. *Aequorea* green-fluorescent protein: expression of the gene and fluorescence characteristics of the recombinant protein. *FEBS Letters* 341:277.

Knutson, J.C. and D. Yee. 1987. Electroporation: parameters affecting transfer of DNA into mammalian cells. *Anal. Biochem.* 164:44.

Kroll, K. and E. Amaya. 1996. Transgenic *Xenopus* embryos from sperm nuclear transplantations reveal FGF signaling requirements during gastrulation. *Development* 122: 3173.

Kuspa, A. and W.F. Loomis. 1996. Ordered yeast artificial chromosome clones representing the *Dictyostelium discoideum* genome. *Proc. Natl. Acad. Sci. USA* 93:5562.

Lavitrano, M., A. Camaioni, V.M. Fazio, S. Dolci, M.G. Farace, and C. Spadafora. 1989. Sperm cells as vectors for introducing foreign DNA into eggs: genetic transformation of mice. *Cell* 57:717.

Maione, B., C. Pittoggi, L. Achene, R. Lorenzini, and C. Spadafora. 1997. Activation of endogenous nucleases in mature sperm cells upon interaction with exogenous DNA. *DNA Cell Biol.* 16:1087.

Moore, K. 2001. Cloning and the beef cattle industry. In: M.J. Fields, J.V. Yelich, and R.S. Sand. (Eds.) *Factors Affecting Calf Crop: Biotechnology of Reproduction.* CRC Press, Boca Raton, FL.

Nakanishi, A. and A. Iritani. 1993. Gene transfer in the chicken by sperm mediated methods. *Mol. Reprod. Dev.* 36:258.

Palmiter, R.D., G. Norstedt, R.E. Gelinas, R.E. Hammer, and R.L. Brinster. 1983. Metallothionin human GH fusion genes stimulate growth in mice. *Science* 222:809.

Perry, A.C.F., T. Wakayama, H. Kishikawa, T. Kasai, M. Okabe, Y. Toyoda, and R. Yanagimachi. 1999. Mammalian transgenesis by intracytoplasmic sperm injection. *Science* 284: 1180.

Schellander, K., J. Peli, F. Schmall, and G. Brem. 1995. Artificial insemination in cattle with DNA-treated sperm. *Anim. Biotech.* 6:41.

Schiestl, R.H. and T.D. Petes. 1991. Integration of DNA fragments by illegitimate recombination in *Saccharomyces cerevisiae. Proc. Nat. Acad. Sci.* 88:7585.

Shemesh, M., M. Gurevich, E. Harel-Markowitz, L. Benvenisti, L.S. Shore, and Y. Stram. 2000. Gene integration into bovine sperm genome and its expression in transgenic offspring. *Mol. Reprod. Devel.* 56:306.

Sperandio, S., V. Lulli, M.L. Bacci, M. Forni, B. Maione, C. Spadafora, and M. Lavitrano. 1996. Sperm-mediated DNA transfer in bovine and swine species. *Anim. Biotech.* 7:59.

Spell, A. and J.M. Robl. 2002. Somatic cell cloning in the beef industry. In: M.J. Fields, J.V. Yelich, and R.S. Sand. (Eds.) *Factors Affecting Calf Crop: Biotechnology of Reproduction.* CRC Press, Boca Raton, FL.

Tsai, H.J., C.H. Lai, and H.S. Yang. 1997. Sperm as a carrier to introduce an exogenous DNA fragment into the oocytes of Japanese abalone (*Haliotis diversicolor suportexta*). *Transgenic Res.* 6:85.

Zelenin, A.V., A.A. Alimov, I.A. Zelenina, M.L. Semenova, M.A. Rodova, B.K. Chernov, and V.A. Kolesnikov. 1993. Transfer of foreign DNA into the cells of developing mouse embryos by microprojectile bombardment. *FEBS Letters* 315:29.

17 Cloning and the Beef Cattle Industry

Karen Moore

CONTENTS

Nuclear transfer or cloning is a very powerful technology for the production of an unlimited number of genetically identical offspring. Cloning has been available for the commercial production of beef cattle, albeit very inefficiently, since the late 1980s, using embryonic donor cells. It was not until the birth of Dolly, the sheep, in 1997 that the true potential of cloning became apparent for enhancing production efficiencies of all livestock species. Dolly was produced by fusing a mammary gland cell from an adult sheep to an enucleated oocyte. This proved that at least some adult cells had not lost their potential to be reprogrammed to an embryonic state and give rise to a new, identical individual. This review will give an overview of nuclear transfer and the potential applications this technology has for improving the beef cattle industry. It will also address the current challenges related to cloning and the research being conducted to make this technology economically feasible to the industry.

INTRODUCTION

Cloning is defined as producing a copy or copies of an individual. This technology has been used for many years to propagate plants. It occurs in animals either naturally or artificially, when a single embryo is split to produce identical twins. The word clone has also been used to describe animals produced by nuclear transfer. The procedure of nuclear transfer involves several steps, which are outlined in Figure 17.1.

0-8493-1117-9/02/$0.00+$1.50

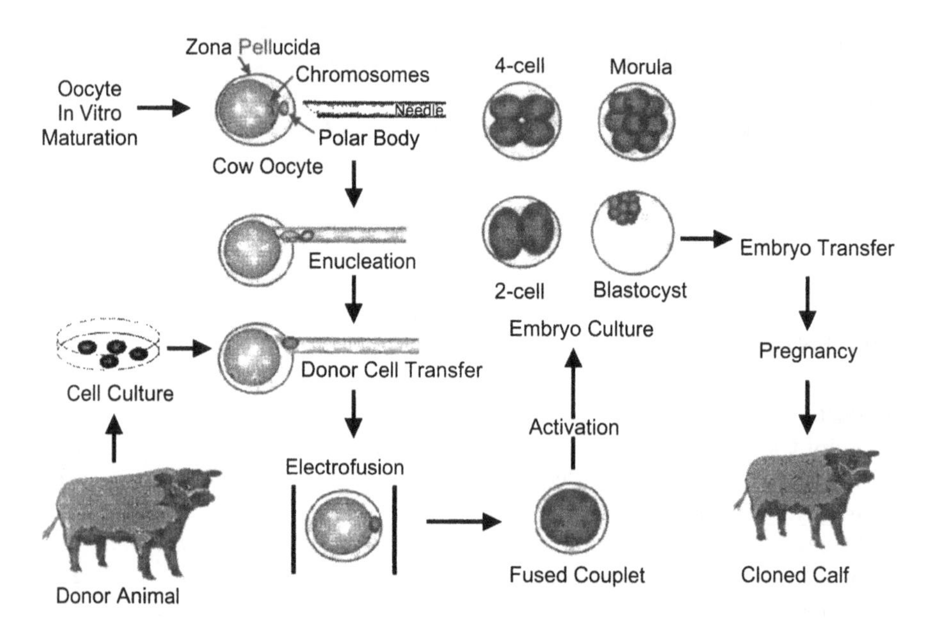

FIGURE 17.1 General overview of procedures for producing a cloned calf.

Briefly, immature eggs, also called oocytes, are collected from slaughterhouse ovaries and matured *in vitro*. After 20 to 22 hours, oocytes are stripped of their surrounding cumulus cells and those with visible polar bodies (mature metaphase II oocytes) are selected for further manipulations. Using micromanipulators, each oocyte is enucleated by piercing the zona pellucida with a glass needle and removing the polar body with a small amount of the adjacent cytoplasm. This should contain all of the oocyte's chromosomal material, which is then discarded. A donor cell from the animal to be cloned is then transferred through the same hole in the zona of each enucleated oocyte to produce couplets. Couplets are fused by alignment in a fusion chamber and applying one or two direct current pulses of 2.25 Kvolts/cm, 15 microseconds each, although voltage and timing may be varied. Fused couplets are later activated by either chemical or electrical stimulation in order to mimic sperm penetration and finish the reprogramming and maturation process. Reconstructed cloned embryos are then cultured *in vitro* for 7 to 9 days and evaluated for development. Viable blastocyst-stage embryos are transferred to synchronized recipient cows via nonsurgical embryo transfer and carried to term in order to produce live cloned calves.

HISTORY OF CLONING

The first successes in cloning were established in amphibians in the 1950s by Briggs and King (1952), when they transplanted embryonic nuclei from frogs (*Rana pipiens*), into enucleated eggs and obtained tadpoles (see Gurdon, 1999, for review). However, it wasn't until 1983, that the first nuclear transfers in mammals were successful, when mice were produced by transferring pronuclei between fertilized zygotes (McGrath and Soulter, 1983). The first successes in livestock followed, with the production of cloned sheep by fusing a cell from a 16-cell embryo to an enucleated oocyte (Willadsen, 1986). From that time through the early 1990s, several groups began developing this technology for the commercial production of cloned beef cattle (Prather et al., 1987; Willadsen, 1989; Bondioli et al., 1990; Bondioli, 1993; Stice and Keefer, 1993; Yang et al., 1993). Embryonic cells of different stages were predominantly used as nuclear donors because they were young and undifferentiated. However, this limited the possible number of identical clones to the number of donor cells within a single embryo. Several groups also tried multigenerational cloning by removing embryonic donor cells from a previously cloned embryo for use in a second, or third round of cloning. This was still quite inefficient and did not allow for genetic modifications. Cloning was thought not only to have tremendous applications for making large numbers of identical offspring, but also a means for making improvements through genetic engineering, or the addition of economically important genes.

Many groups were convinced that embryonic stem cells would need to be established for the production of genetically engineered offspring. Embryonic stem cells are established from the inner cell mass of a developing embryo, which can be maintained in an undifferentiated state indefinitely, in culture. That is, they stay young and remain in an embryonic state, which was thought to be essential for success in nuclear transfer. Stable cell lines would allow genetic alterations to be made, and then those cells could be used in nuclear transfer. While many attempts have been made over the years to isolate embryonic stem cells from cattle, none have been successful (Delhaise et al., 1995; Modlinski et al., 1996, Moens et al., 1996; Stice et al., 1996). True embryonic stem cells have only been isolated from mice (Evans and Kaufman, 1981; Martin 1981), and maybe humans to date (Shamblott et al., 1998; Thomson et al., 1998).

The biggest breakthrough in nuclear transplantation came, when Campbell et al. (1996) demonstrated for the first time that viable offspring could be produced by fusing cultured fetal cells with enucleated oocytes. In 1997, they extended this observation to include the ability to use adult somatic cells (Wilmut et al., 1997). It was then that Dolly, the sheep, was produced from a mammary gland cell taken from an adult sheep. This established for the first time that a differentiated adult cell could be fully reprogrammed into the embryonic state, and could give rise to a new, identical individual. This proved that at least some adult cells remain totipotent, which means they have retained all of the necessary information and abilities to generate a new individual. Cloning with adult cells also offers the advantage of cloning genetically proven animals.

Shortly after the birth of Dolly, it was demonstrated that genetically engineered offspring could also be produced using this same somatic cell cloning technology. Transgenic sheep were cloned from fetal cells that had been genetically altered to contain the human blood clotting factor IX gene (Schnieke et al., 1997). Somatic cell cloning has now been shown to be successful for cloning cattle (Cibelli et al., 1998; Kato et al., 1998; Shiga et al., 1999; Wells et al., 1999; Zakhartchenko et al., 1999a,b,c; Kubota et al., 2000), mice (Wakayama et al., 1998), goats (Baguisi et al., 1999) and most recently pigs (Polejaeva et al., 2000; Betthauser et al., 2000; Onishi et al., 2000).

APPLICATIONS OF CLONING TO THE BEEF CATTLE INDUSTRY

Traditional nuclear transfer involves the transfer and fusion of a donor cell to an enucleated oocyte. This methodology will be useful for numerous applications to benefit the beef industry. Its main application is for expanding superior genetics. Animals of high genetic merit, male or female, can be cloned an unlimited number of times to produce larger numbers of these valuable animals. Animals can be selected for cloning based on their superior genetics for growth, feed efficiency, disease resistance, temperament, or any desirable trait. Producing numerous animals of exceptional genetic merit will allow for more rapid genetic progress and economic benefits.

Cloning will also be useful for producing uniform calf crops of the desired sex. Producers interested in stockers or feedlot animals will have a tremendous advantage, since a uniform group of identical steers can be more effectively managed and marketed. Others may wish to produce a large set of superior replacement heifers. The nutrition, reproduction, and health of these animals should be easier to manage because of animal uniformity. It may also put a new twist on the club calf industry. If all exhibitors had identical clones to raise, it would be a true test of each individual's feeding, exercising, grooming, and handling skills.

Nuclear transplantation has also been useful for salvaging old genetics and exotic breeds. Wells et al. (1998) were able to preserve a breed of cattle from Enderby Island that had essentially gone extinct, using somatic cells from a remaining cow. In another report, Westhusin and Hill were able to clone a 21-year-old Brahman steer named Chance from his skin cells to produce a bull calf clone called Second Chance (Westhusin et al., 2001). Additionally, we may use this technology to import and then propagate new or exotic breeds of cattle that may offer other desirable genetic traits not currently available for use in crossbreeding schemes.

CLONING WITH GENETICALLY ENGINEERED DONOR CELLS

Another way that cloning will have a major impact on the beef industry is cloning with genetically engineered donor cells. Through the various genome projects, the genomes of different organisms are being sequenced, and the genes that impact different traits are being identified. As this information is made available, we will be able to use it to improve cattle by adding, removing, or modifying genes that are of economic importance to the beef cattle industry. Donor cells will be grown in culture where genetic modifications can be made. Only those cells that have the genetic modification will be selected for cloning, so that 100% of the cloned calves will have the desired genetic improvement. This technology is now a reality, with several groups having reported the births of live transgenic calves (Cibelli et al., 1998).

Possibilities for genetic modifications include adding genes for improved heat tolerance and disease resistance. This will have a major impact on reducing animal stress and disease, and therefore will reduce related production costs. Altering growth characteristics can also be accomplished through genetic engineering and cloning. Genes affecting feed efficiency and growth rate can be added for improving beef production. Additionally, those genes that enhance muscling and tenderness may be added for improving beef quality. Moreover, cloned cows could be produced that were genetically modified to produce more milk, or with a reduced incidence of mastitis. All of these examples would result in reduced production costs, improved animal production, and thus, increased profits.

The research and medical communities are also interested in producing cloned cattle. Using a uniform set of identical animals is ideal for conducting experiments. It removes all genetic variability so that only the effects of the treatments are measured. Therefore, it would be faster and easier to test, for example, different feedstuffs or new vaccines to determine the best treatment, and reduce the number of animals needed to come to these conclusions. Additionally, researchers have found that livestock in some cases are better models than mice for studying human and animal diseases. Clones could be genetically engineered to mimic a particular genetic disorder and be compared to nonengineered clones so that only the gene causing the disease would be different. New treatments for the disease would then be more easily tested to determine which is the most effective.

Biopharmaceuticals are another possibility that have been created from genetic engineering and cloning. It is now possible, through genetic engineering, to produce a cow that produces proteins of economic importance in her milk. The cow can be milked and the protein of interest can be harvested in very large quantities much more inexpensively than is currently possible, and without harm to the animal. Pharmaceuticals that are in great demand and very expensive to produce, such as blood clotting factors, could be generated from the milk supply of a single genetically engineered cow. This would provide safer, more affordable medications for both humans and animals.

The medical community is also interested in cloning for production of new cell therapies and organ transplants. Research with cattle oocytes are currently being conducted and could some day lead to the development of human stem cells (Dominko et al., 1999; Kind and Colman, 1999). Diseases such as Parkinson's, diabetes, or even spinal cord injuries could be successfully treated if a viable source of stem cells was available and was not prone to rejection by the immune system. Other groups are working on production of tissue and organ transplants from livestock. These researchers are genetically altering livestock cells so that they are no longer recognized as foreign by the human body. If they succeed, there will no longer be shortages of organs for donation. Such shortages account for more than 3000 deaths annually. Both of these new fields of study are quite controversial and will have major technical and ethical hurdles to cross before they will be accepted.

CHALLENGES

Nuclear transfer has obvious advantages for improving beef cattle production and profits to the industry. Currently, however, the technology is quite inefficient. Table 17.1 gives an overview of some of the latest results from groups from all over the world that are producing cloned cattle. It is apparent that with inefficiencies of manipulation, (an average fusion rate of 64% and an average blastocyst development rate of 33%), production of cloned embryos can still be improved upon. It is not clear yet which

TABLE 17.1
Efficiencies of Nuclear Transfer: A Summary of Results from Several Research Groups Performing Nuclear Transfer in Cattle

Donor Cell	Fusion[a]	Blastocysts[b]	Calves Born[c]	Total Survived[d]	Reference
Fetal	NA	33/276 (12)	4/28 (14)	3/28 (11)	Cibelli et al. (1998)
Cumulus	47/99 (47)	18/37 (49)	5/6 (83)	2/6 (33)	Kato et al. (1998)
Oviductal	94/150 (63)	20/88 (23)	3/4 (75)	2/4 (50)	Kato et al. (1998)
Muscle	222/358 (62)	69/222 (31)	4/26 (15)	2/26 (8)	Shiga et al. (1999)[e]
Granulosa	552/713 (77)	282/552 (51)	10/100 (10)	10/100 (10)	Wells et al. (1999)
Fetal	379/496 (76)	115/379 (30)	2/23 (9)	1/23 (4)	Zakhartchenko et al. (1999a)[e]
Primordial Germ cell	428/507 (84)	115/428 (27)	1/32 (3)	0/32 (0)	Zakhartchenko et al. (1999b)[e]
Mammary	140/223 (63)	36/140 (26)	1/4 (25)	1/4 (25)	Zakhartchenko et al. (1999c)
Skin	82/92 (89)	49/82 (60)	1/16 (6)	0/16 (0)	Zakhartchenko et al. (1999c)
Skin	440/1103 (40)	131/440 (30)	6/54 (11)	4/54 (7)	Kubota et al. (2000)[e]

[a] Total number (percentage) of couplets that fused of total attempted.
[b] Total number (percentage) of viable blastocyst-stage embryos of those cultured *in vitro*.
[c] Total number (percentage) of live calves born of those transferred.
[d] Total number (percentage) of calves surviving of those transferred.
[e] Data from treatments were grouped, giving overall averages.

donor cell types give the best results. Additionally, there are other factors that have an impact on the production of a live calf that need to be addressed. These include early embryonic and fetal losses, which can occur at any time during gestation, but occur most often between days 30 and 90. On average, only 10 to 20% of embryos are carried to term. Furthermore, for those calves that do go to term, gestation is usually extended and calves are born much larger than average (8 to 50% larger). This results in problems with dystocia, resulting in many calves having to be delivered by Caesarian section. This large calf syndrome is not specific to the nuclear transfer calf, however, as this phenomenon has been seen in embryos subjected to *in vitro* culture, or with cows on high urea diets during a critical period of gestation. Nuclear transfer calves may also be born in a weakened state and die shortly thereafter due most commonly to pulmonary insufficiency (see review by Young et al., 1998).

Another challenge when dealing with nuclear transfer calves is an adverse condition seen in some recipient cows, called "hydrops," hydroallantois or hydroamnios. This is a condition that causes an abnormal fluid accumulation within the placental tissues that can result in premature death of the recipient cow and/or calf. These problems all contribute to the current low efficiency of nuclear transfer, with only about 9% of the embryos transferred surviving after calving.

Finally, a challenge of nuclear transfer is the potential loss of genetic diversity. While unlimited numbers of identical cattle can be produced with cloning, it should not replace natural breeding and artificial insemination. If it did, it could result in loss of genetic variation and inbreeding, which are not desirable. This same concern was addressed when artificial insemination became available. Producers are aware of these possibilities, and with proper management of breeding schemes, problems such as these will not occur.

RESEARCH FOR IMPROVED EFFICIENCY

Tremendous interest has been generated in the field of somatic cell nuclear transfer for both the replication of elite beef cattle and for the production of genetically engineered cattle for various agricultural and medical applications. To date, the process is still quite inefficient and due to the long gestation period of the cow, the progress of improving efficiency is slow. However, the potential benefits available from this technology have generated a lot of enthusiasm all over the world, with millions of dollars from both public and private sources being invested in research both in academia and industry. Table 17.2 is a partial listing of the companies currently conducting research in livestock nuclear transfer. Furthermore, there are countless groups in academia attempting to optimize cloning procedures.

It was once thought that production of blastocyst-stage embryos was a good indicator of viability and ability to produce a live calf, but this has not held true. At this time, the only true end point for optimizing nuclear transfer protocols is the production of a live calf. Until other markers or end points are determined that are indicative of live offspring, progress will remain slow.

Several areas of research are being conducted in order to increase the efficiency of nuclear transplantation. Most have worked on optimizing the parameters during *in vitro* manipulations, such as *in vitro* maturation, cell fusion, oocyte activation, and embryo culture. It has been shown that cell cycle synchrony between the oocyte and the donor cell is critical, and a tremendous amount of effort is being placed on optimizing this (see review by Campbell, 1999). Moreover, culture conditions are being highly scrutinized. It is thought that serum in medium may play a role in large

TABLE 17.2
Companies Currently Involved in Cloning Livestock

Company	Location	Species
Advanced Cell Technologies	Massachusetts	Cattle, pigs
Alexion	New York	Pigs
Genzyme Transgenics Corp.	Massachusetts	Cattle, goats
Roslin Institute/Geron Corp.	UK; California	Cattle, pigs
Infigen/ABS	Wisconsin	Cattle
Nexia	Canada	Goats
PPL Therapeutics	Virginia; UK	Cattle, pigs
Pharming	UK	Cattle
Prolinia	Georgia	Cattle
Trans Ova	Iowa	Cattle
Ultimate Genetics	Texas	Cattle

calf syndrome, but this is yet to be determined. Furthermore, many of the calves that have been aborted late in gestation have also had abnormal placentation. The placenta of these calves have fewer cotyledons, which are quite enlarged. Cotyledons attach the placenta to the uterus, and fewer attachments may indicate reduced or altered blood and nutrient flow. Additionally, the umbilical cords of these calves are enlarged. Other parameters also being considered are conditions at embryo transfer, quality of recipients, and other environmental factors. There are many variables to research that may affect the developing clone, resulting in these abnormalities.

Two other areas of concern have garnered attention in nuclear transfer research, imprinting and mitochondrial heteroplasmy. Both deal with the problems of asexual reproduction, or having only one parent. Normally, an embryo arises from the joining of two gametes, one from each sex. The new embryo gets one copy of its DNA from each parent. Certain genes have been imprinted such that only the father's copy of a particular gene functions, or only the mother's copy works. However, imprints differ between gametes and somatic cells, such that some imprints may be erased or changed in somatic cells. This may result in some of the problems associated with nuclear transfer—early embryonic loss and large calf syndrome (Winger et al., 1997; Kono, 1998; Surani, 1999).

The second area is mitochondrial heteroplasmy (Steinborn et al., 1998a,b; Hiendleder et al., 1999; Smith et al., 1999). When a normal embryo is created from a sperm and an egg, the sperm contributes few, if any, mitochondria to the developing embryo. Mitochondria are believed to be solely derived from the maternal oocyte. When we produce nuclear transfer embryos we are creating a different scenario in which there are mitochondria from two sources, the oocyte and the donor cell. This may result in some abnormalities, and only research will determine to what extent this affects the development of the cloned animal.

CONCLUSIONS

It is apparent that cloning will have a great impact on all livestock industries. Unfortunately, this technology is still in its infancy and is not at a point of being economically feasible for the average beef producer. For progress to continue, it is critical for cattlemen to rally for government and industry support of research efforts. Additionally, it is important for cattle producers and researchers to interact in order to identify and direct research to key areas of importance to the beef cattle industry. As efficiencies are improved through both academic and industry research, cattlemen will see the benefits of cloning through increased animal productivity, disease resistance, and increased profits.

REFERENCES

Baguisi, A., et al. 1999. Production of goats by somatic cell nuclear transfer. *Nat. Biotech.* 17:456.

Betthauser, J., et al. 2000. Production of cloned pigs from *in vitro* systems. *Nat. Biotech.* 18:1055.

Bondioli, K.R., M.E. Westhusin, and C.R. Looney. 1990. Production of identical bovine offspring by nuclear transfer. *Theriogenology* 33:165.

Bondioli, K.R. 1993. Nuclear transfer in cattle. *Mol. Reprod. Dev.* 36:274.

Briggs, R. and T.J. King. 1952. Transplantation of living nuclei from blastula cells into enucleated frogs' eggs. *Proc. Natl. Acad. Sci. USA* 38:455.

Campbell, K.H.S., J. McWhir, W.A. Ritchie, and I. Wilmut. 1996. Sheep cloned by nuclear transfer from a cultured cell line. *Nature* 380:64.

Campbell, K.H. 1999. Nuclear transfer in farm animal species. *Semin. Cell. Dev. Biol.* 10:245.

Cibelli, J.B., S.L. Stice, P.J. Golueke, J.J. Kane, J. Jerry, C. Blackwell, F.A. Ponce de Leon, and J.M. Robl. 1998. Cloned transgenic calves produced from nonquiescent fetal fibroblasts. *Science* 280:1256.

Delhaise, F., F.J. Ectors, R. De Roover, F. Ectors, and F. Dessy. 1995. Nuclear transplantation using bovine primordial germ cells from male fetuses. *Reprod. Fertil. Dev.* 7:1217.

Dominko, T., M. Mitalipova, B. Haley, Z. Beyhan, E. Memili, B. McKusick, and N.L. First. 1999. Bovine oocyte cytoplasm supports development of embryos produced by nuclear transfer of somatic cell nuclei from various mammalian species. *Biol. Reprod.* 60:1496.

Evans, M.J. and M.H. Kaufman. 1981. Establishment in culture of pluripotential cells from mouse embryos. *Nature* 292:154.

Gurdon, J.B. 1999. Genetic reprogramming following nuclear transplantation in amphibia. *Cell Dev. Biol.* 10:239.

Hiendleder, S., S.M. Schmutz, G. Erhardt, and R.D. Green. 1999. Transmitochondrial differences and varying levels of heteroplasmy in nuclear transfer cloned cattle. *Mol. Reprod. Dev.* 54:24.

Kato, Y., T. Tani, Y. Sotomaru, K. Kurukawa, J. Kato, H. Doguchi, H. Yasue, and Y. Tsunoda. 1998. Eight calves cloned from somatic cells of a single adult. *Science* 282:2095.

Kind, A. and A. Colman. 1999. Therapeutic cloning: needs and prospects. *Cell Dev. Biol.* 10:279.

Kono, T. 1998. Influence of epigenetic changes during oocyte growth on nuclear reprogramming after nuclear transfer. *Reprod. Fertil. Dev.* 10:593.

Kubota, C., H. Yamakuchi, J. Todoroki, K. Mizoshita, N. Tabara, M. Barber, and X. Yang. 2000. Six cloned calves produced from adult fibroblast cells after long-term culture. *Proc. Natl. Acad. Sci. USA* 97:990.

Martin, G.R. 1981. Isolation of a pluripotent cell line from early mouse embryos cultured in medium conditioned by teratocarcinoma stem cells. *Proc. Natl. Acad. Sci. USA* 78:7634.

McGrath, J. and D. Solter. 1983. Nuclear transplantation in the mouse embryo by microsurgery and cell fusion. *Science* 220:1300.

Modlinski, J.A., M.A. Reed, T.E. Wagner, and J. Karasiewicz. 1996. Embryonic stem cell development capabilities and their possible use in mammalian embryo cloning. *Anim. Reprod. Sci.* 42:437.

Moens, A., P. Chesné, F. Delhaise, A. Delval, F.-J. Ectors, F. Dessy, J.-P. Renard, and Y. Heyman. 1996. Assessment of nuclear totipotency of fetal bovine diploid germ cells by nuclear transfer. *Theriogenology* 46:871.

Onishi, A., M. Iwamotoa, T. Akita, S. Mikawa, K. Takeda, T. Awata, H. Hanada, and A.C. Perry. 2000. Pig cloning by microinjection of fetal fibroblast nuclei. *Science* 289:1118.

Polejaeva, I.A., S.H. Chen, T.D. Vaught, R.L. Page, J. Mullins, S. Ball, Y. Dai, J. Boone, S. Walker, D.L. Ayares, A. Colman, and K.H. Campbell. 2000. Cloned pigs produced by nuclear transfer from adult somatic cells. *Nature* 407:27.

Prather, R.S., F.L. Barnes, M.M. Sims, J.M. Robl, W.H. Eyestone, and N.L. First. 1987. Nuclear transplantation in the bovine embryo: assessment of donor nuclei and recipient oocyte. *Biol. Reprod.* 37:859.

Shamblott, M.J., J. Axelman, S. Wang, E.M. Bugg, J.W. Littlefield, P.J. Donovan, P.D. Blumenthal, G.R. Huggins, and J.D. Gearhart. 1998. Derivation of pluripotent stem cells from cultured human primordial germ cells. *Proc. Natl. Acad. Sci. USA* 95:13726.

Schnieke, A.E., A.J. Kind, W.A. Ritchie, K. Mycock, A.R. Scott, M. Ritchie, I. Wilmut, A. Colman, and K.H. Campbell. 1997. Human factor IX transgenic sheep produced by transfer of nuclei from transfected fetal fibroblasts. *Science* 278:2130.

Shiga, K., T. Fujita, K. Hirose, Y. Sasae, and T. Nagai. 1999. Production of calves by transfer of nuclei from cultured somatic cells obtained from Japanese black bulls. *Theriogenology* 52:527.

Smith, L.C., V. Bordignon, J.M. Garcia, and F.V. Meirelles. 1999. Mitochondrial genotype segregation and effects during mammalian development: applications to biotechnology. *Theriogenology* 53:35.

Steinborn, R., V. Zakhartchenko, J. Jalyazko, D. Klein, E. Wolf, M. Müller, and G. Brem. 1998a. Composition of parental mitochondrial DNA in cloned bovine embryos. *FEBS Letters* 426:352.

Steinborn, R., V. Zakhartchenko, E. Wolf, M. Müller, and G. Brem. 1998b. Non-balanced mix of mitochondrial DNA in cloned cattle produced by cytoplast–blastomere fusion. *FEBS Letters* 426:357.

Stice, S.L. and C.L. Keefer. 1993. Multiple generational bovine embryo cloning. *Biol. Reprod.* 48:715.

Stice, S.L., N.S. Strelchenko, C.L. Keefer, and L. Matthews. 1996. Pluripotent bovine embryonic cell lines direct embryonic development following nuclear transfer. *Biol. Reprod.* 54:100.

Surani, M.A. 1999. Reprogramming a somatic cell nucleus by trans-modification activity in germ cells. *Cell Dev. Biol.* 10:273.

Thomson, J.A., J. Itskovitz-Eldor, S.S. Shapiro, M.A. Waknitz, J.J. Swiergiel, V.S. Marshall, and J.M. Jones. 1998. Embryonic stem cell lines derived from human blastocysts. *Science* 282:1145.

Wakayama T., A.C. Perry, M. Zuccotti, K.R. Johnson, and R. Yanagimachi. 1998. Full-term development of mice from enucleated oocytes injected with cumulus cell nuclei. *Nature* 394:369.

Wells, D.N., P.M. Misica, H.R. Tervit, and W.H. Vivanco. 1998. Adult somatic cell nuclear transfer is used to preserve the last surviving cow of the Enderby Island cattle breed. *Reprod. Fertil. Dev.* 10:369.

Wells, D.N., P.M. Misica, and H.R. Tervit. 1999. Production of cloned calves following nuclear transfer with cultured adult mural granulosa cells. *Biol. Reprod.* 60:996.

Westhsuin, M.E., C.R. Long, T. Shin, J.R. Hill, C.R. Looney, J.H. Pryor, and J.A. Piedrahita. 2001. Cloning to reproduce desired genotypes. *Theriogenology* 55:35.

Willadsen, S.M. 1986. Nuclear transplantation in sheep embryos. *Nature* 320:63.

Willadsen, S.M. 1989. Cloning of sheep and cow embryos. *Genome* 31:956.

Wilmut, I., A.E. Schnieke, J. McWhir, A.J. Kind, and K.H.S. Campbell. 1997. Viable offspring derived from fetal and adult mammalian cells. *Nature* 385:810.

Winger, Q.A., R. De La Fuente, W.A. King, D.T. Armstrong, and A.J. Watson. 1997. Bovine parthenogenesis is characterized by abnormal chromosomal complements: implications for maternal and paternal co-dependence during early bovine development. *Dev. Genet.* 21:160.

Yang, X., S. Jiang, P. Farrell, R.H. Foote, and A.B. McGrath. 1993. Nuclear transfer in cattle: effect of nuclear donor cells, cytoplast age, co-culture, and embryo transfer. *Mol. Reprod. Dev.* 35:29.

Young, L.E., K.D. Sinclair, and I. Wilmut. 1998. Large offspring syndrome in cattle and sheep. *Rev. Reprod.* 3:155.

Zakhartchenko, V., G. Durcova-Hills, M. Stojkovic, W. Schernthaner, K. Prelle, R. Steinborn, M. Müller, G. Brem, and E. Wolf. 1999a. Effects of serum starvation and recloning on the efficiency of nuclear transfer using bovine fetal fibroblasts. *J. Reprod. Fertil.* 115:325.

Zakhartchenko, V., G. Durcova-Hills, W. Schernthaner, M. Stojkovic, H.-D. Reichenbach, S. Mueller, R. Steinborn, M. Mueller, H. Wenigerkind, K. Prelle, E. Wolf, and G. Brem. 1999b. Potential of fetal germ cells for nuclear transfer in cattle. *Mol. Reprod. Dev.* 52:421.

Zakhartchenko, V., R. Alberio, M. Stojkovic, K. Prelle, W. Schernthaner, M. Stojkovic, H. Wenigerkind, R. Wanke, M. Düchler, R. Steinborn, M. Mueller, G. Brem, and E. Wolf. 1999c. Adult cloning in cattle: potential of nuclei from a permanent cell line and from primary cultures. *Mol. Reprod. Dev.* 54:264.

18 Reproductive Real-Time Ultrasound Technology: An Application for Improving Calf Crop in Cattle Operations

G. Cliff Lamb

CONTENTS

O. J. Ginther stated: "gray-scale diagnostic ultrasonography is the most profound technological advance in the field of large animal research and clinical reproduction since the introduction of transrectal palpation and radioimmunoassay of circulating hormones" (Ginther, 1986). It is hard to imagine that many discoveries and procedures related to ovarian, uterine, and fetal function that we use today would have been considered without the development of real-time ultrasound. The research and commercial applications of ultrasound developed for reproduction over the last 15 years support Ginther's statement.

Since ultrasound technology for soft tissue evaluation was first reported (Temple et al., 1956), many different commercial ultrasound devices have been developed for medical examination of body tissues, live evaluation of carcass composition, and diagnostic assessment of pregnancy and diseases in domestic species. The three principal modes of ultrasound are amplitude mode (A-mode), brightness mode (B-mode), and time-motion or motion mode (T–M or M-mode). In 1966, A-mode ultrasound was first used as an aid for pregnancy diagnosis in ewes (Lindahl, 1966).

0-8493-1117-9/02/$0.00+$1.50

With the development of gray-scale ultrasound, ultrasound extended to evaluation of organs such as in the abdomen (James et al., 1976; O'Grady et al., 1978, 1982), heart (Pipers and Hamlin, 1977; Pipers et al., 1978), urinary tract (Cartee et al., 1980), thorax (Rantanen, 1981; Mackey, 1983), and eye (Rogers et al., 1986). Additional developments have led to the use of ultrasound to determine carcass characteristics (Herve and Campbell, 1971; Whittaker et al., 1992), ultrasound-guided biopsies of the liver (Hager et al., 1986), kidney (Hager et al., 1986), prostate (Hager et al., 1986), blood pressure measurements (Carter et al., 1981; Garner et al., 1982), and testicular volume and weight (Bailey et al., 1998).

During the 21 years following the report by Lindahl in 1966 (i.e., 1966 to 1986), 492 articles and books addressed the use of diagnostic ultrasound in animals (Lamb et al., 1988). From 1987 until the present, more than 2200 articles and books have addressed the use of diagnostic ultrasound in large animals, alone. The exponential growth in reports citing ultrasound use over the last 15 years indicates the vast array of applications that scientists and veterinarians have developed. In many cases, we now utilize many of these procedures as a standard for many diagnostic determinations and scientific data collection techniques in large animals.

The area that has arguably benefited more from the development of ultrasound technology than any other area is reproduction in large animals. In many cases, rectal palpation has been replaced by transrectal ultrasonography for pregnancy determination, and diagnoses associated with uterine and ovarian infections. In addition, ultrasonography has added benefits such as fetal sexing and early embryonic detection, and it is less invasive than rectal palpation. From a research standpoint, ultrasound has given us the ability to visually characterize the uterus, fetus, ovary, corpus luteum, and follicles. More accurate measurements of the reproductive organs have opened doors to new areas of research and validated or refuted data from past reports. This chapter will address past, present, and future applications of ultrasound as they relate to reproductive management and research in cattle.

BASIC PRINCIPLES OF ULTRASOUND

Ultrasound is defined as any sound frequency above the normal hearing range of the human ear (i.e., greater than 20,000 hertz). Frequencies commonly used in diagnostic ultrasound range from 1.0 to 20 megahertz (MHz), although most commonly used transducers operate between 3.0 and 7.5 MHz. Diagnostic ultrasound is produced by transducers housing crystals with piezoelectric (pressure–electric) properties. When the piezoelectric crystals are deformed by pressure, electricity is produced. Conversely, when an electric current is applied to the crystals, they will deform. This is the process by which ultrasound is generated and received by the transducer. Pulsed electrical deformation of the crystal produces small sound waves, which impart kinetic energy to molecules of soft tissues. When reflected sound returns to the transducer, a slight deformation of the crystal is produced, which generates an electrical current. The current is displayed on a screen as an image of the tissue interfaces.

As the beam passes through body tissue, a portion of the beam is reflected back to the transducer. Reflection occurs at tissue interfaces of differing acoustic impedance (basically, acoustic impedance = density of the tissue plus velocity of sound in that tissue). The amount of beam reflected back to the transducer is directly proportional to the difference in acoustic impedance at the tissue interface. Because of the large difference in acoustic impedance between soft tissue, bone, and air, most of the sound beam is reflected when it strikes bone. Therefore, tissue interfaces deeper than bone or gas are hidden.

The connection of the transducer to the console has a series of encoders, which spatially orient the returning echoes to display accurately the acoustic interfaces of the tissue on the display screen. There are three main display formats: a) Amplitude mode (A-mode) ultrasound images are one-dimensional displays of returning echo amplitude and distance. Each peak represents a returning echo, the height of which is proportional to its amplitude. b) Brightness mode (B-mode) ultrasound images are two-dimensional displays of dots. The transducer is moved across the surface of a tissue, and the cross-sectional anatomy is depicted. The position of the image on the screen is determined by the time it takes for an echo to return to the transducer. Brightness of the image is proportional to the amplitude of the returning echo (i.e., echogenic). In real-time, echoes are recorded continuously on a nonstorage display screen, which can be frozen or recorded on videotape. This explains why fluid on the screen is black, soft tissues are, generally, gray, and bone is white. The B-mode form of display is the most commonly used for reproductive ultrasound purposes. c) Time-motion or motion mode (T–M or M-mode) is held in place over moving organs, and the display can be printed on an oscilloscope or moving strip of light-sensitive paper. M-mode is used primarily in monitoring heart conditions by measuring cardiac wall motion and valve excursions; however, an actual image of the heart is not produced.

B-mode, real-time imaging transducers are divided into either sector scanners or linear array imagers. Sector scanners emit a wedge-shaped beam and image, whereas linear array products present a more rectangular view. Linear array transducers do not contain one crystal, but up to 64 crystals lying side by side, all emitting an ultrasonic beam. Sector scanners have fewer crystals, which mechanically rotate in the probe or are electronically phased. The advantage of sector scanners is that they require a smaller port of entry into the tissue and are used commonly for follicular aspirations, dominant follicle removal, and ovarian biopsies. Most gynecological applications, such as uterine examination, pregnancy diagnosis, and fetal sexing, use linear array imaging.

HOW SAFE IS DIAGNOSTIC ULTRASOUND FOR REPRODUCTION?

The use of B-mode, real-time diagnostic ultrasonography has presented little verified evidence of undesirable effects in large animals. In contrast to therapeutic ultrasound (physiotherapy) and lithotripsy (calculi disintegration), diagnostic ultrasound is considered to be safe for uses associated with reproductive management. Sound pressure waves produce heat, which causes physiological changes at the cellular level, but diagnostic ultrasound is pulsed and has a low average intensity, whereas therapeutic ultrasound and lithotripsy employ greater intensities. However, acoustic outputs continue to increase (Duck and Martin, 1991) and in some cases overlap with those of therapeutic units. The power output of some Doppler ultrasonic devices is capable of heating fetal and other soft tissue close to the bone (Bly et al., 1992; Bosward et al., 1993; Horder et al., 1998). While Doppler ultrasound is not used on a large scale in reproductive ultrasound evaluations, technicians should be aware of potential biological effects induced by diagnostic levels of ultrasound.

The effects of ultrasound are momentary, as the ultrasound that is neither reflected from tissue surfaces nor transmitted through the surfaces is absorbed by the tissues and converted to heat (Horder et al., 1998). During exposures of most current scanning procedures, the production of ultrasound-induced heat is negligible in any given tissue (WFUMB, 1992). In fact, in these procedures the transducer surface is a greater source of heat (WFUMB, 1992).

Studies in which fetuses were exposed to low-intensity, pulsed ultrasound have yielded conflicting results. Shoji et al. (1975) noted an increase in fetal abnormalities in rats; however, when the protocol was repeated, an increase in fetal abnormalities could not be confirmed (Child et al., 1988). B-mode exposure was reported to depress the fetal auditory brain stem response in the fetal lamb (Siddiqi et al., 1988), and to disrupt myelin in the spinal cords of neonatal rats (Ellisman et al., 1987). Mean fetal weights significantly below those of controls have been noted in mice (Kimmel et al., 1983; Carnes et al., 1991; Hande and Devi, 1993), rats (Jensh et al., 1995), and primates (Tarantal et al., 1993).

Several studies have noted significant increases in the rate of resorption and abortion in exposed pregnant animals. Exposing mice to continuous ultrasound waves had an increased rate of resorptions with increasing intensity of the ultrasound beam (Carnes et al., 1991). In mice or rats exposed to real-time ultrasound, significant detrimental effects were noted on litter size, resorption rate, and stillbirths (Iannaccone et al., 1991), and Demoulin et al. (1985) reported decreased pregnancy rates in women or rats exposed to ultrasound prior to ovulation. In contrast, additional reports (Mahadevan et al., 1987; Gates et al., 1988) indicate no detrimental effects of ultrasound on fertility. In a large study of 2834 pregnant women (Newnham et al., 1993), half the group was ultrasonically examined on five occasions by imaging the fetus and by continuous Doppler investigations of the umbilical and placental vessels. The remaining half of the group was scanned once at 18 weeks of gestation. Babies born after multiple exposures to ultrasound were smaller and lighter than babies born to the single exposure to ultrasound. A similar study (Tarantal et al., 1993) in macaques exposed to repeated ultrasound examinations reported differences in neonatal viability scores, crown–rump length, food intake, leukocyte counts, and behavior. The only large animal study (Squires et al., 1983) addressing the effects of ultrasound on fetal development reported no difference in pregnancy rate among groups given scans and rectal examinations, rectal examinations and inactivated scanning probe, and rectal examinations only.

The literature addressing the biological effects of ultrasound on the uterus, fetus, and ovary is limited, but diagnostic ultrasound can have adverse effects on mammalian tissues. With many conflicting results and unexplained mechanisms, work clearly needs to continue in this area so that ultrasound parameters can be established to avoid adverse effects from excessive exposure to ultrasound waves. However,

B-mode, real-time ultrasound has low acoustic output and presents negligible risk with much benefit to veterinarians and scientists. Technicians should be aware of the possibility of the bioeffects of different modes of ultrasound, remain cognizant of current developments in the field, and adopt a prudent approach by minimizing the number and length of examinations to obtain the required information.

ULTRASONIC EVALUATION OF THE OVARIAN STRUCTURES

The use of ultrasound technology to evaluate ovarian activity has been reviewed in great detail (Pierson and Ginther, 1988; Beal et al., 1992). Ovarian stroma, ovarian vessels, follicles, cysts, corpora hemorrhagica (CH), and corpora lutea (CL) are all structures that have been previously identified by real-time ultrasonography (Cochran et al., 1988; Pierson and Ginther, 1988; Kastelic et al., 1990a,b; Beal et al., 1992; Stewart et al., 1996; Sunderland et al., 1996; Singh et al., 1997). The most distinguishable ovarian structures are antral follicles. Because follicles are fluid-filled structures, they absorb ultrasound waves and are displayed as black on the screen (i.e., anechoic or nonechogenic). In contrast, the ovarian stroma, CH, and CL all contain varying degrees of dense cells, which reflect the ultrasound waves and result in a gray image on the screen.

Ultrasonography has been used to monitor the growth and atresia of individual antral follicles, and it has been established that growth of follicles usually takes place in two or three waves during the estrous cycle (Pierson and Ginther, 1986; 1987a; 1988; Savio et al., 1988; Sirois and Fortune, 1988; Ginther et al., 1989a,b). Sequential ultrasonography also was used to identify whether follicles were growing (increasing in size), static (no change in size), or regressing (decreasing in size; Ginther, 1989b). Ultrasonographically, classified follicle stages have also shown to correlate closely with the ability of follicles to produce hormones (estrogen:progesterone and estrogen:androstenedione ratios, inhibin concentrations, and IGF-binding proteins) indicative of follicular health (Badinga et al., 1992; Guilbault et al., 1993; Price et al., 1995; Carrière et al., 1996; Stewart et al., 1996; Sunderland et al., 1996).

Gray-scale densitometry has been used for quantitative analysis in a number of different fields such as gel electrophoresis, microspectrophotometry, and immunohistochemistry (Sternberger and Sternberger, 1986; Fritz et al., 1989; 1992; Renucci et al., 1991; Ferrandi et al., 1993). Not until recently did computer-assisted analyses focus on ultrasound image attributes of bovine ovarian follicles. Echotexture characteristics of high-resolution ultrasound images of the follicle antrum and wall were correlated with the functional and endocrine status of the follicle (Singh et al., 1998; Tom et al., 1998b). Specific correlations were noted between estradiol or the estradiol:progesterone ratio and echotexture (pixel heterogeneity) characteristics of the follicle antrum and wall (Singh et al., 1998). The mean pixel value and heterogeneity of the antrum and wall were low in preovulatory follicles as well as in growing and early-static phase anovulatory dominant follicles. In late-static and regressing phase subordinate follicles pixel heterogeneity was high (Singh et al., 1998; Tom et al., 1998b). Briefly, mean pixel value is obtained by spot analysis (a randomly selected spherical area) of the antrum and is a measure of the overall gray-scale value of pixels falling under the measuring circle, whereas pixel heterogeneity is a measure of the variation in gray-scale value of pixels falling under the measuring circle. In light of these recent developments, further computer-assisted echotexture analysis may be developed into a diagnostic and prognostic tool to assess physiological and pathological status of bovine ovarian follicles.

Manual palpation or ultrasonographic examination of the cow's genital tract are currently used by veterinarians involved in reproductive management. A recent report (Aslan et al., 2000) evaluating the difference in detection of follicles by ultrasound or rectal palpation concluded that ultrasound was more effective in identifying follicles greater than 10 mm in diameter than rectal palpation. Follicles 10 to 15 mm in diameter were detected in 90% of cases using ultrasonography vs. 62% of the cases using rectal palpation. Follicles greater than 15 mm were detected in 100% of the cases for both ultrasonography and rectal palpation. In a similar review (Hanzen et al., 2000), manual diagnosis of follicles <10 mm was inaccurate, but ultrasound offered the possibility to diagnose follicles <5 mm and to measure the diameter of those follicles. Figure 18.1 demonstrates the appearance of the ovary at various stages of follicular development prior to emergence of a follicular wave, during proestrus, and after development of a follicular cyst.

The ultrasound appearance of luteal structures varies with their stage of development. Corpora hemorrhagica, present from ovulation until day 3 after ovulation, are echogenic, but are less dense than the mature CL. The CH is distinguished from the ovarian stroma by a fine nonechogenic center. Previous reports have been unable to identify the CH before day 3 after ovulation (Pierson and Ginther, 1984), but with sequential daily ultrasound analysis, the CH was distinguished from the stroma (Cochran et al., 1988). The mature CL is easily distinguished from the stroma by a uniform, spherical texture ("salt and pepper" appearance), which is less echogenic than the stroma. Lacunae and a lumen within the CL are easily distinguished as spherical nonechogenic spaces and are easily differentiated from a follicle by the presence of a rim of luteal tissue surrounding the lumen.

The appearance of the CL may be used to estimate the stage of the bovine estrous cycle (Ireland et al., 1980; Pierson and Ginther, 1987b; Kastelic and Ginther, 1989; Kastelic et al., 1990a,b; Singh et al., 1997), yet differences in CL development decrease the accuracy of estimates. A higher percentage of CL in early diestrus tend to have a fluid-filled lumen vs. the CL during late diestrus and advanced stages of pregnancy. We (Spell et al., 2001) determined that luteal diameter was not associated with concentrations of progesterone on day 7 of the estrous cycle, but area and volume were correlated to concentrations of progesterone. Similarly, Marciel et al. (1992) reported a high correlation between progesterone concentration and CL mass and volume. Singh et al. (1997) and Tom et al. (1998a) also found that pixel values of ultrasound images, obtained by quantitative echotexture analysis, decreased from metestrus to mid-diestrus, and increased during proestrus. Pixel values of ultrasound images were highly correlated to plasma and luteal tissue progesterone concentrations and volume densities of luteal cells and stroma. In addition, Perry et al. (1991) determined, with the use of transrectal ultrasonography, that cows with visible luteal cells had serum concentrations of progesterone of greater than 0.5 ng/ml. As with follicles, changes in ultrasound images occur concurrent with structural and hormonal

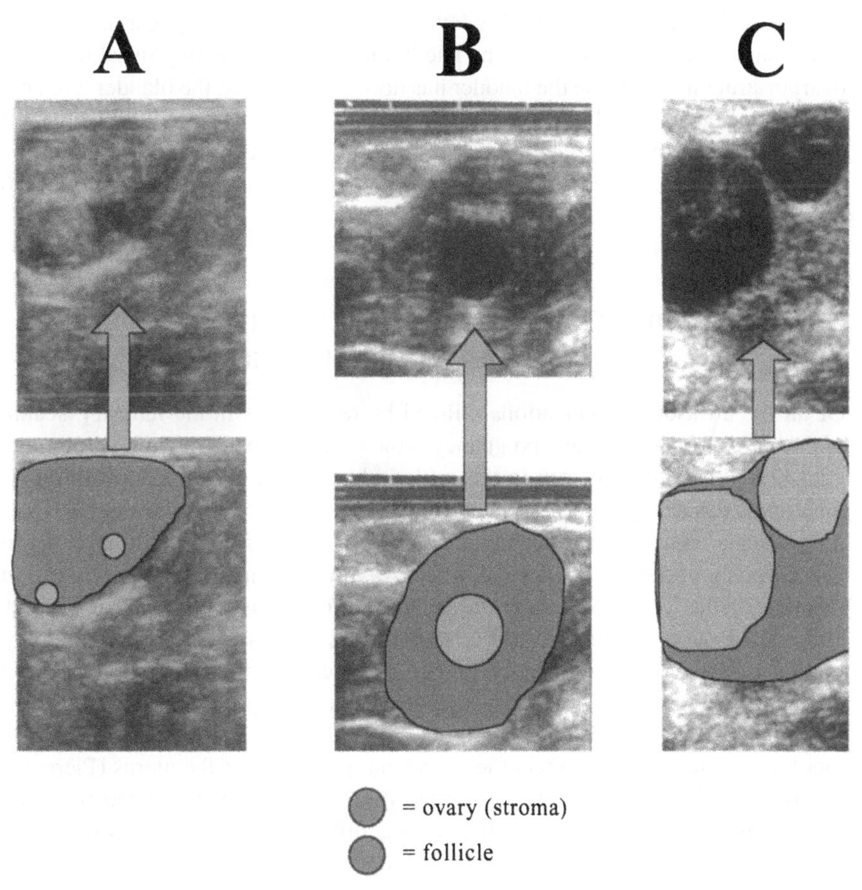

FIGURE 18.1 Ultrasound image of bovine ovaries prior to emergence of a follicular wave (note two small follicles [<5 mm]; Panel A), during proestrus (note preovulatory follicle [13 mm]; Panel B), and after development of a follicular cyst (note delamination of granulosa layer into the antrum; Panel C). Images were taken using a 7.5 MHz transducer.

characteristics of the bovine CL. Further refinements in ultrasound image analysis could aid in distinguishing between the growing and regressing phases of luteal gland development.

Ovarian cysts, blood vessels, and the bladder also are common nonechogenic images on the screen. Luteal cysts can be distinguished from follicular cysts by noting a rim of luteinized cells surrounding the cavity. Follicular cysts tend to be enlarged follicles surrounded by stromal tissue. The simple definition for a follicular cyst in our laboratory is when the mean of the vertical and horizontal diameter exceeds 25 mm. The antral wall of follicular cysts tends to be echogenic, and a sharp border is often present. Blood vessels tend to be confused with follicles, especially with the untrained eye. Altering the plane of scanning by rotating the transducer relative to the ovary elongates blood vessels, or shifting the transducer in the same plane will move the blood vessel across the screen, and it will maintain its spherical shape and size. In this latter case, a follicle will appear and disappear in the same location on the screen. The bladder is easily distinguishable from the ovarian structures because the bladder has no shape. Usually, the bladder will appear wedge-shaped and will often contain sharp echogenic crystals within the fluid. Without question, accuracy and reliability of ultrasonic measurements are achieved by the experience of the operator, a reliable machine, and a cooperative animal.

MONITORING UTERINE AND FETAL CHARACTERISTICS BY ULTRASONIC IMAGING

Of all the ultrasound applications utilized by technicians in the industry, scanning of the uterus for infection and pregnancy is the most commonly practiced commercial application that we have seen in the cattle industry. In a nonpregnant, cycling cow, the uterine tissue appears as a somewhat echogenic structure on the screen. Because the uterus is comprised of soft tissue, it absorbs a portion of the ultrasound waves and reflects a portion of the waves. In this way, we can identify the uterus as a gray structure on the screen. A cross-sectional view of the uterus is displayed as a "rosette" and is easily distinguished from other peripheral tissues, whereas the longitudinal section is less recognizable, yet a trained technician can differentiate between the elongated view of the uterus and other tissues that may appear similar (Figure 18.2). Physiological changes during the estrous cycle lead to physical changes (such as tone) in the uterus, which alters the echogenic properties of the uterus (Pierson and Ginther, 1987a). Even though a scoring system has been developed to describe change in uterine echogenic ability during different stages of the estrous cycle (Pierson and Ginther, 1987a), predicting the stage of the estrous cycle remains inconsistent.

Pathological applications for ultrasound technology have extended to identifying endometritis, pyometra, mucometra, and hydrometra (Perry et al., 1990). With the aid of ultrasound, researchers have determined that uterine infections were related to delayed postpartum folliculogenesis, to the occurrence of short luteal phases after the first postpartum ovulation, and to the development of follicular cysts on the ovaries (Peter and Bosu, 1988).

Reports have indicated the detection of an embryonic vesicle in cattle as early as 9 (Boyd et al., 1988), 10 (Curran et al., 1986a), or 12 days (Pierson and Ginther, 1984) of gestation. In these situations, the exact date of insemination was known and ultrasonography simply was used as a confirmation of pregnancy or to validate that detection of an embryo was possible within the first 2 weeks of pregnancy. In contrast, Kastelic et al. (1989) monitored pregnancy in pregnant and nonpregnant yearling heifers that were all inseminated. Diagnosis of pregnancy in heifers on day 10 through day 16 of gestation resulted in a positive diagnosis for pregnant or nonpregnant of less than 50%. On days 18, 20, and 22 of gestation, accuracy of pregnancy diagnosis improved to 85%, 100%, and 100%, respectively. Although evidence of a pregnancy via ultrasound during days 18 to 22 of gestation yields excellent results, a technician needs to ensure that fluid accumulation in the chorioallantois

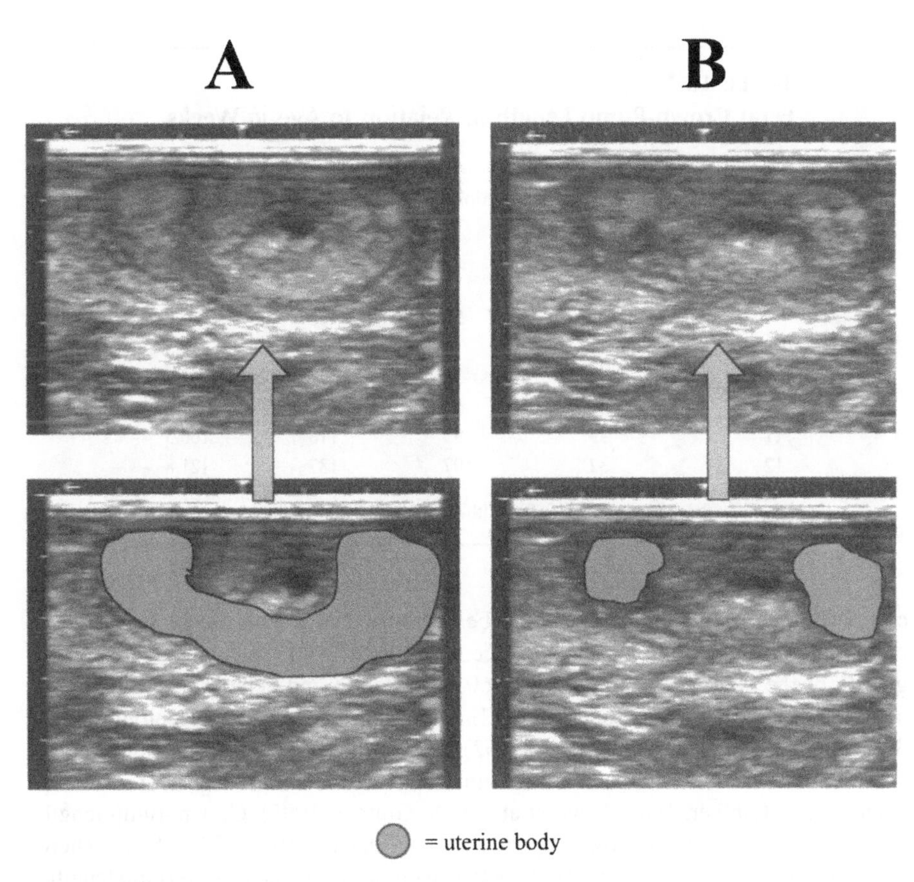

FIGURE 18.2 Ultrasound image depicting an elongated (Panel A) and cross-sectional (Panel B) view of the nonpregnant uterus. Images were taken using a 5.0 MHz transducer.

during early pregnancy (Kastelic et al., 1989) and uterine fluid within the uterus during proestrus and estrus are not confused when making the diagnosis.

Several further reports (Taverne et al., 1985; Hanzen and Delsaux, 1987; Pieterse et al., 1990, Badtram et al., 1991) also indicate the presence of an embryonic vesicle as early as day 25 of gestation. Although Hanzen and Delsaux (1987) utilized a 3.0 MHz transducer for pregnancy diagnosis, they concluded that by day 40 of gestation a positive diagnosis of pregnancy was 100% accurate, whereas overall diagnosis of pregnancy and absence of pregnancy from day 25 of gestation proved to be correct in 94% and 90% of cases, respectively. In 148 dairy cows, pregnancy diagnosis from day 21 to day 25 was 65% accurate, whereas diagnosis of pregnancy from day 26 to 33 was 93% accurate (Pieterse et al., 1990). In their conclusions, the authors state that probable causes of misdiagnosis from day 21 to day 26 were either an accumulation of proestrus or estrus uterine fluid, or the accumulation of pathological fluid in the uterus, or were diagnosed pregnant but experienced early embryonic loss.

Although we have indicated that an embryonic vesicle is detectable by ultrasound as early as 9 days of gestation, accuracy of detection approaches 100% after day 25 of gestation. For practical purposes, the efficiency (i.e., speed and accuracy) of a

TABLE 18.1
Fetal Crown-Rump Length in Relation to Age in Weeks

Fetal Age, weeks	No. of Observations	Crown-Rump Length, mm Minimum	Maximum	Mean
4	25	6	11	8.9
5	35	8	19	12.8
6	50	16	26	20.2
7	47	23	36	27.7
8	41	36	52	45.5
9	48	39	71	62.4
10	43	61	101	87.4
11	39	95	118	106.5
12	32	107	137	121.8

From Hughes and Davies (1989). With permission.

correct diagnosis of pregnancy should be performed in females expected to have embryos that are at least 26 days of age. Curran et al. (1986b) characterized the growth of the embryo from days 20 to 60 of gestation and determined when characteristics such as the heartbeat (day 22), spinal cord (day 28), placentomes (day 35), split hooves (day 44), and ribs (day 52) first became detectable. This information can be used to determine the age of bovine fetuses with a high degree of accuracy (Pierson and Ginther, 1984; Boyd et al., 1988; Ginther, 1995). Crown–rump length measurements were summarized by Hughes and Davies (1989; Table 18.1). There was a significant correlation ($r = 0.98$) between embryo age and crown–rump length.

Prior to the development of ultrasound for pregnancy diagnosis in cattle, technicians were unable to accurately determine the viability or number of embryos or fetuses. Because the heartbeat of a fetus can be detected at approximately 22 days of age, we can accurately assess whether or not the pregnancy is viable. Studies in beef (Diskin and Sreenan, 1980; Beal et al., 1992; Lamb et al., 1997) and dairy (Smith and Stevenson, 1995; Vasconcelos et al., 1997; Fricke et al., 1998; Szenci et al., 1998) cattle have used ultrasound to assess the incidence of embryonic loss. The number of fetuses can most accurately be assessed at between 49 and 55 days of gestation (Davis and Haibel, 1993).

Table 18.2 summarizes the incidence of embryonic loss by study in beef and dairy females. The fertilization rate after AI in beef cows is 90%, whereas the embryonic survival rate is 93% by day 8 and only 56% by day 12 post AI (Diskin and Sreenan, 1980). The incidence of embryonic loss in beef cattle appears to be significantly less than in dairy cattle. Beal et al. (1992) report a 6.5% incidence of embryonic loss in beef cows from day 25 of gestation to day 45. Similarly, Lamb et al. (1997) noted a 4.2% incidence of embryonic loss in beef heifers initially ultrasounded at day 30 of gestation and subsequently palpated rectally at between day 60 and 90 after insemination. In dairy cattle, pregnancy loss from 28 to 56 days after AI was 13.5%, or 0.5% per day (Fricke et al., 1998). This rate of pregnancy

TABLE 18.2
Incidence of Embryonic/Fetal Loss in Cows after an Initial Diagnosis of Pregnancy by Ultrasound, Followed by a Second Diagnosis Prior to or at Calving

Reference	No. Pregnant, Days of Gestation	No. Pregnant, Days of Gestation	No. Embryos Lost	Embryonic Mortality %
		Beef Cattle		
Beal et al. (1992)	138	129	9	6.5
(Cows)	25 days	45 days		
	129	127	2	1.5
	45 days	65 days		
	138	127	11	8.0
	25 days	65 days		
Lamb et al. (1997)	149	143	6	4.0
(Heifers)	30 days	60 days		
	271	260	11	4.1
	35 days	75 days		
	105	100	5	4.8
	30 days	90 days		
		Dairy Cattle		
Smith and Stevenson (1995)	129	113	16	12.4
(Cows and heifers)	28 to 30 days	40 to 54 days		
Vasconcelos et al. (1997)	488	437	51	10.5
(Cows)	28 days	42 days		
	437	409	28	6.3
	42 days	56 days		
	409	402	7	1.7
	56 days	70 days		
	402	395	7	1.7
	56 days	70 days		
	488	395	93	19.1
	28 days	98 days		
Fricke et al. (1998)	89	77	12	13.5
(Cows)	28 days	56 days		
Szenci et al. (1998)	64	52	12	8.6
(Cows)	26 to 58 days	Full term		

loss is similar to the 12.4% reported by Smith and Stevenson (1995) and the 19.1% reported by Vasconcelos et al. (1997) during a comparable stage of pregnancy in lactating dairy cows. The greatest occurrence of pregnancy loss was between day 28 and 42 of gestation (10.5%) and between day 42 and 56 of gestation (6.3%). After day 56 of pregnancy, embryonic losses were reduced to 3.4% from 56 to 98 days of pregnancy and 5.5% from 98 days to calving (Vasconcelos et al., 1997). Specific physiologic mechanisms responsible for pregnancy loss in dairy cattle may include lactational stress associated with increased milk production (Nebel and McGilliard, 1993), negative energy balance (Butler and Smith, 1989), toxic effects of urea and nitrogen (Butler et al., 1995) or reduced ability to respond to increased environmental temperature (Stevenson et al., 1984; Hansen et al., 1992). These studies indicate the usefulness of ultrasonography as a tool to monitor the success of a breeding program, by determining pregnancy rates and embryonic death.

Additional investigators have reported a range of embryonic mortality from day 21 to 60 to be 8% (Boyd et al., 1969) to 35% (Beghelli et al., 1986). In those reports, embryo mortality was determined by the presence of high blood concentrations of progesterone at day 21 to 23 after breeding (presuming a high concentration of progesterone was caused by the embryonic signal to prevent luteal regression) but the absence of an embryo or fetus by rectal palpation at 40 to 60 days after insemination. The authors' rationale assumed that the embryo was lost between day 21 of gestation and the time of palpation; however, there was no positive identification of a viable embryo at day 21 to 23 of gestation. Therefore, ultrasonography provides a tool to accurately differentiate between the failure of a female to conceive or the incidence of embryonic mortality because a heartbeat is detectable at 22 days of gestation.

In a study comparing the efficacy of ultrasonography, bovine pregnancy-specific protein B, and bovine pregnancy-associated glycoprotein 1 tests for pregnancy detection in 138 dairy cows results were mixed (Szenci et al., 1998). When recognition of an embryo proper with a beating heart was used as the criterion for positive ultrasonographic diagnosis, significantly fewer ($P < 0.001$) pregnant cows were correctly identified than by the other two tests. When compared with the noncalving cows, significantly fewer ($P < 0.001$) false positive diagnoses were made by the ultrasonographic tests than by the PSPB and bPAG 1 tests.

Many cattle operations are developing strategies to use fetal sexing as either a marketing or purchasing tool. At approximately day 50 of gestation, male and female fetuses can be differentiated by the relative location of the genital tubercle and development of the genital swellings into the scrotum in male fetuses (Jost, 1971). Fetuses at 48 to 119 days of age have been successfully sexed (Müller and Wittkowski, 1986; Curran et al., 1989; Wideman et al., 1989; Beal et al., 1992). The procedure is reliable, and accuracy has ranged from 92 to 100% (Müller and Wittkowski, 1986; Wideman et al., 1989; Beal et al., 1992). Beal et al. (1992) noted that, of 85 fetuses predicted to be male, 84 were confirmed correct, resulting in 99% accuracy. In addition, of 101 fetuses predicted to be female, 98 were confirmed correct, resulting in 97% accuracy. Recently, we (Lamb et al., unpublished data) determined the sex of 112 fetuses in Angus heifers with 100% accuracy.

For optimal results, the ultrasound transducer should be manipulated to produce a frontal, cross-sectional, or sagittal image of the ventral body surface of the fetus. In larger-framed cows (i.e., Holsteins and Continental beef breeds) or older cows, the optimum window for fetal sexing is usually between days 55 and 70 of gestation, whereas for smaller-framed cows (Jerseys and English beef breeds), the ideal window is usually between days 55 and 80 of gestation. There are two limitations that could inhibit the ability of a technician to determine the sex of a fetus: 1) as the fetus increases in size, it becomes more difficult to move the transducer relative to

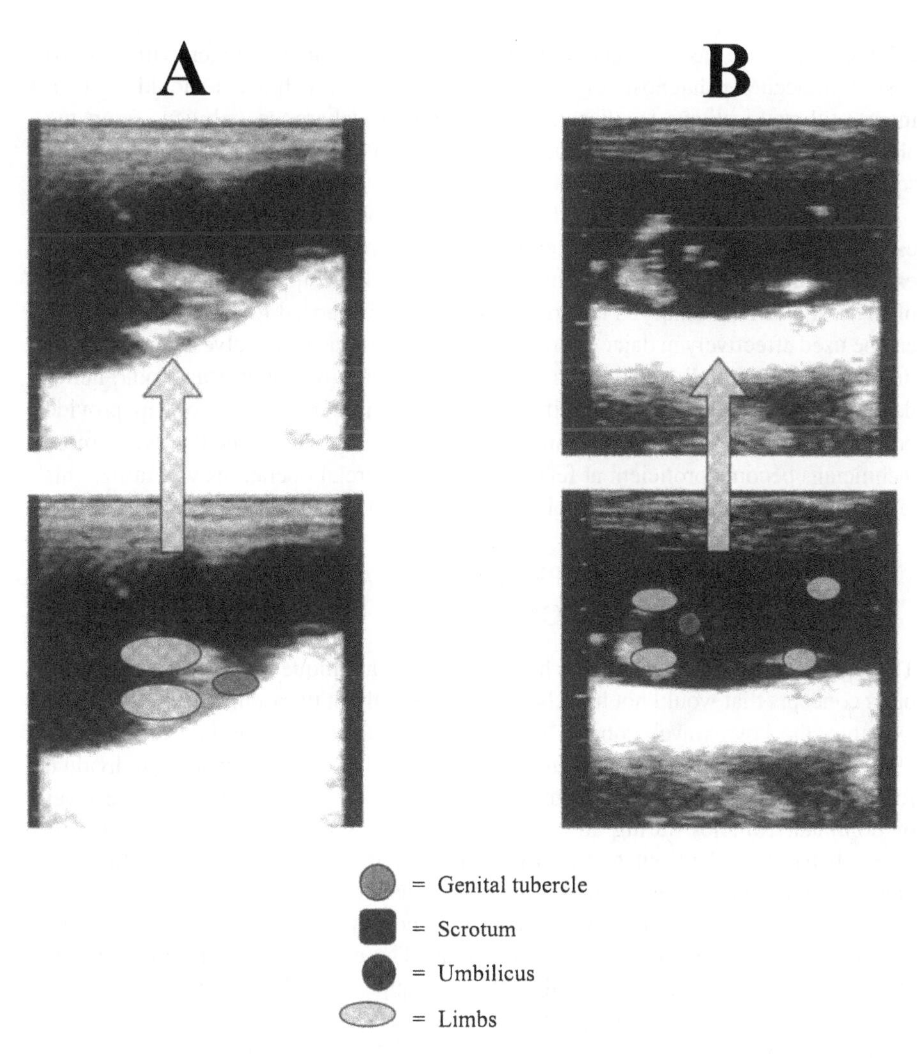

FIGURE 18.3 Ultrasound image of a female bovine fetus (65 days of gestation; Panel A) and a sagittal view of a male fetus (65 days of gestation; Panel B). Images were taken using a 5.0 MHz transducer.

the fetus to obtain the desired image; and, 2) the gravid horn is more likely to descend ventrally into the abdominal cavity in larger or older cows, making fetal sexing virtually impossible without retracting the gravid horn.

Figure 18.3 illustrates the cross-sectional image of a female fetus (65 days of gestation; Panel A) and a sagittal view of a male fetus (65 days of gestation; Panel B). The umbilicus can be used as an excellent landmark when determining the location of the genital tubercle or presence of a scrotum in males. In the male, the genital tubercle is located adjacent to and caudal to the umbilicus, whereas the genital tubercle in the female is located just ventral to the tail. The scrotum is detectable between the hind legs of the male fetus. The genital tubercle and scrotum are echogenic and are easily detected on an ultrasound screen as echogenic images. To ensure an accurate diagnosis of sex, for each patient, a technician should view an image at three locations: 1) adjacent to the umbilicus, where the umbilicus enters the abdomen (possible male genital tubercle); 2) the area between the back legs (possible scrotum); and, 3) ventral to the tail (possible female genital tubercle).

In beef cattle operations, fetal sexing remains limited to purebred operations especially in conjunction with an embryo transfer program. Determination of sex, especially after the successful transfer of embryos to recipients, allows marketing of male and female embryos before the pregnancy is carried to term. This strategy can be used effectively in dairy operations trying to produce bull calves of a particular mating for sale to bull studs. From a commercial cattle operation standpoint, heifer development operations are utilizing fetal sexing as a marketing tool to provide potential buyers with females that are pregnant with fetuses of a specific sex. As more technicians become proficient at fetal sexing, commercial operations will utilize this technology to enhance the marketability and efficiency of their cattle operations.

ADDITIONAL REPRODUCTIVE APPLICATIONS FOR ULTRASONOGRAPHY IN CATTLE

Through the use of ultrasound, we have made several unique discoveries and developed concepts that would not have been possible without ultrasound technology. For example, the "two-wave" concept of follicular development was first proposed in 1960, but remained controversial until 1988 when ultrasound analysis of individual follicles demonstrated that heifers usually had two or three distinct waves of turnover of dominant follicles during an estrous cycle (Sirois and Fortune, 1988). Today, research has branched out to focus on follicular dynamics associated with wave emergence, follicular atresia, and follicular dominance. Just as important, new reproductive uses have been established since the discovery that ultrasound could impact reproductive management in cattle operations. The embryo transfer community is utilizing ultrasound to understand the effects of superovulation on follicular growth and to adjust FSH regimens to optimize embryo recovery. In addition, researchers and technicians now use ultrasound as a valuable tool for transvaginal oocyte recovery, follicular aspiration, and ovarian biopsies.

An efficient exploitation of the female gamete pool and shortening of the generation interval are important aims in modern reproductive management systems. To maximize the use of genetically valuable cows without compromising donor fertility requires an abundant availability of developmentally competent cumulus oocyte complexes (COC) for *in vitro* fertilization. Aspirating follicles from ovaries of slaughtered animals is a procedure that is commonly used, but it is a nonrepeatable approach and the genetic origin of the oocyte is unknown. Laparoscopy has been used for retrieval of oocytes from living animals (Lambert et al., 1983, 1986). Although laparoscopy is a repeatable procedure, it can be traumatic to the oocyte donor. Therefore, ultrasound-guided technology for the aspiration of small and preovulatory follicles provides an excellent alternative. This technology has been successfully employed in humans (Dellenbach et al., 1985; Kemeter and Feichtinger, 1986; Lenz et al., 1987), horses (Brück et al., 1992; Cook et al., 1992; Brück and

Greve, 1994; Meintjes et al., 1995b), and cattle (Pieterse et al., 1989, 1991, 1992; Simon et al., 1993; Walton et al., 1993; Bungartz et al., 1995; Santl et al., 1998; Carlin et al., 1999; Hashimoto et al., 1999a,b). Ultrasound-guided aspiration of oocytes allows for repeated recovery of oocytes from the same animal without causing a reduction in fertility attributed to the adhesions and scars formed with laparoscopical techniques. In addition, oocytes can be recovered biweekly from cows stimulated with PMSG or FSH to enhance follicle numbers (Pieterse et al., 1989, 1992; Walton et al., 1993) and from clinically infertile animals (Looney et al., 1994). Ultrasound-guided aspirations were also performed on cows in their first trimester of pregnancy with no observed detrimental effects on the fetus (Meintjes et al., 1995a).

Hashimoto et al. (1999b) recently tested the effect of the frequency (i.e., 5.0 or 7.5 MHz) of the ultrasound transducer on the collection rate of oocytes and COCs. In this study, the authors concluded that the number of oocytes and COCs per donor obtained with the 7.5 MHz transducer (11.2 and 9.0, respectively) was greater than those obtained with the 5.0 MHz transducer (4.3 and 3.5, respectively). The efficacy of the 7.5 MHz probe over the 5.0 MHz probe was attributed to the greater clarity of images obtained with the higher-frequency transducer. Ultrasound technology also has been used to assess multiple embryo production following interfollicular transfer of oocytes (i.e., transfer of multiple oocytes from donor follicles to a single recipient preovulatory follicle; Bergfelt et al., 1998). Using transvaginal ultrasonography, the preovulatory follicle was punctured with a 25-gauge needle between the exposed and nonexposed portions of the follicular wall, and five or six oocytes with transfer medium were infused into the antrum. Sixteen of a potential 43 (37%) oocytes/embryos were recovered from the oviduct, of which eight fertilized embryos were recovered from three recipients.

An ultrasound-guided transvaginal technique for CL biopsy has been developed and tested in cattle (Kot et al., 1999). The biopsy needle set consisted of an inner needle with a 20-mm-long specimen notch, an outer cannula with a cutting edge, and an automated spring-loaded handle with a trigger. The biopsy needle set was inserted into the channel guide of the handle of a convex-array transvaginal ultrasound probe. The transducer was positioned in the vaginal fornix, and the ovary was manipulated transrectally against the vaginal wall and transducer face. During monitoring on the ultrasound screen, the inner needle was pushed through the vaginal wall into the CL, and the cutting cannula was fired, cutting and trapping luteal tissue in the specimen notch. This is a practical procedure that may be useful for experimental and diagnostic purposes with little trauma to the donor.

When superstimulation of follicles is used for embryo transfer, the most effective protocol is to initiate FSH treatments at the beginning of a new follicular wave. This usually results in increased ovulations and an increase in transferable embryos. In most cases, the emergence of a new follicle can be detected by sequential daily ultrasound scanning of the ovary. However, eliminating the dominant follicle will initiate the emergence of a new wave. Using transvaginal ultrasonography, dominant follicle ablation can be performed to initiate a new wave (Baracaldo et al., 2000; Shaw and Good, 2000). Shaw and Good (2000) reported that greater numbers of total ova/embryos were recovered from cows whose dominant follicle was ablated, and ablation also accounted for a greater number of nontransferable embryos. The authors concluded that synchronization of follicular waves following dominant follicle ablation increases total ova/embryo output. Ablation of the two largest follicles also was as efficacious as ablating all follicles greater than 5 mm and was as effective as estradiol plus progesterone in synchronizing follicular wave emergence for superstimulation in cattle (Baracaldo et al., 2000).

Computer-assisted evaluation of ultrasonographic image attributes indicative of viability and atresia of ovarian follicles has the potential to become an integral part of ovarian superstimulation protocols. However, in many cases, animal handling facilities, laboratories providing image analysis services, and the individual making clinical decisions are geographically separated. The feasibility of remote assessment of follicular status and ovarian response to superstimulation was demonstrated using Internet and videoconferencing techniques (Pierson and Adams, 1999). Ultrasound images of the ovarian responses from heifers were digitally acquired and transmitted to a distant laboratory for quantitative assessment. Images from follicles which ovulated in response to luteolysis and GnRH treatments were visually and quantitatively different from atretic follicles. The use of the Internet for data transfer, image analysis, and clinical evaluation can provide useful information to practitioners who wish to have access to the biological information, but do not have the resources or technical skills to invest in the equipment required to make quantitative assessments of visual data.

SUMMARY

The impact of real-time ultrasound on the study of reproduction has been dramatic, and the further development of portable ultrasound machines has given clinicians an added tool for diagnostic reproductive management. Ultrasound is commonly used to monitor uterine anatomy, involution, and pathology. In addition, it has been used to detect pregnancy, study embryonic mortality, monitor fetal development, and determine fetal sex. The applications of ultrasound used by scientists include the ability to monitor follicular characteristics, ovarian function, and aid in follicular aspirations and oocyte retrieval. In the future, as technology improves, technicians will have an opportunity to use the Internet or videoconferencing for ultrasound image analyses. With every new technological development, scientists, veterinarians, and producers discover new possibilities for the use of reproductive ultrasound to enhance the scientific merit of research or improve reproductive efficiency in cattle operations.

REFERENCES

Aslan, S., M. Findik, N. Erunal-Maral, H. Kalender, M. Celebi, and E. Saban. 2000. Comparison of various examination methods used in ovarian diagnostics in cattle. *Dtsch. Tierarztl. Wochenschr.* 107:227.

Badinga, L., M.A. Driancourt, J.D. Savio, D. Wolfenson, M. Drost, R.L. de la Sota, and W.W. Thatcher. 1992. Endocrine and ovarian responses associated with the first-wave dominant follicle in cattle. *Biol. Reprod.* 47:871.

Badtram, G.A., J.D. Gaines, C.B. Thomas, and W.T.K. Bosu. 1991. Factors influencing the accuracy of early pregnancy detection in cattle by real-time ultrasound scanning of the uterus. *Theriogenology* 35:1153.

Bailey, T.L., R.S. Hudson, T.A. Powe, M.G. Riddell, D.F. Wolfe, and R.L. Carson. 1998. Caliper and ultrasonographic measurements of bovine testicles and a mathematical formula for determining testicular volume and weight *in vivo*. *Theriogenology* 49:581.

Baracaldo, M.I., M.F. Martinez, G.P. Adams, and R.J. Mapletoft. 2000. Superovulatory response following transvaginal follicle ablation in cattle. *Theriogenology* 53:1239.

Beal, W.E., R.C. Perry, and L.R. Corah. 1992. The use of ultrasound in monitoring reproductive physiology of beef cattle. *J. Anim. Sci.* 70:924.

Beghelli, V., C. Boiti, E. Parmigiani, and S. Barbacini. 1986. Pregnancy diagnosis and embryonic mortality in the cow. In: J.M. Sreenan and M.G. Diskin (Eds.) *Embryonic Mortality in Farm Animals.* Martinus Nijhoff, Boston, pp. 159.

Bergfelt, D.R., G.M. Brogliatti, and G.P. Adams. 1998. Gamete recovery and follicular transfer (graft) using transvaginal ultrasonography in cattle. *Theriogenology* 50:15.

Bly, S.H.P., P.R. Vlahovich, P.R. Mabee, and R.G. Hussey. 1992. Computed estimates of maximum temperature elevations in fetal tissues during transabdominal pulsed Doppler examinations. *Ultrasound Med. Biol.* 18:389.

Bosward, K.L., S.B. Barnett, A.K.W. Wood, M.J. Edwards, and G. Kossoff. 1993. Heating of guinea-pig fetal brain during exposure to pulsed ultrasound. *Ultrasound Med. Biol.* 19:415.

Boyd, J., P. Bacisch, A. Young, and J.A. McCracken. 1969. Fertilization and embryonic survival in dairy cattle. *Br. Vet. J.* 125:87.

Boyd, J.S., S.N. Omran, and T.R. Ayliffe. 1988. Use of a high frequency transducer with real time B-mode ultrasound scanning to identify early pregnancy in cows. *Vet. Rec.* 123:8.

Brück, I.K. Raun, B. Synnestvedt, and T. Greve. 1992. Follicle aspiration in the mare using a transvaginal ultrasound-guided technique. *Equine Vet. J.* 24:58.

Brück, I. and T. Greve. 1994. Transvaginal ultrasound-guided aspiration of follicular fluid in the mare. *Theriogenology* 41:170 (Abstr.).

Bungartz, L., A. Lucas-Hahn, D. Rath, and H. Niemann. 1995. Collection of oocytes from cattle via follicular aspiration aided by ultrasound with or without gonadotropin pretreatment and in different reproductive stages. *Theriogenology* 43:667.

Butler, W.R and R.D. Smith. 1989. Interrelationships between energy balance and postpartum reproductive function in diary cattle. *J. Dairy Sci.* 72:767.

Butler, W.R., D.J.R. Cherney, and C.C. Elrod. 1995. Milk urea nitrogen (MUN) analysis: field trial results on conception rates and dietary inputs. *Proc. Cornell Nutr. Conf.* pp. 89.

Carlin, S.K., A.S. Garst, C.G. Tarraf, T.L. Bailey, M.L. McGilliard, J.R. Gibbons, A. Ahmadzadeh, and F.C. Gwazdauskas. 1999. Effects of ultrasound-guided transvaginal follicular aspiration on oocyte recovery and hormonal profiles before and after GnRH treatment. *Theriogenology* 51:1489.

Carnes, K., R. Hess, and F. Dunn. 1991. Effects of *in utero* ultrasound exposure on the development of the fetal mouse testis. *Biol. Reprod.* 45:432.

Carrière, P.D., D. Harvey, and G.M. Cooke. 1996. The role of pregnenolone-metabolizing enzymes in the regulation of oestradiol biosynthesis during development of the first wave dominate follicle in the cow. *J. Endocrinol.* 149:233.

Cartee, R.E., B.A. Selcer, and C.S. Patton. 1980. Ultrasonic diagnosis of renal disease in small animals. *J. Am. Vet. Med. Assoc.* 176:426.

Carter, J., J.A. Reynoldson, G.D. Thorburn, and W.A. Bates. 1981. Blood flow measurement during exercise in sheep using Doppler ultrasonic method. *Med. Biol. Eng. Comput.* 19:373.

Child, S.Z., E.L. Carstensen, A.H. Gates, and W.J. Hall. 1988. Testing for the teratogenicity of pulsed ultrasound in mice. *Ultrasound Med. Biol.* 14:493.

Cochran, S.N., T.R. Ayliffe, and J.S. Boyd. 1988. Preliminary observations of bovine ovarian structures using B-mode real time ultrasound. *Vet. Rec.* 122:465.

Cook, N.L., E.L. Squires, B.S. Ray, V.M. Cook, and D.J. Jasko. 1992. Transvaginal ultrasonically guided follicular aspiration of equine oocytes. *J. Equine Vet. Sci.* 12:204.

Curran, S., R.A. Pierson, and O.J. Ginther. 1986a. Ultrasonographic appearance of the bovine conceptus from days 10 through 20. *J. Am. Vet. Med. Assoc.* 189:1289.

Curran, S., R.A. Pierson, and O.J. Ginther. 1986b. Ultrasonographic appearance of the bovine conceptus from days 20 through 60. *J. Am. Vet. Med. Assoc.* 189:1295.

Curran, S., J.P. Kastelic, and O.J. Ginther. 1989. Determining sex of the bovine fetus by ultrasonic assessment of the relative location of the genital tubercle. *Anim. Reprod. Sci.* 19:217.

Davis, M.E. and G.K. Haibel. 1993. Use of real-time ultrasound to identify multiple fetuses in beef cattle. *Theriogenology* 40:373.

Demoulin, A., R. Bologne, J. Hustin, and R. Lambotte. 1985. Is ultrasound monitoring of follicular growth harmless? *Ann. NY Acad. Sci.* 442:146.

Dellenbach, P., I. Nisand, L. Morreau, B. Feger, C. Plumere, and P. Gerlinger. 1985. Transvaginal sonographically controlled follicle puncture for oocyte retrieval. *Fertil. Steril.* 44:656.

Diskin, M.G. and J.M. Sreenan. 1980. Fertilization and embryonic mortality rates in beef heifers after artificial insemination. *J. Reprod. Fertil.* 59:463.

Duck, F.A. and K. Martin. 1991. Trends in diagnostic ultrasound exposure. *Phys. Med. Biol.* 36:1423.

Ellisman, M.H., D.E. Palmer, and M.P. Andre. 1987. Diagnostic levels of ultrasound may disrupt myelination. *Exp. Neurol.* 98:78.

Ferrandi, B., F. Cremonesti, R. Gieger, A.L. Consiglio, A. Carnevali, and F. Porcelli. 1993. Quantitative cytochemical study of some enzymatic activities in preovulatory bovine oocytes after *in vitro* maturation. *Acta Histochem.* 95:89.

Fricke, P.M., J.N. Guenther, and M.C. Wiltbank. 1998. Efficacy of decreasing the dose of GnRH used in a protocol for synchronization of ovulation and timed AI in lactating dairy cows. *Theriogenology* 50:1275.

Fritz, P., J. Hoenes, D. Lutz, H. Multhaupt, A. Mischinski, R. Dorrer, P. Schwarzmann, H.V. Tuczek, and W. Muller. 1989. Quantitative immunohistochemistry: standardization and possible application in research and surgical pathology. *Acta Histochem.* (Suppl. XXXVII):213.

Fritz, P., H. Multhaupt, J. Hoenes, D. Lutz, R. Doerrer, P. Schwarzmann, and H.V. Tuczek. 1992. Quantitative histochemistry. *Prog. Histochem. Cytochem.* 24:1.

Garner, H.E., J.R. Coffman, A.W. Hahn, and J. Hartley. 1982. Indirect blood pressure measurement in the horse. *Proc. Am. Assoc. Equine Pract.* 18:343.

Gates, A.H., E.L. Carstensen, S.Z. Child, W.J. Hall, and C.L. Maczynski. 1988. Murine ovulatory response to ultrasound exposure and its gynecological relevance. *Ultrasound Med. Biol.* 14:485.

Ginther, O.J. 1986. Ultrasonic imaging and reproductive event in the mare. Cross Plains, WI, Equiservices, Inc., pp. 1.

Ginther, O.J., J.P. Kastelic, and L. Knopf. 1989a. Composition and characteristics of follicular waves during the bovine oestrous cycle. *Anim. Reprod. Sci.* 20:187.

Ginther, O.J., J.P. Kastelic, and L. Knopf. 1989b. Intraovarian relationship among dominate and subordinate follicles and corpus luteum in heifers. *Theriogenology.* 32:787.

Ginther, O.J. 1995. Ultrasonic Imaging and Animal Reproduction: Fundamentals Book 1 pp. 7 and 147, Equiservices Publishing, Cross Plains, WI.

Guilbault, L.A., P. Rouillier, P. Matton, R.G. Glencross, A.J. Beard, and P.G. Knight. 1993. Relationships between the levels of atresia and inhibin contents (subunit and – dimer) in morphologically dominant follicles during their growing and regressing phases of development in cattle. *Biol. Reprod.* 48:268.

Hager, D., T. Nyland, and P. Fisher. 1986. Ultrasound guided biopsy of the canine liver, kidney, and prostate. *Vet. Radiol. Ultrasound* 26:82.

Hande, M.P. and P.U. Devi. 1993. Effect of *in utero* exposure to diagnostic ultrasound on the postnatal survival and growth of mouse. *Teratology* 48:405.

Hansen, P.J., W.W. Thatcher, and A.D. Ealy. 1992. Methods for reducing effects of heat stress on pregnancy. In: VanHorn H.H. and Wilcox C.J. (Eds.). *Large Dairy Herd Management.* Am. Dairy Sci. Assoc. Champaign IL. pp. 116.

Hanzen, C. and B. Delsaux. 1987. Use of transrectal B-mode ultrasound imaging in bovine pregnancy diagnosis. *Vet. Rec.* 121:200.

Hanzen, C., M. Pieterse, O. Scenczi, and M. Drost. 2000. Relative accuracy of the identification of ovarian structures in the cow by ultrasonography and palpation per rectum. *Vet. J.* 159:161.

Hashimoto, S., R. Takakura, M. Kishi, T. Sudo, N. Minami, and M. Yamada. 1999a. Ultrasound-guided follicle aspiration: the collection of bovine cumulus-oocyte complexes from ovaries of slaughtered or live cows. *Theriogenology* 51:757.

Hashimoto, S., R. Takakura, N. Minami, and M. Yamada. 1999b. Ultrasound-guided follicle aspiration: effect of the frequency of a linear transvaginal probe on the collection of bovine oocytes. *Theriogenology* 52:131.

Herve, M.P. and E.A. Campbell. 1971. Prediction of the weight of total muscle by ultrasonic measurements in steers and calves. *Res. Vet. Sci.* 12:427.

Horder, M.M., S.B. Barnett, G.J. Vella, M.J. Edwards, and A.K Wood. 1998. Ultrasound-induced temperature increase in guinea-pig fetal brain *in utero:* third-trimester gestation. *Ultrasound Med. Biol.* 24:1501.

Hughes, E.A. and D.A.R. Davies. 1989. Practical uses of ultrasound in early pregnancy in cattle. *Vet. Rec.* 124:456.

Iannaccone, P.M., M. Van Gorder, E.L. Madsen, A.O. Martin, and R. Garton. 1991. The role of preimplantation sonographic exposure in postimplantation development and pregnancy outcome. *J. Ultrasound Med.* 10:659.

Ireland, J. J., R. L. Murphy, and P. B. Coalson. 1980. Accuracy of predicting stages of bovine estrous cycle by gross appearance of the corpus luteum. *J. Dairy Sci.* 663:155.

James, A.E. Jr., J.B. Brayton, G. Novak, D. Wight, T.K. Shehan, R.M. Bush, and R.C. Sanders. 1976. The use of diagnostic ultrasound in evaluation of the abdomen in primates with emphasis on the rhesus monkey (*Macaca mulatta*). *J. Med. Primatol.* 5:160.

Jensh, R., P.A. Lewin, M.T. Poczobutt, B.B. Goldberg, J. Oler, M. Goldman, and R. Brent. 1995. Effects of prenatal ultrasound exposure on adult offspring behavior in the Wistar rat. *Proc. Soc. Exp. Biol. Med.* 210:171.

Jost, A. 1971. Embryonic sexual differentiation. In: H.W. Jones and W.W. Scott (Eds.) *Hermaphroditism, Genital Anomalies and Related Endocrine Disorders* (2nd Ed.) Williams and Wilkin, Baltimore, MD. pp. 16.

Kastelic, J.P. and O.J. Ginther. 1989. Fate of conceptus and corpus luteum after induced embryonic loss in heifers. *J. Am. Vet. Med. Assoc.* 194:922.

Kastelic, J.P., S. Curran, and O.J. Ginther. 1989. Accuracy of ultrasonography for pregnancy diagnosis on days 10 to 22 in heifers. *Theriogenology* 31:813 (Abstr.).

Kastelic, J.P., D.R. Bergfelt, and O.J. Ginther. 1990a. Relationship between ultrasonic assessment of the corpus luteum and plasma progesterone concentration in heifers. *Theriogenology* 33:1269.

Kastelic, J.P., R.A. Pierson, and O.J. Ginther. 1990b. Ultrasonic morphology of corpora lutea and central cavities during the estrous cycle and early pregnancy in heifers. *Theriogenology* 34:487.

Kemeter, P. and W. Feichtinger. 1986. Transvaginal oocyte retrieval using a transvaginal sector probe combined with an automated puncture device. *Hum. Reprod.* 1:21.

Kimmel, C.A., M.E. Stratmeyer, W.D. Galloway, J.B. Laborde, N. Brown, and F. Pinkavitch. 1983. The embryotoxic effects of ultrasound exposure in pregnant ICR mice. *Teratology* 27:245.

Kot, K., L.E. Anderson, S.J. Tsai, M.C. Wiltbank, and O.J. Ginther. 1999. Transvaginal, ultrasound-guided biopsy of the corpus luteum in cattle. *Theriogenology* 52:987.

Lamb, C.R., J.L. Stowater, and F.S. Pipers. 1988. The first twenty-one years of veterinary diagnostic ultrasound: a bibliography. *Vet. Radiol. Ultrasound* 29:37.

Lamb, G.C., B.L. Miller, V. Traffas, and L.R. Corah. 1997. Estrus detection, first service conception, and embryonic death in beef heifers synchronized with MGA and prostaglandin. *Kansas AES Rep. Prog.* 783:97.

Lambert, R.D., C. Bernard, J.E. Rioux, R. Beland, D. D'Amours, and A. Montreuil. 1983. Endoscopy in cattle by the paralumbar route: technique for ovarian examination and follicular aspiration. *Theriogenology* 20:149.

Lambert, R.D., M.A. Sirard, C. Bernard, J. Beland, J.E. Rioux, P. Leclerc, D.P. Menard, and M. Bedoya. 1986. *In vitro* fertilization of bovine oocytes matured *in vivo* and collected at laparoscopy. *Theriogenology* 25:117 (Abstr.).

Lenz, S., J. Leeton, and P. Renou. 1987. Transvaginal recovery of oocytes for *in vitro* fertilization using vaginal ultrasound. *J. In Vitro Fertil. Embryo Trans.* 4:51.

Lindahl, I.L. 1966. Detection of pregnancy in sheep by means of ultrasound. *Nature* 212: 642.

Looney, C.R., B.R. Lindsey, C.L. Gonzeth, and D.L. Johnson. 1994. Commercial aspects of oocyte retrieval and *in vitro* fertilization (IVF) for embryo production in problem cows. *Theriogenology* 41:67.

Mackey, V.S. 1983. Equine pleuropneumonia: radiology–diagnostic ultrasound–pleuroscopy. *Proc. Am. Assoc. Equine Pract.* 29:75.

Mahadevan, M., K. Chalder, D. Wiseman, A. Leader, and P.J. Taylor. 1987. Evidence for an absence of deleterious effects of ultrasound on human oocytes. *J. In Vitro Fert. Embryo Trans.* 4:277.

Marciel, M., H. Rodriguez-Martinez, and H. Gustafsson. 1992. Fine structure of corpora lutea in superovulated heifers. *Zentralbl. Veterinärmed.* 39:89.

Meintjes, M., M.S. Bellow, J.B. Broussard, J.B. Paul, and R.A. Godke. 1995a. Transvaginal aspiration of bovine oocytes from hormone-treated pregnant beef cattle for *in vitro* fertilization. *J. Anim. Sci.* 73:967.

Meintjes, M., M.S. Bellow, J.B. Paul, J.R. Broossard, L.Y. Li, D. Paccamonti, B.E. Eilts, and R.A. Godke. 1995b. Transvaginal ultrasound-guided oocyte retrieval in cyclic and pregnant horse and pony mares for *in vitro* fertilization. *Biol. Reprod.* 1:355.

Müller, E. and G. Wittkowski. 1986. Visualization of male and female characteristics of bovine fetuses by real-time ultrasonics. *Theriogenology* 25:571 (Abstr.)

Nebel, R.L. and M.L. McGilliard. 1993. Interactions of high milk yield and reproductive performance in dairy cows. *J. Dairy Sci.* 76:3257.

Newnham, J.P., S.F. Evans, C.A. Michael, F.J. Stanley, and L.I. Landau. 1993. Effects of frequent ultrasound during pregnancy: a randomized controlled trial. *Lancet* 342:887.

O'Grady, J.P., E.C. Davidson, Jr., W.D. Thomas, G.N. Esra, L. Gluck, and M.V. Kulovich. 1978. Cesarean delivery in a gorilla. *J. Am. Vet. Med. Assoc.* 173:1137.

O'Grady, J.P., C.H. Yaeger, G.N. Esra, and W. Thomas. 1982. Ultrasonic evaluation of echinococcosis in four lowland gorillas. *J. Am. Vet. Med. Assoc.* 181:1348.

Perry, R.C., W.E. Beal, and L.R. Corah. 1990. Reproductive applications of ultrasound in cattle part 2: monitoring uterine characteristics and pregnancy. *Agri-practice* 11:31.

Perry, R.C., L.R. Corah, G.H. Kiracofe, J.S. Stevenson, and W.E. Beal. 1991. Endocrine changes and ultrasonography of ovaries in suckled beef cows during resumption of postpartum estrous cycles. *J. Anim. Sci.* 69:2548.

Peter, A.T. and W.T.K. Bosu. 1988. Influence of intrauterine infections and follicular development on the response to GnRH administration in postpartum dairy cows. *Theriogenology* 29:1163.

Pierson, R.A. and G.P. Adams. 1999. Remote assessment of ovarian response and follicular status using visual analysis of ultrasound images. *Theriogenology* 51:47.

Pierson, R.A. and O.J. Ginther. 1984. Ultrasonography for detection of pregnancy and study of embryonic development in heifers. *Theriogenology* 22:225 (Abstr.).

Pierson, R.A. and O.J. Ginther. 1986. Ovarian follicular populations during early pregnancy in heifers. *Theriogenology* 26:649.

Pierson, R.A. and O.J. Ginther. 1987a. Follicular populations during the estrous cycle in heifers I. Influence of day. *Anim. Reprod. Sci.* 14:165.

Pierson, R.A. and O.J. Ginther. 1987b. Reliability of diagnostic ultrasonography for identification and measurement of follicles and detecting the corpus luteum in heifers. *Theriogenology* 28:929.

Pierson, R.A. and O.J. Ginther. 1988. Follicular populations during the estrous cycle in heifers III. Time of selection of ovulatory follicle. *Anim. Reprod. Sci.* 16:81.

Pieterse, M.C., K.A. Kappen, T.A.M. Kruip, and M.A.M. Taverne. 1989. Aspiration of bovine oocytes during transvaginal ultrasound scanning of ovaries. *Theriogenology* 30:751.

Pieterse, M.C., O. Szenci, A.H. Willemse, C.S.A. Bajcsy, S.J. Dieleman, and M.A.M. Taverne. 1990. Early pregnancy diagnosis in cattle by means of linear-array real-time ultrasound scanning of the uterus and a qualitative and quantitative milk progesterone test. *Theriogenology* 33:697.

Pieterse, M.C., P.L.A.M. Vos, T.A.M. Kruip, Y.A. Wurth, T.H. Van Beneden, A.H. Willemse, and M.A.M. Taverne. 1991. Transvaginal ultrasound-guided follicular aspiration of bovine oocytes. *Theriogenology* 35:19.

Pieterse, M.C., P.L.A.M. Vos, T.A.M. Kruip, Y.A. Wurth, T.H. Van Beneden, A. H. Willemse, and M.A.M. Taverne. 1992. Repeated transvaginal ultrasound-guided ovum pick-up in ECG-treated cows. *Theriogenology* 37:273 (Abstr.).

Pipers, F.S. and R.L. Hamlin. 1977. Echocardiography in the horse. *J. Am. Vet. Med. Assoc.* 170:815.

Pipers, F.S., V. Reef, R.L. Hamlin, and D.M. Rings. 1978. Echocardiography in the bovine animal. *Bov. Pract.* 13:114.

Price, C.A., P.D. Carrière, B. Bhatia, and N.P. Groome. 1995. Comparison of hormonal and histological changes during follicular growth, as measured by ultrasonography, in cattle. *J. Reprod. Fertil.* 103:63.

Rantanen, N.W. 1981. Ultrasound appearance of normal lung borders and adjacent viscera of the horse. *Vet. Radiol. Ultrasound* 22:217.

Renucci, R.P., C. Feuerstein, M. Manier, P. Lorimier, M. Savasta, and J. Thibault. 1991. Quantitative image analysis with densitometry for immunohistochemistry and autoradiography of receptor binding sites—methodological considerations. *J. Neurosci. Res.* 28:583.

Rogers, M., R.E. Cartee, W. Miller, and A.K. Ibrahim. 1986. Evaluation of the extirpated equine eye using B-mode ultrasonography. *Vet. Radio. Ultrasound* 27:24.

Santl, B., H. Wenigerkind, W. Schernthaner, J. Modl, M. Stojkovic, K. Prelle, W. Holtz, G. Brem, and E. Wolf. 1998. Comparison of ultrasound-guided vs laparoscopic transvaginal ovum pick-up (OPU) in Simmental heifers. *Theriogenology* 50:89.

Savio, J.D., L. Keenan, M.P. Boland, and J.F. Roche. 1988. Pattern of growth of dominant follicles during the oestrous cycle in heifers. *J. Reprod. Fertil.* 83:663.

Shaw, D.W. and T.E. Good. 2000. Recovery rates and embryo quality following dominant follicle ablation in superovulated cattle. *Theriogenology* 53:1521.

Shoji, R., U. Murakami, and T. Shimizu. 1975. Influence of low-intensity ultrasonic irradiation on prenatal development of two inbred mouse strains. *Teratology* 12:227.

Siddiqi, T.A., R.A. Meyer, J.R. Woods, and M.A. Piessinger. 1988. Ultrasound effects on fetal auditory brain stem responses. *Obstet. Gynecol.* 72:752.

Simon, L., L.Bungartz, D. Rath, and H. Niemann. 1993. Repeated bovine oocyte collection by means of a permanently rinsed ultrasound-guided aspiration unit. *Theriogenology* 39:312 (Abstr.).

Singh, J., R.A. Pierson, and G.P. Adams. 1997. Ultrasound image attributes of the bovine corpus luteum: structural and functional correlates. *J. Reprod. Fertil.* 109:35.

Singh, J., R.A. Pierson, and G.P. Adams. 1998. Ultrasound image attributes of bovine ovarian follicles and endocrine and functional correlates. *J. Reprod. Fertil.* 112:19.

Sirois, J. and J.E. Fortune. 1988. Ovarian follicular dynamics during the estrous cycle in heifers monitored by real time ultrasonography. *Biol. Reprod.* 39:308.

Smith, M.W. and J.S. Stevenson. 1995. Fate of the dominant follicle, embryonal survival, and pregnancy rates in dairy cattle treated with prostaglandin $F_{2\alpha}$ and progestins in the absence or presence of a functional corpus luteum. *J. Anim. Sci.* 73:3743.

Spell, A.R., W.R. Beal, L.R. Corah, and G.C. Lamb. 2001. Evaluating recipient and embryo factors that effect pregnancy rates of embryo transfer in beef cattle. *Theriogenology* 56:287.

Squires, E.L., J.L. Voss, M.D. Villahoz, and R.K. Shideler. 1983. *Proc. Twenty-ninth Am. Assoc. Equine Pract.* pp. 31.

Sternberger, L.A. and N.H. Sternberger. 1986. The unlabeled antibody method: comparison of peroxidase–antiperoxidase with avidin–biotin complex by a new method of quantification. *J. Histochem. Cytochem.* 34:599.

Stevenson, J.S., M.K. Schmidt, and E.P. Call. 1984. Stage of estrous cycle, time of insemination, and seasonal effects on estrus and fertility of Holstein heifers after prostaglandin $F_{2\alpha}$. *J. Dairy Sci.* 67:1798.

Stewart, R.E., L.J. Spicer, T.D. Hamilton, B.E. Keefer, L.J. Dawson, G.L. Morgan, and S.E. Echternkamp. 1996. Levels of insulin-like growth factor (IGF) binding proteins, luteinizing hormone and IGF-I receptors, and steroids in dominant follicles during the first follicular wave in cattle exhibiting regular estrous cycles. *Endocrinology* 137:2842.

Sunderland, S.J., P.G. Knight, M.P. Boland, J.F. Roche, and J.J. Ireland. 1996. Alterations in intrafollicular levels of different molecular mass forms of inhibin during development of follicular- and luteal-phase dominant follicles in heifers. *Biol. Reprod.* 54:453.

Szenci, O., J.F. Beckers, P. Humblot, J. Sulon, G. Sasser, M.A. Taverne, J. Varga, R. Baltusen, and G. Schekk. 1998. Comparison of ultrasonography, bovine pregnancy-specific protein B, and bovine pregnancy-associated glycoprotein 1 tests for pregnancy detection in dairy cows. *Theriogenology* 50:77.

Tarantal, A.F., W.D. O'Brien, and A.G. Hendrickx. 1993. Evaluation of the bioeffects of prenatal ultrasound exposure in the cynomolgus macaque (*Macaca fascicularis*): III. Developmental and hematologic studies. *Teratology* 47:159.

Taverne, M.A.M., O. Szenci, J. Szetag, and A. Piros. 1985. Pregnancy diagnosis in cows with linear-array real-time ultrasound scanning: a preliminary note. *Vet. Quarterly* 7:264.

Temple, R.S., H.H. Stonaker, and D. Howry. 1956. Ultrasonic and conductive methods for estimating fat thickness in live cattle. *Am. Soc. Anim. Prod. West. Sec. Proc.* 7:477 (Abstr.).

Tom, J.W., R.A. Pierson, and G.P. Adams. 1998a. Quantitative echotexture analysis of bovine corpora lutea. *Theriogenology* 49:1345.

Tom, J.W., R.A. Pierson, and G.P. Adams. 1998b. Quantitative echotexture analysis of bovine ovarian follicles. *Theriogenology* 50:339.

Vasconcelos, J.L.M., R.W. Silcox, J.A. Lacerda, J.R. Pursley, and M.C. Wiltbank. 1997. Pregnancy rate, pregnancy loss, and response to heat stress after AI at 2 different times from ovulation in dairy cows. *Biol. Reprod.* 56 (Suppl. 1):140.

Walton, S.J., K.A. Christie, and R.B. Stubbings. 1993. Evaluation of frequency of ultrasound-guided follicle aspiration on bovine ovarian dynamics. *Theriogenology* 39:336 (Abstr.).

WFUMB (World Federation for Ultrasound in Medicine and Biology). 1992. WFUMB Symposium on Safety and Standardization in Medical Ultrasound. Issues and Recommendations Regarding Thermal Mechanisms for Biological Effects of Ultrasound. Hornbaek, Denmark, 1991. *Ultrasound Med. Biol.* 18:731.

Whittaker, A.D., B. Park, B.R. Thane, R.K. Miller, and J.W. Savell. 1992. Principles of ultrasound and measurement of intramuscular fat. *J. Anim. Sci.* 70:942.

Wideman, D., C.G. Dorn, and D.C. Kraemer. 1989. Sex detection of the bovine fetus using linear array real-time ultrasonography. *Theriogenology* 31:272 (Abstr.).

19 Factors that Affect Embryonic Survival in the Cow: Application of Technology to Improve Calf Crop

E. Keith Inskeep

CONTENTS

Luteal progesterone is essential for the preparation of the uterus and oocyte before breeding, as well as for the maintenance of uterine quiescence and the embryo/fetus during most of gestation in the cow. Prior to and immediately after estrus, progesterone regulates the mechanisms necessary for maternal recognition of pregnancy. Cows in which short estrous cycles (short luteal phases) were produced by early weaning (day 30 postpartum) or normal estrous cycles were produced by pretreatment with progestogen before early weaning have been used to elucidate specific points at which fertility is compromised in the early postpartum beef cow.

In mated cows with short luteal phases, daily supplementation with progestogen, beginning on day 3, failed to maintain pregnancy, despite the fact that fertilization,

0-8493-1117-9/02/$0.00+$1.50

early embryonic development to day 3, and transport of the embryo into the uterus appeared normal. When normal embryos were transferred on day 7 after estrus into cows with short luteal phases that received daily supplementation with progestogen, or embryos from cows with short luteal phases were transferred on day 6 after mating into cows with normal cycles, pregnancy rates in each case were about half those achieved using progestogen-pretreated cows with normal luteal phases. In cows with short luteal phases supplemented with progestogen, pregnancy rates were improved dramatically when the regressing corpus luteum was removed on day 4 or 5 after mating. Thus the early regressing corpus luteum produces an embryotoxic effect in addition to depriving the uterus and embryo of progesterone.

In cows with normal estrous cycles, fertility has been lower if concentrations of estradiol were elevated for more than 3 days before ovulation. Persistent follicles that produce more estradiol have developed when progesterone was low and frequency of tonic pulses of LH was high. In cows with such follicles, oocytes appeared to resume maturation well before the surge of LH, and embryos died in the oviduct before the 16-cell stage. In early postpartum cows with transferred embryos and cows with normal cycles, excessive follicular development and high estradiol-17β during the luteal phase, specifically on days 14 through 17 after insemination, were detrimental to embryo survival.

Finally, losses occur during the late embryonic period, between days 30 and 40 of pregnancy. A local relationship between uterine and luteal function remains important during this time. In preliminary studies using cows with replacement corpora lutea induced on days 28 to 31, continuation of the pregnancy was associated with higher, but not excessive, concentrations of prostaglandin (PG) $F_{2\alpha}$ and lower concentrations of estradiol. If the replacement corpora lutea were induced after day 36 on the ovary adjacent to the embryo, all pregnancies continued.

To maximize fertility in the cow, treatments for synchronizing estrus must provide high progesterone and low estradiol before estrus and mating. They must reduce luteolytic influences such as excesses of $PGF_{2\alpha}$ or estradiol, early after mating and during maternal recognition of pregnancy and the late embryonic period.

INTRODUCTION

Secretion of progesterone by the corpus luteum is essential throughout gestation to achieve 100% success in the maintenance of uterine quiescence and survival of the embryo/fetus, and for normal parturition, in the cow (McDonald et al., 1952). Prior to and immediately after estrus, progesterone regulates the establishment and timing of mechanisms necessary for luteal regression in the nonpregnant cow and for maternal recognition of pregnancy. Utilizing this knowledge, researchers and breeders can transfer embryos into ovariectomized cows and pregnancy can be completed successfully by providing exogenous progestogen on a continuous daily basis (Inskeep and Baker, 1985). Given this success, why are some embryos lost in the apparently normal cow?

Much of the pregnancy loss in cattle is due to early embryonic death (Sreenan and Diskin, 1986) after either natural or artificial insemination or embryo transfer.

According to the review by Thatcher et al. (1994), approximately 30% of repeat breeder cows experience embryonic loss by day 7 of pregnancy. Additional embryonic losses occur gradually from days 8 to 17 (approximately 40% of total losses) and between days 17 and 24 (approximately 24% of total losses). Losses between days 17 and 24 were estimated at 6 to 12% of pregnancies in two studies in which dairy heifers were bred at synchronized estrus after two treatments with $PGF_{2\alpha}$ 11 days apart (Van Cleeff et al., 1991). Published estimates of late embryonic death rate (days 27 to 42) average 10 to 12% (Thatcher et al., 1994; Smith and Stevenson, 1995; Vasconcelos et al., 1997). These figures must be considered in conjunction with estimates of fertilization failure (12%; Kidder et al., 1954) and fetal losses that may range as high as 8% of animals pregnant at 42 days (Vasconcelos et al., 1997). Thus, wastage occurs throughout pregnancy, but is concentrated in the embryonic period, or the first 40 days after breeding.

Although more of the studies have been done in dairy than in beef cows, embryonic deaths occur and have been documented and studied at four times:

1. The period from days 4 through 9 after mating, when excessive secretion of $PGF_{2\alpha}$ may occur, as shown in postpartum cows with short luteal phases
2. The early postovulatory period, before day 6 after mating, in cows in which persistent follicles developed during lowered concentrations of progesterone and produced oocytes that were fertilizable; in this scenario, most embryos failed to reach the 16-cell stage
3. The period of maternal recognition of pregnancy, on days 14 through 17, when lower pregnancy rates have been associated with large follicles and high concentrations of estradiol
4. The late embryonic period, from days 28 to 40

Endocrine imbalances, specifically involving excessive concentrations of hormones, have been implicated as causes of embryonic death in the first three of these four periods. Recent data indicate that insufficient amounts of $PGF_{2\alpha}$ and high concentrations of estrogen may contribute in the fourth period.

THE PATTERN OF RETURN OF FERTILITY POSTPARTUM

During the postpartum period, the cow undergoes a transition from an anovulatory, anestrous, infertile animal into a fertile animal with normal ovulatory estrous cycles and a higher conception rate. As early as 1953, Asdell listed "breeding back too soon after calving" as a fault of management contributing to infertility, citing the work of Shannon et al. (1952). In their classic research bulletin, Wisconsin workers (Casida et al., 1968) compiled unweighted means of fertility in postpartum dairy and beef cows from the earlier literature (Table 19.1) and summarized results of studies in which they had observed similar patterns of increasing fertility with increasing postpartum interval.

TABLE 19.1
Unweighted Means for Conception Rates (%) to First Service Following Parturition in Cattle, Summarized by 30-Day Periods

	Intervals Following Parturition (days)					
Type of Cattle	<30	31–60	61–90	91–120	121–150	151–180
Dairy[a]	39	53	62	62	65	64
Beef[b]	33	58	69	74		

[a] Compiled by Casida et al. (1968) from five published studies.
[b] Compiled by Casida et al. (1968) from three published studies.

Conception rates in dairy cows inseminated at first estrus, measured by rectal palpation at 39 to 45 days after breeding were 5, 52, 54, and 75% for cows bred at <21, 21 to 40, 41 to 60, and >61 days postpartum, respectively (Casida et al., 1968). Cows that were not inseminated until the first estrus after 74 days postpartum had a conception rate of 69%, compared to a mean of 46.5% for all cows inseminated at first estrus. In contrast, Olds and Cooper (1970) reported that breeding at the first estrus after 40 days postpartum increased services per cow by only 0.08 and mean time from first breeding to conception by 5 days compared to breeding at first estrus after 60 days postpartum. Thus, calving interval was shortened by an average of 15 days by the earlier breeding.

Conception rates for beef cows studied by Casida et al. (1968), calculated from fertilization rates determined on day 3 after insemination and embryos palpated per rectum on days 38 to 44, averaged 15, 55, 64, and 86%, respectively, for cows inseminated at <31, 31 to 50, 51 to 90, and >91 days postpartum. The interaction of day postpartum with whether fertility was measured at 3 or 38 to 44 days after insemination had no effect on conception rate. From this observation, the Wisconsin workers concluded that embryonic death was a minimal contributor to early postpartum infertility in beef cows. However, if one combines their data from slaughter at 3 days after insemination over all postpartum intervals examined, fertilization rate was 52%, while pregnancy rate for cows slaughtered at 15 days or palpated at 38 to 44 days was 35%. Thus, estimated embryonic death would be 17 percentage points or nearly 33% of the embryos present at day 3 after insemination. Holness et al. (1980) reported that only 12% of first postpartum ovulations were fertile in beef cows, and conception rates were particularly low in the early postpartum period in *Bos indicus* cattle (Wells et al., 1985).

Factors identified as causes of reduced fertility in the Wisconsin studies included ovulation adjacent to the previously pregnant uterine horn and diameter of that horn (an indicator of degree of uterine involution). Failure of ovulation (formation of a palpable corpus luteum) at first estrus (in 10.5% of calving intervals) and ovulation without observed estrus (detected in 46.5% of calving intervals) also reduced fertility. In a more recent study, beef cows were inseminated at 25 to 29 or 35 to 39 days postpartum, at a fixed time in relation to estrous synchronization. No negative effect of the uterine horn of previous pregnancy on conception rate was detected in cows that ovulated (Bridges et al., 2000a). However, ovulation failure was a significant factor limiting pregnancy rate in these early postpartum cows.

EARLY EMBRYONIC DEATH ASSOCIATED WITH SHORT DURATION OF THE LUTEAL PHASE: LESSONS FROM THE POSTPARTUM COW

Much can be learned by consideration of the physiologic points at which fertility is compromised during the postpartum transition. Specifically, the restoration of cyclic ovulation, either naturally or by induction, provides an experimental situation in which one can determine roles of selected endocrine events. The occurrence of a short luteal phase following first ovulation or first estrus was reported by Menge et al. (1962) and Morrow et al. (1966). Smaller corpora lutea on day 15 were noted in cows bred at first postpartum estrus (Casida et al., 1968). Short-lived corpora lutea occurred in the pubertal heifer and ewe lamb, the postpartum ewe and cow returning to ovulatory activity, the ewe emerging from seasonal anestrus (Garverick and Smith, 1986; Lauderdale, 1986) and in postpartum water buffaloes (Usmani et al., 1990). Obviously, a corpus luteum that had regressed before day 14 could not support pregnancy; maternal recognition of pregnancy in the cow occurs on days 14 to 17.

Variables such as follicular development, pre- and postovulatory concentrations of gonadotropins, and luteal receptors for LH were shown to be responsible for some variations in level of luteal function (reviewed by Lishman and Inskeep, 1991), but not for its duration. In a series of studies, Copelin et al. (1987, 1989) at Missouri, Peter et al. (1989) at Wisconsin, and Cooper et al. (1991) in West Virginia showed clearly that premature uterine secretion of $PGF_{2\alpha}$ was responsible for the short luteal phase.

Several research groups, beginning with Ramirez-Godinez et al. (1981, 1982b) and Sheffel et al. (1982), observed that pretreatment with a progestogen usually resulted in formation of a corpus luteum with a normal functional lifespan, in response to weaning or injection of gonadotropins. Secretion of $PGF_{2\alpha}$ rose during treatment of anestrous cows with progestogen, in the same manner as during a short luteal phase after injection of hCG in control cows (Cooper et al., 1991). Thus, if the uterus had not been exposed previously to progestogen, secretion of $PGF_{2\alpha}$ increased prematurely when the first corpus luteum began to secrete progesterone. In subsequent studies, Johnson et al. (1992) showed that secretion of $PGF_{2\alpha}$ during treatment was not necessary for the effect of progestogen to normalize the subsequent luteal phase. Zollers et al. (1993) demonstrated that pretreatment with progestogen increased numbers of receptors for progesterone in the uterus on day 5 after estrus. Upregulation of uterine progesterone receptors appears to be essential to the timing of secretion of $PGF_{2\alpha}$.

Bellows et al. (1974) found that beef cows from which calves were weaned at about 35 days postpartum would consistently exhibit estrus in 4 to 5 days and form corpora lutea. Casida et al. (1968) and Ramirez-Godinez et al. (1982a) obtained evidence that ovulation and fertilization occurred at the expected time after estrus preceding a short luteal phase in early weaned cows. Logically, fertility should be improved by pretreatment of the postpartum cow with progestogen because of the prevention of the shortened luteal phase (Ramirez-Godinez et al., 1981). At West Virginia, an experimental model was designed in which to test where fertility fails in the postpartum beef cow. Calves were weaned at about 30 days postpartum from cows that had not formed corpora lutea before their calves were weaned. Half of the cows received progestogen treatment (6 mg norgestomet implants for 9 days, ending 2 days after early weaning). Control cows were expected to have short luteal phases/estrous cycles in all cases. Cows pretreated with progestogen were expected to have normal luteal phases/estrous cycles in an average of at least 80% of cases, based upon the data cited in the previous paragraph. Importantly, cows in both groups were at the same stage postpartum when studied.

Using this model, Breuel et al. (1993a) began a series of studies comparing components of fertility in cows with short or normal luteal phases. First, they removed and flushed the oviducts from cows in each group at day 3 after breeding. Corpora lutea were formed in 100% of the cows, and recovery of oocytes/embryos (86%), fertilization (68%), development of fertilized oocytes to the 4- to 8-cell stage (100%), and embryo quality did not differ between cows with short or normal luteal phases. When uteri were flushed nonsurgically on day 6, recovery of oocytes/embryos (79%), fertilization rate (82%), and development to at least the 4-cell stage (90%) again did not differ (Breuel et al., 1993a). It appeared that loss of the embryo must be a consequence of early luteal regression, so it was expected that supplemental treatment with progestogen would maintain pregnancy.

Progestogen therapy was tested by providing a daily supplement of melengestrol acetate (MGA) in feed beginning on day 4 after breeding (Breuel et al., 1993a). No pregnancies were maintained in cows with short luteal phases. In contrast, 41% of all norgestomet-pretreated cows and 50% of those cows that had normal luteal phases maintained pregnancy regardless of whether or not they received MGA. Supplemental injections of 200 mg progesterone daily gave the same result; no pregnancies occurred in cows with short-lived corpora lutea. Surprisingly, 12 of the 13 control cows that were deleted from these experiments because they had shown a spontaneous short luteal phase before breeding conceived at the postweaning estrus, even though they were bred at an average of only 33 days postpartum.

Whether the oocytes in cows with short luteal phases were inherently defective or the uteri of such cows were hostile to embryo survival was addressed in two experiments utilizing embryo transfer. First, two good-quality frozen–thawed embryos were transferred on day 7 after estrus into the uteri of postpartum cows expected to have short (control) or normal (norgestomet pretreated) luteal phases. All cows received 200 mg per day of supplemental progesterone, subcutaneously, beginning on day 4 after estrus. Pregnancies were maintained in 28% of control cows compared to 58% of norgestomet-pretreated cows ($P < 0.05$; Butcher et al., 1992). So part of the loss occurred after day 7 and in spite of therapy with progestogen.

Second, oocytes/embryos were flushed from the uteri of control and norgestomet-pretreated cows on day 6 after breeding, and if viable, transferred into the uteri of nonlactating, cycling recipients on day 6. Survival rates for embryos deemed fit to transfer did not differ with source (50 and 73% for cows with short and normal luteal phases, respectively; Schrick et al., 1993). However, pregnancy rate, determined as the number of recipients pregnant divided by the number of experimental cows from which an embryo or oocyte was recovered on day 6, was 13% for cows with a short luteal phase compared to 32% for cows with a normal luteal phase ($P = 0.06$). Likewise, embryonic survival for all fertilized oocytes found on day 6 was 23% for cows with a short luteal phase and 47% for cows with a normal luteal phase ($P = 0.08$). The values for cows with a normal luteal phase were comparable to pregnancy rates obtained in early weaned, mated cows with normal luteal phases (33%, Ramirez-Godinez et al., 1981; 50%, Breuel et al., 1993a).

Combining the above results, it was concluded that about half of the difference in ability to maintain pregnancy between cows with short and normal luteal phases (when supplemental progestogen was provided) could be attributed to effects on the oocyte or embryo before day 7 after estrus and the other half to a hostile uterine environment on or after day 7. The apparent timing of embryo loss was strikingly similar to the timing of increased uterine secretion of $PGF_{2\alpha}$ on days 4 through 9 after estrus in cows with short luteal phases (Cooper et al., 1991). Moreover, Schrick et al. (1993) had observed that concentrations of $PGF_{2\alpha}$ in flushings from the uterine lumen of cows with short luteal phases were more than double those from cows with normal luteal phases (636 ± 82 and 288 ± 90 pg/ml, respectively). Embryo quality tended to be correlated negatively with concentrations of $PGF_{2\alpha}$ in flushings from the uterine lumen ($r = -0.42$; $P = 0.07$). Because embryo quality was lower on day 6 (Schrick et al., 1993) than on day 3 (Breuel et al., 1993a), it was proposed that the specific problem in short luteal phase cows was likely to have occurred after the embryo entered the uterus. A direct embryotoxic effect of $PGF_{2\alpha}$ seemed possible because that had been suggested for mouse (Harper and Skarnes, 1972) and shown for rabbit (Maurer and Beier, 1976) and rat (Breuel et al., 1993b) embryos.

Effects of $PGF_{2\alpha}$ on embryo survival have been examined in several studies in cows in which daily supplemental progestogen was provided to replace the regressed corpus luteum. Buford et al. (1996) showed that $PGF_{2\alpha}$ was detrimental to embryos when given to normally cycling beef cows during days 4 to 7 after estrus and insemination, an interval similar to that during which high embryo mortality had been observed in cows with short luteal phases. On the other hand, the prostaglandin endoperoxide synthase inhibitor, flunixin meglumine, given by intramuscular injections in a dosage of 1 g/cow, three times/day at 8-hour intervals did not reduce pregnancy rates. Therefore, flunixin meglumine was considered safe to use in an experiment in postpartum beef cows.

Buford et al. (1996) tested whether embryonic survival was improved when the rise in $PGF_{2\alpha}$ that caused luteal regression was reduced by treatment with flunixin meglumine. Calves were weaned at ~28 days postpartum to induce estrous behavior and ovulation in cows with expected short luteal phases. Cows were bred at observed estrus by natural service and 12 hours later by artificial insemination to high-fertility bulls. All cows received 300 mg/day of progesterone in corn oil, subcutaneous 14, from day 3.5 after mating until pregnancy determination at day 30. Cows were allotted at random among three treatments: saline, flunixin meglumine (to inhibit synthesis of $PGF_{2\alpha}$), and flunixin meglumine plus removal of the corpus luteum (lutectomy). The latter treatment was intended to ask the secondary question, whether luteal maintenance *per se*, if it should occur in the group treated with flunixin meglumine, affected embryonic survival. Flunixin meglumine was given at 1 g every 8 hours on days 4 through 9, and lutectomy was performed on day 7. Pregnancy rates for all cows treated were 21, 27, and 53%, respectively ($P < 0.05$ for saline or flunixin meglumine only vs. flunixin meglumine plus lutectomy). The surprise finding was that pregnancy rate was not increased by flunixin meglumine alone, but was increased by the combination with lutectomy. Therefore, the regressing or partially regressing corpus luteum appeared to be a component of the embryotoxic effect of $PGF_{2\alpha}$.

To confirm that the corpus luteum was indeed required for the embryotoxic effect of $PGF_{2\alpha}$ in the early postbreeding period, another experiment in nonlactating, cycling cows was designed to test whether lutectomy would overcome this effect (Buford et al., 1996). Twenty-eight nonlactating beef cows were mated, supplemented with progestogen and assigned at random to three groups. Cows in group one received saline; those in groups two and three received 15 mg $PGF_{2\alpha}$, every 8 hours on days 5 through 8 after estrus, and those in group three were lutectomized on day 5. Pregnancy rates were 67, 22, and 80%, respectively, at ultrasonography on day 30 ($P < 0.05$). Thus, the combination of lutectomy and flunixin meglumine was required to prevent the toxic effects of $PGF_{2\alpha}$ on the early embryo in early postpartum cows (endogenous $PGF_{2\alpha}$), and lutectomy alone was effective in nonlactating, cycling cows (exogenous $PGF_{2\alpha}$). From these data, it seemed possible that even subluteolytic concentrations of $PGF_{2\alpha}$ (Schramm et al., 1983) could play a role in embryonic loss during early development via release of an embryotoxin from the corpus luteum. Shelton et al. (1990) observed in subfertile dairy cows that peripheral concentrations of progesterone increased more slowly after estrus than in heifers. They also saw that luteal cells from those cows in culture had a less favorable steroidogenic response to LH and PGE_2 than luteal cells from heifers.

Following estrus, concentrations of $PGF_{2\alpha}$ fall to basal values; slight increases on day 5 (as determined by concentrations of 15-keto, 13,14-dihydro-$PGF_{2\alpha}$) were associated with metestrous bleeding (Kindahl et al., 1976). Schallenberger et al. (1989) observed an increase in concentrations of $PGF_{2\alpha}$ until day 6 after estrus and artificial insemination. In light of these data and the findings in postpartum cows, the incidence of naturally occurring early increases in concentrations of $PGF_{2\alpha}$ and effects of such increases on development of the embryo should be determined in cycling cows. Secretion of $PGF_{2\alpha}$ may be especially important during cycles which are of relatively normal duration but have lowered concentrations of progesterone (Lishman and Inskeep, 1991). The majority of embryonic mortality in subfertile dairy cows occurred 6 to 7 days after estrus (Ayalon, 1978), when the morula was developing into the blastocyst. Maurer and Chenault (1983) observed that 67% of embryonic mortality had occurred or was occurring by day 8 of gestation in beef cows. Seals et al. (1998) showed that premature luteal regression by $PGF_{2\alpha}$ on days 5 through 8 caused embryonic death in cows supplemented with progestogen (confirming the results of Buford et al., 1996), but treatment on either days 10 through 13 or 15 through 18 of pregnancy was not effective. Either the embryo is susceptible only until about day 8, or older regressing corpora lutea do not produce or promote production of the embryotoxic factor.

Many products of luteolysis or partial luteolysis could play a role in embryonic loss (Buford et al., 1996). Involvement of luteal $PGF_{2\alpha}$ is worthy of further evaluation, because Hu et al. (1990) observed that short-lived corpora lutea produced more $PGF_{2\alpha}$ than did corpora lutea with a normal life span. Hernandez-Fonseca et al. (2000) used transfer of an embryo to each uterine horn to test whether an embryotoxin might be delivered locally to the uterine horn adjacent to the regressing corpus luteum. The reduction in survival of embryos in ipsilateral and contralateral uterine horns did not differ, so they concluded that the effect was systemic or through the uterine lumen.

Oxytocin, which is released from the corpus luteum in response to $PGF_{2\alpha}$ (Schallenberger et al., 1984), can increase secretion of $PGF_{2\alpha}$ from the uterus (Newcomb et al., 1977; Milvae and Hansel, 1980). In a study discussed above, Buford et al. (1996) observed that injections of $PGF_{2\alpha}$ increased concentrations of oxytocin in serum of intact, but not lutectomized, nonlactating cows. Lemaster et al. (1999) treated cows with oxytocin rather than $PGF_{2\alpha}$ in the same type of experiment as that done by Buford et al. (1996). Treatment on days 5 through 8 reduced pregnancy rate to 33% (compared to 80% in control cows), but in this case, the effect of oxytocin was blocked by concurrent treatment with flunixin meglumine (80% pregnancy rate). Thus, if oxytocin is involved in the embryotoxic effect, its role is to release more $PGF_{2\alpha}$, rather than to have a direct effect on the embryo.

On the other hand, oxytocin, injected to cause milk letdown or released by uterine manipulation (Roberts et al., 1975) associated with embryo transfer, could very well play a significant role in early embryonic death in cattle, with or without causing complete luteolysis. Other factors that increase secretion of $PGF_{2\alpha}$, such as heat stress (Malayer et al., 1990), uterine infection (Manns et al., 1985), or mastitis (Cullor, 1990; Barker et al., 1998, 1999) could cause embryonic death through this mechanism (see review by Zavy, 1994 for detailed discussion).

FOLLICULAR AND HORMONAL PATTERNS IN THE PREOVULATORY PERIOD

Our interest in effects of follicular development on fertility was piqued during the course of the studies in postpartum cows presented above. Breuel et al. (1993a) examined data on fertility for cows with normal luteal phases, regardless of treatment. Cows with larger preovulatory follicles (12.9 ± 0.9 mm) 5 days before the surge of LH had a lower conception rate (36%) and higher preovulatory concentrations of estradiol associated with that estrus than those with smaller follicles (7.5 ± 1.8 mm) at that time, which averaged 91% conception. From the literature covering nearly a half century of attempts to synchronize estrus in cattle, it is apparent that under conditions of low progesterone or progestogen, a largest ("dominant") follicle often becomes persistent. The larger size of follicles in animals completing treatment with low dosages of progesterone was recognized many years ago (Ulberg et al., 1951). However, this knowledge was not fully appreciated or utilized until the last decade.

As summarized by Kinder et al. (1996), circulating concentrations of progesterone during the estrous cycle determine frequency of secretion of pulses of gonadotropin-releasing hormone (GnRH) from the hypothalamus. Frequency of pulses of GnRH, in turn, determines frequency of secretion of pulses of luteinizing hormone (LH) from the anterior pituitary. A high frequency of pulses of LH stimulates continued growth of, and increased secretion of, estradiol-17β by the largest ovarian follicle. A low frequency of pulses of LH fails to support continued follicular growth and allows earlier degeneration or more frequent replacement of the largest follicle. Each time that the largest follicle of a wave stops growing, an increase in secretion of FSH recruits a new cohort of follicles (Adams et al., 1992). Ginther et al. (1996) reviewed the process of selection of the largest (usually referred to as dominant) follicle, the roles of loss of dominance by the largest follicle and increased secretion of FSH in the selection process, and the acquisition of dependency on LH by the largest follicle of a cohort.

Data in support of the above sequence of hormonal events are extensive. We analyzed data pooled from nine studies (Ireland and Roche, 1982; Roberson et al., 1989; Kojima et al., 1992; Stock and Fortune, 1993; Stumpf et al., 1993; Custer et al., 1994; Cupp et al., 1995; Bergfeld et al., 1996; Cooperative Regional Research Project NE-161, 1996) in which progesterone, estradiol-17β, and frequency of pulses of LH were measured during steady-state conditions. Concentrations of progesterone in peripheral circulation accounted for 37% of the variation in frequency of LH pulses and 38% of the variation in concentrations of estradiol. In turn, LH pulse frequency accounted for 50% of the variation in concentrations of estradiol.

When a persistent follicle ovulates, the oocyte is likely to be at a later stage of maturation (Mihm et al., 1994b; Revah and Butler, 1996). Although the oocyte is fertilizable, development of the resultant zygote is retarded, and early embryonic death ensues before the 16-cell stage in a majority of cases (Wishart, 1977; Ahmad et al., 1995). This scenario provides an explanation for the lowered fertility seen with low dosages of progesterone or progestogens as well as in naturally occurring cases of low progesterone during the estrous cycle before breeding (Folman et al., 1973; Meisterling and Dailey, 1987). Importantly, the presence of a large dominant follicle did not decrease the developmental competence of oocytes from smaller antral follicles (Smith et al., 1996). Thus, if a persistent follicle regresses, normal fertility can be expected for the oocyte from the next follicle that develops and ovulates.

Workers utilizing repeated ultrasonic imaging to monitor ovarian follicles in different size categories (Pierson and Ginther, 1987) or individual follicles (Sirois and Fortune, 1988) over time have confirmed the postulate by Rajakoski (1960) that follicular growth in cattle occurs in a wave-like pattern. A "wave" begins with the emergence of a group or cohort of follicles >4 mm in diameter (Knopf et al., 1989) and is characterized by development of a single large follicle and regression of several subordinates (Ginther et al., 1989a). The largest follicle, although often referred to as "dominant," is anovulatory and becomes atretic during a luteal phase, because the corpus luteum is truly the dominant structure in the ovary and its restriction of pulses of LH stops growth of the largest follicle (Taft et al., 1996). The largest follicle present at the onset of luteolysis may become dominant and ovulate during the ensuing follicular phase. Most often two (Ginther et al., 1989c; Knopf et al., 1989; Rajamahendran and Taylor, 1991), but sometimes three (Savio et al. 1988; Sirois and Fortune, 1988) or even four (more frequently in Brahman cattle;

Rhodes et al., 1995) large follicles (or waves) occur in sequence during an estrous cycle. Successive waves emerged on days 0 (day of ovulation) and 10 of estrous cycles with two waves, and on days 0, 9, and 16 of cycles with three waves (Ginther et al., 1989c); thus a new follicular wave occurred about every 7 to 10 days.

The ovulatory follicle in cows with two waves of follicular development is older and larger than the ovulatory follicle in cows with three waves of follicular development during an estrous cycle (Ginther et al., 1989c). Given the greater secretion of estradiol-17β from the ovulatory follicle, discussed above, it was suspected that patterns of secretion of estradiol, before breeding, contributed to embryonic losses during days 1 to 4 after breeding, before the 16-cell stage (Mihm et al., 1994 a,b; Ahmad et al., 1995; Cooperative Regional Research Project, NE-161, 1996; Revah and Butler, 1996).

If exposure of the preovulatory oocyte to a longer duration of high concentrations of estrogen can compromise embryo survival, then conception rates might be expected to be lower in those cows with two, rather than three, waves of follicular development. Indeed, in data collected at four research stations, conception rate to first service was reduced in lactating dairy cows in which the ovulatory follicle came from the second (37/64; 58%) compared to the third (20/21; 95%; $P < 0.05$) wave of follicular development during the estrous cycle before insemination (NE-161 Regional Research Project Annual Report, 1996, unpublished). In one study, Ahmad et al. (1997) found a similar trend in beef animals. Conception rates were 82% in 44 heifers and lactating cows in which the ovulatory follicle came from the second wave and 100% in eight heifers and cows in which the ovulatory follicle came from the third wave of follicular development during the estrous cycle before insemination. In a second replicate of that study, even fewer cows had three waves of follicular development, and it was not possible to detect any difference in fertility due to number of follicular waves (H. Hernandez-Fonseca, unpublished data, West Virginia Agricultural and Forestry Experiment Station).

SOURCES OF VARIATION IN WAVE PATTERNS OF FOLLICULAR DEVELOPMENT

Although concentrations of progesterone during the luteal phase clearly influence the duration of persistence of a follicle and the number of follicular waves during an estrous cycle (Richards et al., 1990; Sanchez et al., 1993, 1995; Smith and Stevenson, 1995), it is obvious that other factors must play a role. In the study by Ahmad et al. (1997), for example, peripheral concentrations of progesterone and estradiol-17β, in samples collected every other day from beef animals with two and three waves, differed only in relation to the time that luteal regression occurred, not in mean concentrations during the luteal phase.

Ginther et al. (1996) pointed out, and Ahmad et al. (1997) also discussed, the variation among laboratories (herds) in whether two or three follicular waves occurred during an estrous cycle in the majority of animals (Sirois and Fortune, 1988; Ginther et al., 1989a; Knopf et al., 1989; Rajamahendran and Taylor, 1991). More often two waves, as in the studies of fertility cited above (Ahmad et al., 1997;

Cooperative Regional Research Project, NE-161, unpublished), seemed to be the predominant pattern. There is evidence that the proportion of animals with two vs. three waves varies with nutrition (Murphy et al., 1991) and body condition (Burke et al., 1995).

Ginther et al. (1989b) showed that follicular waves continue to occur during early pregnancy in the cow, at average intervals of 9 days. They reported that days of origin of the largest follicles of the first and second follicular waves after estrus did not differ between pregnant and nonbred heifers. Likewise, the largest follicle of the first wave did not differ in maximum diameter. Based upon a lack of difference in interval from emergence of one follicular wave to the next during the first 70 days of pregnancy, and significant differences among individual heifers in magnitude of that interval, Ginther et al. (1989b) hypothesized that patterns of follicular development were repeatable and would favor greater proportions of two-wave or three-wave cycles as a characteristic of the individual. Rhodes et al. (1995) found some evidence of repeatability of number of follicular waves from one cycle to the next in Brahman cattle, although they studied only a small number of animals. Ahmad et al. (1997) observed that eight of nine nonpregnant beef animals in their study had two follicular waves in both the estrous cycle before insemination and the equivalent period after breeding. In more extensive observations of cows for two consecutive cycles before breeding, Hernandez-Fonseca (unpublished data, West Virginia) did not find evidence of repeatability from one cycle to the next in uninseminated animals.

HOW DO PERSISTENT FOLLICLES CAUSE LOW FERTILITY?

The sequential relationship of low progesterone, increased frequency of pulses of LH, a persistent largest follicle, increased secretion of estradiol-17β, and decreased fertility (Savio et al., 1993a,b; Stock and Fortune, 1993; Wehrman et al., 1993) is widely accepted as one of causes and effects. However, it is not clear whether the reduction in fertility in a cow with a persistent follicle is due to effects of estrogens, LH, or both. Differences in concentrations of estradiol that are associated with subsequent changes in fertility may be confusing. For example, fertility was reduced equally (55 and 59%, compared to 84% in controls) in beef cows and heifers in which the estrous cycle was extended by approximately 7 days with either low (one controlled internal drug releasing device, CIDR-B; 3.7 ng/ml) or high (two CIDR-B; 5.9 ng/ml) progesterone in a study by Washburn and Keller (1992). Estradiol-17β was quite a bit higher (11 vs. 4 pg/ml) on the last day of treatment in the animals on the lower dosage of progesterone. In another group of animals in that study, in which a norgestomet implant (6 mg) was used to delay estrus for 7 days, estradiol was only slightly higher in absolute value (14 pg/ml), yet fertility was further reduced to 32%.

In one experiment in the study by Revah and Butler (1996), cows were treated with FSH to produce multiple follicles, which then persisted during low progesterone. However, concentrations of estradiol-17β declined to very low values (<1 pg/ml) during the 7 days immediately prior to estrus. Even so, follicular oocytes were at a later stage of maturation (meiosis had resumed), just as in animals with high concentrations of estradiol-17β from a single persistent follicle in another of their experiments and in the study by Mihm et al. (1994b). Revah and Butler (1996) reviewed studies that provided evidence that either LH or estrogens could be responsible for advances in oocyte maturation, and decreases in rates of fertilization, implantation, and embryo survival, as well as increased rates of embryonic and congenital anomalies in rats.

In a recent study, Shaham-Albalancy et al. (1996) showed that concentration of progesterone before estrus altered endometrial morphology during the subsequent estrous cycle and that a low concentration of progesterone (2.1 to 2.3 ng/ml) during that period increased subsequent secretion of $PGF_{2\alpha}$, as measured by its major metabolite. These effects could lead to a decrease in fertility even though the original oocyte was healthy. However, it is important that all negative effects of low concentrations of progesterone before ovulation appear to be lethal before day 25 of pregnancy in cattle (Thatcher et al., 1994). In fact, Wehrman et al. (1996) have shown that development of a persistent follicle before synchronized estrus did not alter rate of survival of embryos transferred on day 7 after that estrus.

Treatment regimens designed to avoid ovulation of persistent follicles now abound. Several of these methods have been discussed in detail by others in this book, so they will not be treated in depth here. One such regimen that we have found very effective for fixed-time insemination in beef cows was presented by Bridges et al. (1999).

ROLE OF FOLLICULAR SECRETION OF ESTRADIOL IN EMBRYONIC MORTALITY DURING MATERNAL RECOGNITION OF PREGNANCY

Patterns of follicular development, and resultant secretion of estradiol-17β, during days 14 to 17 after breeding may be an important factor in embryonic loss during maternal recognition of pregnancy. This concept was originally presented by Macmillan et al. (1986) and has been supported in subsequent studies by Thatcher et al. (1989) and others, in which ovulation or atresia of the largest follicle during the mid-luteal phase sometimes increased pregnancy rate. Again, our attention was brought to this time frame as a result of observations made during preliminary studies of the survival of embryos when transferred into postpartum cows with short luteal phases that were supplemented with progestogens (Butcher et al., 1992). Attempts were made to provide supplemental progestogen by silastic implants containing up to 25 mg norgestomet. However, large follicles developed in the ovaries during days 12 to 20 and embryos transferred on day 7 failed to survive. Therefore, high dosages of injected progesterone or flurogestone acetate were used in subsequent studies in which embryo survival was increased (Butcher et al., 1992). Even short periods of deprivation of progesterone can decrease embryo survival during the maternal recognition period. Lulai et al. (1994b) studied the effects of initiation of luteal regression on day 15, either 24 or 36 hours before beginning replacement therapy with norgestomet. Embryo survival was 84% in control heifers and cows, but was reduced to 45 and 13%, respectively, when replacement therapy was delayed for 24 or 36 hours.

Evidence for association of embryonic loss with excessive secretion of estrogen during maternal recognition of pregnancy in beef cows was obtained by Pritchard et al. (1994). They sampled concentrations of progesterone and estradiol in peripheral blood during days 14 to 17 after breeding in over 100 lactating beef cows. Cows were divided into three groups according to concentrations of estradiol, the lower quarter, middle half, and upper quarter, which averaged 1.6, 2.1, and 3.1 pg/ml of estradiol, respectively, during the 4-day sampling period. Conception rate to first service by artificial insemination declined as concentration of estradiol increased. Mean conception rates were 77, 60, and 42%, respectively. Numerous studies have been done in which gonadotropin-releasing hormone (GnRH) or human chorionic gonadotropin (hCG) have been used to ovulate or luteinize large follicles during this stage after breeding, with considerable variation in response. Based upon the review by Lewis et al. (1990), one cannot conclude that these treatments are routinely valuable.

Given the above information, workers in the NE-161 regional research project proposed that secretion of estrogen from a large follicle during days 14 through 17 (or beyond) after breeding can compromise embryo survival, either directly or through interference with mechanisms of maternal recognition of pregnancy/luteal maintenance. Further, they proposed that cows with two waves of follicular development during the equivalent of an estrous cycle after breeding would have such a follicle. Ahmad et al. (1997) found that fewer animals conceived among those that had two (70%) rather than three (96%; $P < 0.05$) waves of follicular development during the equivalent of one estrous cycle after insemination. Surprisingly however, concentrations of estrogen in peripheral blood were not greater on day 14 after estrus and insemination in animals with two waves than in those with three waves.

LATE EMBRYONIC MORTALITY: THE PERI-ATTACHMENT PERIOD

Attachment of the embryo in the uterus is initiated around day 30 in the cow, with marked development of the placentomes between days 30 and 40 (Melton et al., 1951; King et al., 1982). Recent authors have illustrated pregnancy losses occurring both before (Van Cleeff et al., 1991) and after day 25 of gestation (Schallenberger et al., 1989; Kastelic et al., 1991; Van Cleeff et al., 1991; Wolff, 1992; Smith and Stevenson, 1995). As pointed out earlier, based upon return intervals exceeding 27 days after breeding, Thatcher et al. (1994) estimated that late embryonic death rate was 10.6% in heifers bred at estrus after synchronization with two injections of $PGF_{2\alpha}$ 11 days apart. Vasconcelos et al. (1997) found that 10.5% of lactating dairy cows that were pregnant at 28 days had lost the pregnancy by day 42. Those losses constituted the majority of losses between days 22 and 98 (Table 19.2).

It is disturbing that losses of pregnancy during the late embryonic and early fetal period do not appear to be reduced by treatments that have been designed to avoid the early embryonic death associated with persistent follicles and to allow timed insemination in the dairy cow. The data in Table 19.2 illustrate the wide variation among different studies (unpublished data graciously provided by Drs. J. S. Stevenson and W. W. Thatcher). Data on comparable control cows are lacking, and within experiment there usually have not been sufficient numbers of cows to detect meaningful differences among regimens for synchronization of estrus. In one study in Kansas, treatment with progesterone during the Ov-Synch protocol increased pregnancy rate at 56 days, in part by increasing embryo survival between days 28 and 56 (Table 19.2).

Unfortunately, few workers have collected data on embryonic death in beef cows after such treatments and fixed-time insemination. In the study reported by Bridges et al. (1999) only one of 71 cows pregnant at day 39 failed to calve. Four studies have been done in animals that were inseminated 12 hours after detection of estrus. In Brahman crossbred heifers, fertilization rate was 93% of intact ova, 78% had intact embryos on day 16, and 72% were pregnant on day 35 (Smith et al., 1982). Beal et al. (1992) diagnosed pregnancy by ultrasonography at 25, 45, and 65 days in 205 beef cows that initially had 138 viable embryos. Losses were 6.5% to day 45 and another 1.5% to day 65. Lamb et al. (1997) measured embryo mortality in *Bos taurus* heifers on three ranches with herds of 169 to 439 head. These heifers had been inseminated 12 hours after they were first detected in estrus in response to an injection of $PGF_{2\alpha}$ 17 days after withdrawal of MGA, which had been fed at 0.5 mg/day for 14 days. Conception rates as determined by ultrasonography at 29 to 33 days after insemination ranged from 44 to 67%. Of 525 pregnant heifers, 4.2% did not have viable embryos at palpation 60 to 90 days after the end of the breeding season. Dunne et al. (2000) measured embryo survival at slaughter on day 14 as 68%. By ultrasonography at day 30, their estimate was 76% pregnant, while at full term, 71.8% calved, so that the late embryonic and fetal loss was 4.2 percentage points. Thus, they concluded that most losses occurred before day 14 and that losses after day 30 were approximately 5.5%. Clearly, late embryonic losses are lower in beef cattle than in dairy cattle.

Drost et al. (1999) have used embryo transfer (ET) to attempt to overcome some of the effects of heat stress in lactating dairy cows in the summer in Florida. Conception rates at day 42 were improved from 21.4% for artificial insemination to 35.4% for ET from superovulated donors. Embryo mortality, as estimated from the difference in pregnancies at day 42 and cows with high progesterone on day 22, was 64.7% in cows bred by artificial insemination and 41.3% in cows given ET.

There is evidence that late embryonic loss precedes luteolysis. In seven of eight heifers in which embryonic death was detected between days 25 and 40 post-breeding, the onset of luteal regression, as detected by ultrasonography, began at least 3 days after embryonic death, as indicated by loss of heartbeat (Kastelic et al., 1991). In another study utilizing 70 pregnant cows, seven pregnancies were lost between days 35 and 42 after breeding; embryo death in each of these seven cows preceded luteal regression, detected by ultrasonography and declining concentrations of progesterone in milk (Wolff, 1992). Schallenberger et al. (1989) observed an increased secretion of $PGF_{2\alpha}$ between days 30 and 36 in pregnant heifers, one of which lost the pregnancy. However, from the magnitude and timing of secretion of $PGF_{2\alpha}$, or the luteal response to it, no firm conclusions could be drawn.

TABLE 19.2
Recent Estimates of Late Embryonic or Early Fetal Loss in Cows Bred by Timed AI at Synchronized Estrus

Research Leader/ Station (Author)	Synchronization Method	Pretreatment Status of Cows	AI Plan	Period Studies		Pregnancy Rate (%)		Number Pregnant	Estimated Loss (%)
				Day 1	Day 2	Day 1	Day 2		
Thatcher/FL (Moreira)	OvSynch	Lact. Cyclic	T[a]	32	74	48.4	40.5	185	14.6
		Lact. Anestrus				22.2	20.5	26	7.7
	OvSynch only	Lact.				36.9	30.0	94	17.4
	$PGF_{2\alpha}$ 12 and 26d before OvSynch	Lact.				48.0	41.8	117	10.7
	+ bST	Lact.				43.7	36.6	143	15.6
	– bST	Lact.				39.5	34.1	68	9.5
	All	Lact.				42.3	35.8	211	13.7
Stevenson/KS	OvSynch	Lact. Cyclic (58%)	T[b]	28	56	35.6	20.0	91	45.9
(El Zarkouny)	CIDR between GnRH and $PGF_{2\alpha}$	Lact. Cyclic (67%)				58.2*	44.0*	90	24.4
	Control	Lact. Cyclic (36%)				—	21.0	80	—
(Cartmill)	OvSynch	Lact. Cyclic	T[c]	28	38–56	—	—	70	30.0

		Lact. Anestrus				—	—	8	12.0
	$PGF_{2\alpha}$ 12d before OvSynch	Lact. Cyclic				—	—	80	24.0
		Lact. Anestrus				—	—	16	31.0
	2 × $PGF_{2\alpha}$ 12d apart	Lact. Cyclic				—	—	73	26.0
		Lact. Anestrus				—	—	9	56.0
	All	Lact. Cyclic				—	—	223	26.0
		Lact. Anestrus				—	—	67	33.0
(*J. Dairy Sci.*)	OvSynch	Lact. Cyclic		27–30/40–50		—	—	173	13.3
		Lact. Anestrus				—	—	30	30.0
	Select Synch	Lact. Cyclic				—	—	113	8.8
		Lact. Anestrus				—	—	13	15.4
	All	Lact. Cyclic				—	—	286	11.5
		Lact. Anestrus				—	—	43	25.6*
	OvSynch	Lact.				—	—	203	15.8
	Select Synch	Lact.				—	—	126	9.5
Wiltbank/WI	OvSynch	Lact.	T[e]	28	98	29.2	22.4	815	22.2
(Vasconcelos[d])		Lact.	T[f]			35.1	28.7	786	17.0

[a] Timed AI 64 hours after $PGF_{2\alpha}$, 16 hours after second GnRH.
[b] Timed AI 65 to 67 hours after $PGF_{2\alpha}$, 17 to 19 hours after second GnRH.
[c] Timed AI 64 to 68 hours after $PGF_{2\alpha}$, 16 to 20 hours after second GnRH.
[d] Vasconcelos et al. (1997).
[e] Timed AI 48 hours after $PGF_{2\alpha}$, 0 hours after second GnRH.
[f] Timed AI 72 hours after $PGF_{2\alpha}$, 24 hours after second GnRH.
* $P < 0.05$.

Lulai et al. (1994a) induced new corpora lutea on days 36 to 40 of pregnancy, during progestogen treatment and after induced regression of the original corpora lutea. When on the ovary adjacent to the pregnant uterine horn, induced corpora lutea were maintained after progestogen withdrawal (and supported the pregnancy) in four of four heifers. Pregnancies continued and induced corpora lutea on the opposite ovary were maintained in only one of six heifers.

In research conducted at West Virginia, maintenance of pregnancy was examined after induction of new corpora lutea between days 27 and 54 post-breeding. This was done in cows in which original corpora lutea had either regressed or been removed earlier and pregnancy had been maintained with an exogenous progestogen until induction of the new corpora lutea. After the presence of one or more new corpora lutea was confirmed, progestogen was withdrawn gradually over a 6-day period. In the preliminary studies by Wright et al. (1994), pregnancy was maintained only when the new corpus luteum was induced on the ovary adjacent to the embryo.

Bridges et al. (2000b) removed original corpora lutea on day 26 of pregnancy, induced new corpora lutea between days 28 and 31, and examined patterns of secretion of $PGF_{2\alpha}$ and estradiol during days 31 through 35. In cows with higher concentrations of $PGF_{2\alpha}$, more progesterone was secreted by the induced corpus luteum, and maintenance of pregnancy tended to be higher. In addition, there was a tendency for more pregnancies to be maintained when concentrations of estradiol were lower.

In the pooled data from the latter two studies, when a new corpus luteum was induced on the ipsilateral ovary later than day 36 after mating, 21 of 21 pregnancies were maintained. However, when the corpus luteum was induced on or before day 36 after mating, only 15 of 30 pregnancies were maintained. Further studies of the timing and nature of embryonic deaths after day 25 of pregnancy and the hormonal patterns with which they are associated are needed.

IMPLICATIONS

From the extensive data reviewed, it is clear that embryonic mortality is a significant limiting factor to the success of establishment and maintenance of pregnancy in cattle, particularly in dairy cattle. However, critical times at which losses may occur have been identified in beef cattle. These times include the period immediately before estrus, days 4 through 8 after estrus, the period of maternal recognition of pregnancy during days 14 through 17 after estrus, and the late embryonic/early fetal period, between days 28 and 42 to 50 (Figure 19.1). Some hormonal patterns or imbalances, with which embryonic mortality is associated, have been identified. Producers and researchers alike must be aware that the management required to maximize fertility in the cow is not simple. Treatments for synchronizing estrus must provide high progesterone, keep LH and estradiol low during treatment, and lead to development of a highly functional corpus luteum after mating (Figure 19.1). Such treatments must reduce luteolytic influences such as excesses of $PGF_{2\alpha}$ or estradiol, early after mating, during maternal recognition of pregnancy, and during the late embryonic period.

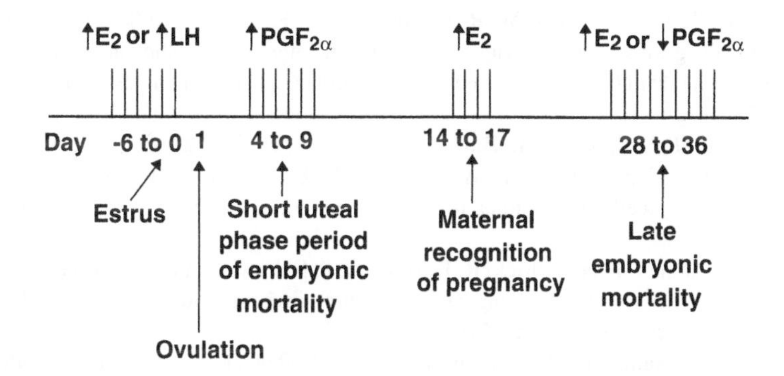

FIGURE 19.1 Points at which fertility may be compromised by abnormal concentrations of hormones in the cow. E_2 = estradiol-17β, LH = luteinizing hormone, $PGF_{2\alpha}$ = prostaglandin $PGF_{2\alpha}$.

REFERENCES

Adams, G.P., R. L. Matteri, and O. J. Ginther. 1992. Effect of progesterone on ovarian follicles, emergence of follicular waves and circulating follicle-stimulating hormone in heifers. *J. Reprod. Fertil.* 96:627.

Ahmad, N., F. N. Schrick, R. L. Butcher, and E. K. Inskeep. 1995. Effect of persistent follicles on early embryonic losses in beef cows. *Biol. Reprod.* 52:1129.

Ahmad, N., E. C. Townsend, R. A. Dailey, and E. K. Inskeep. 1997. Relationships of hormonal patterns and fertility to occurrence of two or three waves of ovarian follicles, before and after breeding, in beef cows and heifers. *Anim. Reprod. Sci.* 49:13.

Asdell, S. A. 1953. Factors involved in sterility of farm animals. *Iowa State College J. Sci.* 28:127.

Ayalon, N. 1978. A review of embryonic mortality in cattle. *J. Reprod. Fertil.* 54:483.

Barker, A. K. Inskeep, E. Townsend, and R. Dailey. 1999. Effects of gram-positive bacterial pathogens on reproductive efficiency of ewes. *Biol. Reprod.* 60 (Suppl. 1):256.

Barker, A. R., F. N. Schrick, M. J. Lewis, H. H. Dowlen, and S. P. Oliver. 1998. Influence of clinical mastitis during early lactation on reproductive performance of Jersey cows. *J. Dairy Sci.* 81:1285.

Bellows, R. A., R. E. Short, J. J. Urick, and O. F. Pahnish. 1974. Effects of early weaning on postpartum reproduction of the dam and growth of calves born as multiples or singles. *J. Anim. Sci.* 39:589.

Bergfeld, E. G., F. N. Kojima, A. S. Cupp, M. E. Wehrman, K. E. Peters, V. Mariscal, T. Sanchez, R. J. Kittok, and J. E. Kinder. 1996. Changing dose of progesterone results in sudden changes in frequency of luteinizing hormone pulses and secretion of 17β-estradiol in bovine females. *Biol. Reprod.* 54:546.

Breuel, K. F., P. E. Lewis, F. N. Schrick, A. W. Lishman, E. K. Inskeep, and R. L. Butcher. 1993a. Factors affecting fertility in the postpartum cow: role of the oocyte and follicle in conception rate. *Biol. Reprod.* 48:655.

Breuel, K. F., A. Fukuda, and F. N. Schrick. 1993b. Effects of prostaglandin $F_{2\alpha}$ on development of 8-cell rat embryos *in vitro*. *Biol. Reprod.* 48 (Suppl. 1):173.

Bridges, P. J., P. E. Lewis, W. R. Wagner, and E. K. Inskeep. 1999. Follicular growth, estrus and pregnancy after fixed-time insemination in beef cows treated with intravaginal progesterone inserts and estradiol benzoate. *Theriogenology* 52:573.

Bridges, P. J., R. Taft, P. E. Lewis, W. R. Wagner, and E. K. Inskeep. 2000a. Effect of the previously gravid uterine horn and postpartum interval on follicular diameter and conception rate in beef cows treated with estradiol benzoate and progesterone. *J. Anim. Sci.* 78:2172.

Bridges, P. J., D. J. Wright, W. I. Buford, N. Ahmad, H. Hernandez-Fonseca, M. L. McCormick, F. N. Schrick, R. A. Dailey, P. E. Lewis, and E. K. Inskeep. 2000b. Ability of induced corpora lutea to maintain pregnancy in beef cows. *J. Anim. Sci.* 78:2942.

Buford, W. I., N. Ahmad. F. N. Schrick, R. L. Butcher, P. E. Lewis, and E. K. Inskeep. 1996. Embryotoxicity of a regressing corpus luteum in beef cows supplemented with progestogen. *Biol. Reprod.* 54:531.

Burke, J. M., J. H. Hampton, C. R. Staples, and W. W. Thatcher. 1995. Body condition influences maintenance of a persistent first wave dominant follicle in dairy cattle. *J. Anim. Sci.* 73 (Suppl. 1):230.

Butcher, R. L., J. E. Reber, A. W. Lishman, K. F. Breuel, F. N. Schrick, J. C. Spitzer, and E. K. Inskeep. 1992. Maintenance of pregnancy in postpartum beef cows that have short-lived corpora lutea. *J. Anim. Sci.* 70: 3831.

Casida, L. E., W. E. Graves, E. R. Hauser, J. W. Lauderdale, J. W. Riesen, S. Saiduddin, and W. J. Tyler. 1968. Studies on the postpartum cow. Univ. Wisconsin, Madison. *Res. Bull.* 270.

Cooperative Regional Research Project, NE-161. 1996. Relationship of fertility to patterns of ovarian follicular development and associated hormonal profiles in dairy cows and heifers. *J. Anim. Sci.* 74:1943.

Cooper, D. A., D. A. Carver, P. Villeneuve, W. J. Silvia, and E. K. Inskeep. 1991. Effects of progestogen treatment on concentrations of prostaglandins and oxytocin in plasma from the posterior vena cava of post-partum beef cows. *J. Reprod. Fertil.* 91:411.

Copelin, J. P., M. F. Smith, H. A. Garverick, and R. S. Youngquist. 1987. Effect of the uterus on subnormal luteal function in anestrous beef cows. *J. Anim. Sci.* 64:1506.

Copelin, J.P., M. F. Smith, D. H. Keisler, and H. A. Garverick. 1989. Effect of active immunization of pre-partum and post-partum cows against prostaglandin $F_{2\alpha}$ on lifespan and progesterone secretion of short-lived corpora lutea. *J. Reprod. Fertil.* 87:199.

Cullor, J. S. 1990. Mastitis and its influence upon reproductive performance in dairy cattle. *International Symp. Bov. Mastitis,* Indianapolis, IN, pp. 176.

Cupp, A., T. T. Stumpf, F. N. Kojima, L. A. Werth, M. S. Roberson, R. J. Kittok, and J. E. Kinder. 1995. Secretion of gonadotrophins change during the luteal phase of the bovine oestrous cycle in the absence of corresponding changes in progesterone or 17β-oestradiol. *Anim. Reprod. Sci.* 37:109.

Custer, E. E., W. E. Beal, S. J. Wilson, A. W. Meadows, J. G. Berardinelli, and R. Adair. 1994. Effect of melengestrol acetate (MGA) or progesterone-releasing intravaginal device (PRID) on follicular development, concentrations of estradiol 17-β and progesterone, and luteinizing hormone release during an artificially lengthened bovine estrous cycle. *J. Anim. Sci.* 72:1282.

Drost, M. J. D. Ambrose, M-J. Thatcher, C. K. Cantrell, K. E. Wolfsdorf, J. F. Hasler, and W. W. Thatcher. 1999. Conception rates after artificial insemination or embryo transfer in lactating dairy cows during summer in Florida. *Theriogenology* 52:1161.

Folman, Y., M. Rosenberg, Z. Herz, and M. Davidson. 1973. The relationship between plasma progesterone concentrations and conception in postpartum dairy cows maintained on two levels of nutrition. *J. Reprod. Fertil.* 34:267.

Garverick, H. A. and M. F. Smith. 1986. Mechanisms associated with subnormal luteal function. *J. Anim. Sci.* 62 (Suppl. 2):2.

Ginther, O. J., J. P. Kastelic, and L. Knopf. 1989a. Composition and characteristics of follicular waves during the bovine estrous cycle. *Anim. Reprod. Sci.* 20:187.

Ginther, O. J., Kastelic, J. P., and Knopf, L. 1989b. Intraovarian relationships among dominant and subordinate follicles and the corpus luteum in heifers. *Theriogenology* 32:787.

Ginther, O. J., L. Knopf, and J. P. Kastelic. 1989c. Ovarian follicular dynamics in heifers during early pregnancy. *Biol. Reprod.* 41:247.

Ginther, O. J., M. C. Wiltbank, P. M. Fricke, J. R. Gibbons, and K. Kot. 1996. Selection of the dominant follicle in cattle. *Biol. Reprod.* 55:1187.

Harper, M. J. K. and R. C. Skarnes. 1972. Inhibition of abortion and fetal death produced by endotoxin or prostaglandin $F_{2\alpha}$. *Prostaglandins* 2:295.

Hernandez-Fonseca, H. J., B. L. Sayre, R. L. Butcher, and E. K. Inskeep. 2000. Embryotoxic effects adjacent and opposite to the early regressing bovine corpus luteum. *Theriogenology* 54:83.

Holness, D. H., D. H. Hale, and J. D. H. Hopley. 1980. Ovarian activity and conception during the post-partum period in Afrikaner and Mashona cows. *Zimbabwe J. Agric. Res.* 18:3.

Hu, Y., J. D. H. Sanders, S. G. Kurz, J. S. Ottobre, and M. L. Day. 1990. *In vitro* prostaglandin production by bovine corpora lutea destined to be normal or short-lived. *Biol. Reprod.* 42:801.

Inskeep, E. K. and R. D. Baker. 1985. Successful transfer of bovine embryos into ovariectomized recipients. *J. Anim. Sci.* 61 (Suppl. 1):409.

Ireland, J. J. and J. F. Roche. 1982. Effect of progesterone on basal LH and episodic LH and FSH in heifers. *J. Reprod. Fertil.* 64:295.

Johnson, S. K., R. P. Del Vecchio, E. C. Townsend, and E. K. Inskeep. 1992. Role of prostaglandin $F_{2\alpha}$ in follicular development and subsequent luteal life span in early postpartum beef cows. *Dom. Anim. Endocrinol.* 9:49.

Kastelic, J. P., D. L. Northey, and O. J. Ginther. 1991. Spontaneous embryonic death on days 20 to 40 in heifers. *Theriogenology* 35:351.

Kidder, H. E., W. G. Black, J. N. Wiltbank, L. C. Ulberg, and L. E. Casida. 1954. Fertilization rates and embryonic death rates in cows bred to bulls of different levels of fertility. *J. Dairy Sci.* 37:691.

Kindahl, H., L-E. Edqvist, A. Bane, and E. Granstrom. 1976. Blood levels of progesterone and 15-keto-13,14-dihydro-prostaglandin $F_{2\alpha}$ during the normal estrous cycle and early pregnancy in heifers. *Acta Endocrinol.* 82:134.

Kinder, J. E., F. N. Kojima, E. G. M. Bergfeld, M. E. Wehrman, and K. E. Fike. 1996. Progestin and estrogen regulation of pulsatile LH release and development of persistent ovarian follicles in cattle. *J. Anim. Sci.* 74:1424.

King, G. J., B. A. Atkinson, and H. A. Robertson. 1982. Implantation and early placentation in domestic ungulates. *J. Reprod. Fertil.* 31 (Suppl.):17.

Knopf, L., J. P. Kastelic, E. Schallenberger, and O. J. Ginther. 1989. Ovarian follicular dynamics in heifers: test of two-wave hypothesis by ultrasonically monitoring individual follicles. *Dom. Anim. Endocrinol.* 6:111.

Kojima, N., T. T. Stumpf, A. S. Cupp, L. A. Werth, M. S. Roberson, M. W. Wolfe, R. J. Kittok, and J. E. Kinder. 1992. Exogenous progesterone and progestins as used in estrous synchrony regimens do not mimic the corpus luteum in regulation of luteinizing hormone and 17β-estradiol in circulation of cows. *Biol. Reprod.* 47:1009.

Lamb, G. C., B. L. Miller, V. Traffas, and L. R. Corah. 1997. Estrus detection, first service conception, and embryonic death in beef heifers synchronized with MGA and prostaglandin. *Cattleman's Day Rep., Kansas State Univ.* pp. 97.

Lauderdale, J. W. 1986. A review of patterns of change in luteal function. *J. Anim. Sci.* 62 (Suppl. 2):79.

Lemaster, J. W., R. C. Seals, F. M. Hopkins, and F. N. Schrick. 1999. Effects of administration of oxytocin on embryonic survival in progestogen-supplemented cattle. *Prostaglandins* 57:259.

Lewis, G. S., D. W. Caldwell, C. E. Rexroad, Jr., H. H. Dowlen, and J. R. Owen. 1990. Effects of gonadotropin-releasing hormone and human chorionic gonadotropin on pregnancy rate in dairy cattle. *J. Dairy Sci.* 73:66.

Lishman, A. W. and E. K. Inskeep. 1991. Deficiencies in luteal function during re-initiation of breeding activity in beef cows and in ewes. *South African J. Anim. Sci.* 21:59.

Lulai, C., I. Dobrinski, J. P. Kastelic, and R. J. Mapletoft. 1994a. Induction of luteal regression, ovulation and development of new luteal tissue during early pregnancy in heifers. *Anim. Reprod. Sci.* 35:163.

Lulai, C., J. P. Kastelic, T. D. Carruthers, and R. J. Mapletoft. 1994b. Role of luteal regression in embryo death in cattle. *Theriogenology* 41:1081.

Macmillan, K. L., V. K. Taufa, and A. M. Day. 1986. Effects of an agonist of gonadotropin releasing hormone (Buserelin) in cattle. III. Pregnancy rates after a post-insemination injection during metoestrus or dioestrus. *Anim. Reprod. Sci.* 11:1.

Malayer, J. R., P. J. Hansen, T. S. Gross, and W. W. Thatcher. 1990. Regulation of heat shock-induced alterations in the release of prostaglandins by the uterine endometrium of cows. *Theriogenology* 34:219.

Manns, J. G., J. R. Nkuuhe, and F. Bristol. 1985. Prostaglandin concentrations in uterine fluid of cows with pyometra. *Can. J. Comp. Med.* 49:436.

Maurer, R. R. and H. M. Beier. 1976. Uterine proteins and development *in vitro* of rabbit preimplantation embryos. *J. Reprod. Fertil.* 48:33.

Maurer, R. R. and J. R. Chenault. 1983. Fertilization failure and embryonic mortality in parous and nonparous beef cattle. *J. Anim. Sci.* 56:1186.

McDonald, L. M., R. E. Nichols, and S. H. McNutt. 1952. Study of corpus luteum ablation and progesterone replacement therapy in the cow. *Am. J. Vet. Res.* 13:446.

Meisterling, E. M. and R. A. Dailey. 1987. Use of concentrations of progesterone and estradiol-17β in milk in monitoring postpartum ovarian function in dairy cows. *J. Dairy Sci.* 70:2154.

Melton, A. A., R. O. Berry, and O. D. Butler. 1951. The interval between the time of ovulation and attachment of the bovine embryo. *J. Anim. Sci.* 10:993.

Menge, A. C., S. E. Mares, W. J. Tyler, and L. E. Casida. 1962. Variation and association among postpartum reproduction and production characteristics in Holstein–Friesian cattle. *J. Dairy Sci.* 45:233.

Mihm, M., A. Baguisi, M. P. Boland, and J. F. Roche. 1994a. Association between the duration of dominance of the ovulatory follicle and pregnancy rate in beef heifers. *J. Reprod. Fertil.* 102:123.

Mihm, M., N. Curran, P. Hyttel, M. P. Boland, and J. F. Roche. 1994b. Resumption of meiosis in cattle oocytes from preovulatory follicles with a short and long duration of dominance. *J. Reprod. Fertil. Abstr. Series* 13:14.

Milvae, R. A. and W. Hansel. 1980. Concurrent uterine venous and ovarian arterial prostaglandin F concentrations in heifers treated with oxytocin. *J. Reprod. Fertil.* 60:7.

Morrow, D. A., S. J. Roberts, K. McEntee, and H. G. Gray. 1966. Postpartum ovarian activity and uterine involution in dairy cattle. *J. Am. Vet. Med. Assoc.* 149:1596.

Murphy, M. G., W. J. Enright, M. A. Crowe, K. McConnell, L. J. Spicer, M. P. Boland, and J. F. Roche. 1991. Effect of dietary intake on pattern of growth of dominant follicles during the oestrus cycle in beef heifers. *J. Reprod. Fertil.* 92:333.

Newcomb, R., W. D. Booth, and L. E. A. Rowson. 1977. The effect of oxytocin treatment on the levels of prostaglandin F in the blood of heifers. *J. Reprod. Fertil.* 49:17.

Olds, D. and T. Cooper. 1970. Effect of postpartum rest period in dairy cattle on the occurrence of breeding abnormalities and on calving intervals. *J. Am. Vet. Med. Assoc.* 157:92.

Peter, A.T., W. K. Bosu, R. M. Liptrap, and E. Cummings. 1989. Temporal changes in serum prostaglandin $F_{2\alpha}$ and oxytocin in dairy cows with short luteal phases after the first postpartum ovulation. *Theriogenology* 32:277.

Pierson, R. A. and O. J. Ginther. 1987. Follicular populations during the estrous cycle in heifers I. Influence of day. *Anim. Reprod. Sci.* 14:165.

Pritchard, J. Y., F. N. Schrick, and E. K. Inskeep. 1994. Relationship of pregnancy rate to peripheral concentrations of progesterone and estradiol in beef cows. *Theriogenology* 42:247.

Rajakoski, E. 1960. The ovarian follicular system in sexually mature heifers with special reference to seasonal, cyclical, and left-right variations. *Acta Endocrinol. Suppl.* 52:7.

Rajamahendran, R. and C. Taylor. 1991. Follicular dynamics and temporal relationships among body temperature, oestrus, the surge of luteinizing hormone and ovulation in Holstein heifers treated with norgestomet. *J. Reprod. Fertil.* 92:461.

Ramirez-Godinez, J. A., G. H. Kiracofe, R. M. McKee, R. R. Schalles, and R. J. Kittok. 1981. Reducing the incidence of short estrous cycles in beef cows with norgestomet. *Theriogenology* 15:623.

Ramirez-Godinez, J. A., G. H. Kiracofe, D. L. Carnahan, M. F. Spire, K. B. Beeman, J. S. Stevenson, and R. R. Schalles. 1982a. Evidence for ovulation and fertilization in beef cows with short estrous cycles. *Theriogenology* 17:409.

Ramirez-Godinez, J. A., G. H. Kiracofe, R. R. Schalles, and G. D. Niswender. 1982b. Endocrine patterns in the postpartum beef cow associated with weaning: a comparison of the short and subsequent normal cycles. *J. Anim. Sci.* 55:153.

Revah, I. and W. R. Butler. 1996. Prolonged dominance of follicles reduces the viability of bovine oocytes. *J. Reprod. Fertil.* 106:39.

Rhodes, F. M., G. De'ath, and K. W. Entwistle. 1995. Animal and temporal effects on ovarian follicular dynamics in Brahman heifers. *Anim. Reprod. Sci.* 38:265.

Richards, M. W., R. D. Geisert, L. J. Dawson, and L. E. Rice. 1990. Pregnancy response after estrus synchronization of cyclic cows with or without a corpus luteum prior to breeding. *Theriogenology* 34:1185.

Roberson, M.S., M. W. Wolfe, T. T. Stumpf, R. J. Kittok, and J. E. Kinder. 1989. Luteinizing hormone secretion and corpus luteum function in cows receiving two levels of progesterone. *Biol. Reprod.* 41:997.

Roberts, J. S., B. Barcikowski, L. Wilson, Jr., R. C. Skarnes, and J. A. McCracken. 1975. Hormonal and related factors affecting the release of prostaglandin $F_{2\alpha}$ from the uterus. *J. Steroid Biochem.* 6:1091.

Sanchez, T., M. E. Wehrman, E. G. Bergfeld, K. E. Peters, F. N. Kojima, A. S. Cupp, V. Mariscal, R. J. Kittok, R. J. Rasby, and J. E. Kinder. 1993. Pregnancy rate is greater when the corpus luteum is present during the period of progestin treatment to synchronize time of estrus in cows and heifers. *Biol. Reprod.* 49:1102.

Sanchez, T., M. E. Wehrman, F. N. Kojima, A. S. Cupp, E. G. Bergfeld, K. E. Peters, V. Mariscal, R. J. Kittok, and J. E. Kinder. 1995. Dosage of the synthetic progestin, norgestomet, influences luteinizing hormone pulse frequency and endogenous secretion of 17β-estradiol in heifers. *Biol. Reprod.* 52:464.

Savio, J. D., L. Keenan, M. P. Boland, and J. F. Roche. 1988. Pattern of growth of dominant follicles during the oestrous cycle in heifers. *J. Reprod. Fertil.* 83:663.

Savio, J. D., W. W. Thatcher, L. Badinga, R. L. de la Sota, and D. Wolfenson. 1993a. Regulation of dominant follicle turnover during the oestrous cycle in cows. *J. Reprod. Fertil.* 97:197.

Savio, J. D., W. W. Thatcher, G. R. Morris, K. Entwistle, M. Drost, and M. R. Mattiacci. 1993b. Effects of induction of low plasma progesterone concentrations with a progesterone-releasing intravaginal device on follicular turnover and fertility in cattle. *J. Reprod. Fertil.* 98:77.

Schallenberger, E., D. Schams, B. Bullermann, and D. L. Walters. 1984. Pulsatile secretion of gonadotrophins, ovarian steroids and ovarian oxytocin during prostaglandin-induced regression of the corpus luteum in the cow. *J. Reprod. Fertil.* 71:493.

Schallenberger, E., D. Schams, and H. H. D. Meyer. 1989. Sequences of pituitary, ovarian and uterine hormone secretion during the first 5 weeks of pregnancy in dairy cattle. *J. Reprod. Fertil.* (Suppl. 37):277.

Schramm, W., L. Bovaird, M. E. Glew, G. Schramm, and J. A. McCracken. 1983. Corpus luteum regression induced by ultra-low pulses of prostaglandin $F_{2\alpha}$. *Prostaglandins* 26:347.

Schrick, F. N., E. K. Inskeep, and R. L. Butcher. 1993. Pregnancy rates for embryos transferred from early postpartum beef cows into recipients with normal estrous cycles. *Biol. Reprod.* 49:617.

Seals, R. C., J. W. Lemaster, F. M. Hopkins, and F. N. Schrick. 1998. Effects of elevated concentrations of prostaglandin $F_{2\alpha}$ on pregnancy rates in progestogen-supplemented cattle. *Prostaglandins* 56:377.

Shaham-Albalancy, A., M. Rosenberg, Y. Folman, A. Nyska, and D. Wolfenson. 1996. The effect of progesterone concentration during the luteal phase of the estrous cycle on uterine endometrial morphology and function during the subsequent estrous cycle of dairy cows. *Thirteenth Int. Cong. Anim. Reprod. A. I.* 19-18 (Abstr.)

Shannon, F. P., G. W. Salisbury, and N. L. Van Demark. 1952. The fertility of cows inseminated at various intervals after calving. *J. Anim. Sci.* 11:355.

Sheffel, C. E., B. R. Pratt, W. L. Ferrell, and E. K. Inskeep. 1982. Induced corpora lutea in the postpartum beef cow. II. Effects of treatment with progestogen and gonadotropins. *J. Anim. Sci.* 54:830.

Shelton, K., M. F. Gayerie de Abreu, M. G. Hunter, T. J. Parkinson, and G. E. Lamming. 1990. Luteal inadequacy during the early luteal phase of subfertile cows. *J. Reprod. Fertil.* 90:1.

Sirois, J. and J. E. Fortune. 1988. Ovarian follicular dynamics during the estrous cycle in heifers monitored by real-time ultrasonography. *Biol. Reprod.* 39:308.

Smith, L. C., M. Olivera-Angel, N. P. Groome, B. Bhatia, and C. A. Price. 1996. Oocyte quality in small antral follicles in the presence or absence of a large dominant follicle in cattle. *J. Reprod. Fertil.* 106:193.

Smith, M. W. and J. S. Stevenson. 1995. Fate of the dominant follicle, embryonal survival, and pregnancy rates in dairy cattle treated with prostaglandin $F_{2\alpha}$ and progestins in the absence or presence of a functional corpus luteum. *J. Anim. Sci.* 73:3743.

Sreenan, J. M. and M. G. Diskin. 1986. The extent and timing of embryonic mortality in cattle. *Embryonic Mortality in Farm Animals.* J. M. Sreenan and M. G. Diskin. (Eds.) Martinus Nijhoff, Dordrecht. pp. 1.

Stock, A. E. and J. E. Fortune. 1993. Ovarian follicular dominance: relationship between prolonged growth of the ovulatory follicle and endocrine parameters. *Endocrinology* 132:1108.

Stumpf, T. T., M.S. Roberson, M. W. Wolfe, D. L. Hamernik, R. J. Kittok, and J. E. Kinder. 1993. Progesterone, 17β-estradiol and opioid neuropeptides modulate pattern of LH in circulation of the cow. *Biol. Reprod.* 49:1096.

Taft, R., N. Ahmad, and E. K. Inskeep. 1996. Exogenous pulses of luteinizing hormone cause persistence of the largest bovine ovarian follicle. *J. Anim. Sci.* 74:2985.

Thatcher, W. W., K. L. Macmillan, P. J. Hansen, and M. Drost. 1989. Concepts for regulation of corpus luteum function by the conceptus and ovarian follicles to improve fertility. *Theriogenology* 31:149.

Thatcher, W. W., C. R. Staples, G. Danet-Desnoyers, B. Oldick, and E-P. Schmitt. 1994. Embryo health and mortality in sheep and cattle. *J. Anim. Sci.* 72 (Suppl. 3):16.

Ulberg, L. C., R. E. Christian, and L. E. Casida. 1951. Ovarian response of heifers to progesterone injections. *J. Anim. Sci.* 10:752.

Usmani, R. H., R. A. Dailey, and E. K. Inskeep. 1990. Effects of limited suckling and varying prepartum nutrition on postpartum reproductive traits of milked buffaloes *J. Dairy Sci.* 73:1564.

Van Cleeff, J., M. Drost, and W. W. Thatcher. 1991. Effects of postinsemination progesterone supplementation on fertility and subsequent estrous responses of dairy heifers. *Theriogenology* 36:795.

Vasconcelos, J. L. M., R. L. Silcox, J. A. Lacerda, J. R. Pursley, and M. C. Wiltbank. 1997. Pregnancy rate, pregnancy loss and response to heat stress after AI at 2 different times from ovulation in dairy cows. *Biol. Reprod.* 56 (Suppl. 1):140.

Washburn, S. P. and M. L. Keller. 1992. Fertility of beef cattle when estrous cycles are extended with progestogens. *J. Anim. Sci.* 70 (Suppl. 1):255.

Wehrman, M. E., K. E Fike, E. J. Melvin, E. G. M. Bergfeld, and J. E. Kinder. 1996. Development of a persistent ovarian follicle during synchronization of estrus does not alter conception rate after embryo transfer in cattle. *Theriogenology* 45:291 (Abstr.).

Wehrman, M. E., M. S. Roberson, A. S. Cupp, F. N. Kojima, T. T. Stumpf, L. A. Werth, M. W. Wolfe, R. J. Kittok, and J. E. Kinder. 1993. Increasing exogenous progesterone during synchronization of estrus decreases endogenous 17-β estradiol and increases conception in cows. *Biol. Reprod.* 49:214.

Wells, P. L., D. H. Holness, P. J. Freymark, C. T. McCabe, and A. W. Lishman. 1985. Fertility in the Afrikaner cow: ovarian recovery and conception in suckled and non-suckled cows postpartum. *Anim. Reprod. Sci.* 8:315.

Wishart, D. F. 1977. Synchronization of oestrus in heifers using steroid (SC 5914, SC 9880, and SC 21009) treatment for 21 days: the effect of treatment on the ovum collection and fertilization rate and the development of the early embryo. *Theriogenology* 8:249.

Wolff, N. 1992. Detection of embryonic mortality in cattle using sonography. *Tierarztl Prax* 20:373.

Wright, D. J., W. I. Buford, F. N. Schrick, and E. K. Inskeep. 1994. Ability of induced secondary corpora lutea to maintain pregnancy in beef cows. *J. Anim. Sci.* 72 (Suppl. 2): 125.

Zavy, M. T. 1994. Embryonic mortality in cattle. In M. T. Zavy and R. D. Geisert, (Eds.). *Embryonic Mortality in Domestic Species.* CRC Press, Boca Raton, FL.

Zollers, W. G., H. A. Garverick, M. F. Smith, R. J. Moffatt, B. E. Salfen, and R. S. Youngquist. 1993. Concentrations of progesterone and oxytocin receptors in endometrium of postpartum cows expected to have a short or normal oestrous cycle. *J. Reprod. Fertil.* 97:329.

20 Sexing Sperm for Beef and Dairy Cattle Breeding

George E. Seidel, Jr.

CONTENTS

It is now possible to sex sperm of most mammalian species with greater than 90% accuracy with an instrument called a flow cytometer/cell sorter (Seidel, 1999). Because bovine X chromosome-bearing sperm have about 4% more DNA than Y chromosome-bearing sperm, measuring DNA content can be used to discriminate between the two types of sperm. Currently about 2500 sperm of each sex can be sexed per second at 90% accuracy, so it would take 2 hours to sort the minimal number of sperm in a typical artificial insemination dose. The sorting process is not

0-8493-1117-9/02/$0.00+$1.50

innocuous; for example, sperm exit the equipment at nearly 60 miles per hour. However, damage to sperm from sorting is less than that caused by cryopreservation.

Despite the limitations, hundreds of heifers have recently calved or become pregnant following artificial insemination with sexed, frozen sperm using lower sperm numbers per dose than are used conventionally (Seidel et al., 1999), and improvements are being made to sperm sexing technology at a rapid pace. To date, there is no evidence that these sexing procedures result in abnormal offspring.

The main deterrent to immediate mass application of this technology is that pregnancy rates are slightly lower than with unsexed semen, and this problem is exacerbated by the relatively low numbers of sperm available for insemination. Nevertheless, it is likely that these problems will be solved or circumvented, so that sexed sperm will be commercially available from selected bulls within the next year. Possibly the first applications will be limited to artificial insemination of heifers, which have higher fertility with fewer sperm than cows. With additional research, use of sexed sperm with cattle could become widespread.

Here, I summarize potential applications of sexed sperm in cattle, assuming that accuracy of sexing is about 90% (already true), that fertility is greater than 90% of fertility of unsexed semen (currently 70 to 80%; Seidel et al., 1999), and that sexed, frozen sperm will be available at a reasonable cost from a broad selection of bulls (sperm from some bulls may not tolerate the trauma of sexing, particularly if also frozen).

APPLICATIONS PRODUCING REPLACEMENT HEIFERS

Increasing the Percentage of Heifer Calves to Expand the Herd or Sell Replacements

Normally, averaged over thousands of animals, 49% of calves born will be heifers, and a few of these will be sterile freemartins. Due to chance alone, it is not unusual to have only 40% heifers from 100 consecutive calvings. For example, there is about a 20% chance that at least 8 of the next dozen calves born in any herd will be bulls. Sperm sexed for females at 90% accuracy would greatly decrease these vagaries of sex ratio; about 90% of calves born would, in the long run, be heifers. A program using artificial insemination of X-selected sperm would enable rapid herd expansion without the risk of introducing diseases that occur with purchased animals. It would greatly increase the practicality of producing replacement heifers for sale.

Note, however, if most herds were using such a program, there would be a large surplus of heifers, and the price of replacement heifers could drop substantially, making some heifer-rearing systems less profitable. This problem likely would be offset in the long run because of the profitability of breeding older beef and dairy cows to terminal-cross beef sires for rapidly growing male calves.

Increasing Selection Intensity by Choosing Genetically Superior Dams of Replacements

At equilibrium, about 40% of beef females and 80% of dairy females must be bred for herd replacements to maintain herd size because over half the calves born are bulls, and some of the heifers born either die, become unthrifty, or do not become pregnant. With accurately sexed semen, only half as many females would need to be bred for replacements, thus increasing selection intensity. Note that even though most genetic progress is made on the bull side of the pedigree, increasing selection intensity on the female side would further improve the cow herd. According to Van Vleck (1981), this could be worth nearly $50/mating in dairy cows.

Breeding Heifers to Have Heifer Calves to Decrease the Incidence of Calving Difficulty

A major problem is dystocia when heifers calve. This can be minimized by breeding only well grown (but not fat) heifers and by using service sires that produce a low percentage of difficult births. The latter course, while reasonably effective, can result in lighter calves that will develop into smaller cows. This strategy could become a problem after several generations of such a program. The majority of dystocias are due to bull calves, which average about 5 lb heavier in birth weight than heifer calves. A large study in New Zealand (Morris et al., 1986) with primiparous beef heifers illustrates this well; death losses from birth to weaning were 10% for heifer calves and 18% for bull calves, mostly due to sequelae of dystocia. To decrease dystocia substantially, one could use bulls that sire a low percentage of calves with difficult birth plus sperm sexed to produce 90% heifer calves. There is the added benefit that these first-calf heifers should be better genetically, on average, than the older cows in the herd. In my opinion, this will be one of the most important uses of sexed sperm, both in dairy and beef cattle production.

Dispensing with the Cow Herd Using an All-Female System for Beef Production

Theoretically, it is possible to have every beef female replace herself with a heifer calf just before she is fattened for slaughter. Heifers would be bred with sexed semen to calve at 21 to 22 months of age with low-birth-weight heifer calves. They would be fed very well while lactating for about 4 months, when weaning would occur. Early-weaned, 4-month-old calves grow very efficiently and could be fed well so they become pregnant at 12 to 13 months of age to start the next cycle. The 25- to 26-month-old newly weaned mothers would be fed another 2 months and sold as fat cattle that still would produce premium carcasses at 27 to 28 months of age.

This system would be more efficient than current systems because there is no cow herd; all heifers would be in one phase or another of growth. A huge cost of conventional beef systems is maintenance of the cow herd, which would not exist with this approach. Obviously such a system would not work on extensive pastures without considerable supplemental feed.

Proving Young Dairy Bulls

To maintain the current rate of genetic progress in most dairy cattle breeding programs, it is necessary to progeny test young bulls. This continues to be problematic, and incentives are used to encourage farmers to produce calves and lactations for these programs. It is particularly frustrating that half of the calves born are bulls, and therefore useless for proving dairy sires. If semen sexed at 90% accuracy for heifer calves were available, only 55% as many cows would need to be bred to get the same number of heifers for proofs. The fringe benefits of having 90% female calves for replacements might make such a program very successful.

A related problem is obtaining young bulls from elite cows to progeny test in the first place; half the time a heifer is produced. This often is dealt with by superovulating the potential bull mother, so by chance at least one bull is produced. This is the reason that over 80% of proven dairy bulls in North America are produced via embryo transfer. Sexed semen without superovulation will be an attractive alternative in some situations.

Circumventing the Shortage of Dairy Heifer Calves Born Due to Lengthened Lactations

The average dairy cow has three calves in her lifetime. For a variety of reasons, calving often is a traumatic time for dairy cows, resulting in various health problems. One management technique that is growing in popularity is to lengthen lactations by propping up the lactation curve with bovine growth hormone. This, however, will have the net effect of fewer calves per lifetime, and if the strategy is taken to the extreme, a herd may not produce sufficient heifer calves to maintain herd size without sexed sperm. At the very least, selection intensity could be near zero on the dam's side of the pedigree. Skewing the sex ratio to favor female offspring would be a very sensible way to circumvent this problem.

Sexing Sperm for *In Vitro* Fertilization, Superovulation, and Embryo Transfer Programs

The first calves produced with accurately sexed semen resulted from *in vitro* fertilization (IVF), which requires many fewer sperm than artificial insemination (Cran et al., 1993). Accuracy of sexing sperm was 90%. Typically, with IVF or superovulation in dairy cattle, one would want heifer calves, since large numbers of full-brother bulls from most dams would be difficult to market at a profit because such programs are relatively costly. Another problem is that sex ratios with most IVF programs without sexed sperm are in the range of 55 to 60% bull calves (Pegoraro et al., 1998). Although it now is possible to sex embryos resulting from IVF or standard superovulation and nonsurgical embryo recovery programs with reasonable accuracy (Seidel, 1999), the sexing process is relatively expensive, and embryos of the less valuable sex are often discarded. It would be much more elegant and likely less expensive to sex sperm so that embryos of the less valuable sex are not even produced.

More research is required for these applications. At the present time, fertilization and embryo development rates in IVF programs with sexed sperm are lower than with unsexed semen. Few data are available from inseminating superovulated cows with sexed sperm. In the long term, however, sexed sperm are likely to be especially useful when coupled with these biotechnologies.

APPLICATIONS PRODUCING BULL CALVES

SEXING SPERM FOR *IN VITRO* FERTILIZATION, SUPEROVULATION, AND EMBRYO TRANSFER PROGRAMS

The above heading was just presented in the context of producing heifer calves in dairy cattle. However, with IVF or superovulation in beef cattle, one would often want bull calves, since large numbers of full-brother bulls would be very marketable for large natural breeding programs.

TERMINAL-CROSS BEEF PROGRAMS

An example of the practical economic value of sexing to produce steers is from my own herd of beef cattle. Most of the cows are Hereford and Angus crossbreds that are inseminated with Charolais semen. With this terminal-cross program, both steers and heifers are sold for fattening at weaning at about 7 to 8 months of age. Over a 4-year period, the heifer calves averaged 491 lb and sold at an average price of $.85/lb (US $417 per calf). Under identical management, the steers averaged 519 lb at an average of $.92/lb (US $477 per calf), a $60 advantage due to sex (average of $30 per calf). The reason steer calves command higher prices is that they grow more efficiently.

I would manage this herd quite differently if sexed semen were used. Replacement heifers sired by bulls that transmit good maternal traits would be obtained from 20% of the cows (those with the best maternal traits), and the remaining cows would produce terminal-cross male calves. Note that it would work quite well to have most of the heifers have female calves for replacements and most of the cows have terminal-cross bull calves. Another attractive option would be to have an all male terminal-cross program and purchase all replacement heifers.

MALE OFFSPRING FROM DAIRY COWS

Over the past decade, the value of 3-day-old Holstein bulls in the U.S. has varied between less than $10 and greater than $100. Recently, it has been on the low side, especially for the "colored" breeds. Over the past few years, this has been even more problematic in the U.K. The majority of dairy calves end up being used for veal or beef, and this will continue. Those calves destined for veal or beef production systems would be more valuable if sired by beef breeds such as Angus or Charolais. They would be even more valuable if male, on the order of a $60 advantage for beef, because males are larger and grow more efficiently. Dystocia and resulting death with bull calves is only slightly higher than with heifer calves from multiparous cows (Morris et al., 1986).

Programs for producing male calves for beef from dairy cows using sires that transmit good carcass qualities mesh especially well with programs to have female calves from heifers and from the best cows genetically. Such integrated programs would make the $10 dairy calf for meat a thing of the past, but would also compete with more traditional beef production systems.

OTHER CONSIDERATIONS

In addition to being financially successful if sexing costs are low, accuracy is high, fertility is normal, and calves are normal, sexed sperm programs will result in more efficient milk and meat production. Fewer animals will be required per unit of product, making the use of this technology ecologically sound. Less feed will be required, and less manure will be produced than without sexed semen.

Sexing technology will not be totally benign. There could be some dislocations as beef–dairy cross calves become more difficult to distinguish from beef breeds. Systems of raising replacements could change substantially. Increased efficiency translates into still fewer cows. Despite these side effects, sexed sperm likely would be considered very beneficial to the long-term health of the cattle industry, primarily because it would enable providing better products for consumers in a shorter time frame, and at lower cost than not using sexed sperm.

REFERENCES

Cran, D.G., L.A. Johnson, N.G.A. Miller, D. Cochrane, and C. Polge. 1993. Production of bovine calves following separation of X- and Y-chromosome bearing sperm and *in vitro* fertilization. *Vet. Rec.* 132:40.

Morris, C.A., G.L. Bennett, R.L. Baker, and A.H. Carter. 1986. Birthweight, dystocia and calf mortality in some New Zealand beef breeding herds. *J. Anim. Sci.* 62:327.

Pegoraro, L.M.C., J.M. Thuard, N. Delalleau, B. Guerin, J.C. Deschamps, B. Marquant-LeGuienne, and P. Humblot. 1998. Comparison of sex ratio and cell number of IVM-IVF bovine blastocysts co-cultured with bovine oviduct epithelial cells or with Vero cells. *Theriogenology* 49:1579.

Seidel, G.E., Jr. 1999. Sexing mammalian sperm and embryos—state of the art. *J. Reprod. Fertil.* (Suppl. 54):475.

Seidel, G.E., Jr., J.L., Schenk, L.A. Herickhoff, S.P. Doyle, Z. Brink, R.D. Green, and D.G. Cran. 1999. Insemination of heifers with sexed sperm. *Theriogenology* 52:1407.

Van Vleck, L.D. 1981. Potential genetic impact of artificial insemination, sex selection, embryo transfer, cloning, and sexing in dairy cattle. In: B.G. Brackett, G.E. Seidel, Jr., and S.M. Seidel (Eds.) *New Technologies in Animal Breeding.* Academic Press, New York, NY. pp. 221.

21 New Developments in Managing the Bull

R. L. Ax, H. E. Hawkins, S. K. DeNise, T. R. Holm, H. M. Zhang, J. N. Oyarzo, and M. E. Bellin

CONTENTS

Fertility as a trait is five to ten times more important to profitability of a beef herd than any other trait we can measure. Unfortunately, heritability estimates for fertility put the value at only about 10% (Taylor et al., 1985). How, then, can we manage bulls to optimize overall herd fertility? Breeding soundness evaluations (BSE) can be performed to qualify bulls as potential satisfactory breeders. However, as we enter the new millennium, the majority of bulls in the U.S. are still not routinely subjected to a BSE.

Fertility *per se* of individual bulls cannot be predicted from a BSE. Bulls with identical physical quality semen characteristics obtained with a BSE can vary substantially in actual fertility. Biochemical tests are being developed which enable screening of specific proteins in semen samples which correspond to higher fertility.

Finally, in multiple-sire pastures, all bulls are not created equal. From DNA-based parentage tests, results indicate that approximately one half of the bulls will sire 80% of the calves that are subsequently born. The purpose of this chapter is to provide a perspective of how we need to manage our bull power to maximize fertility in the cow herd.

0-8493-1117-9/02/$0.00+$1.50

BREEDING SOUNDNESS EVALUATIONS

The American Society for Theriogenology publishes guidelines that list minimal acceptable thresholds for bulls to qualify on the basis of a breeding soundness evaluation (BSE) (Chenoweth et al., 1992). Bulls are qualified as "potential satisfactory breeders" from the outcome of that exam. A BSE includes a general physical exam, review of vaccinations, measurement of scrotal circumference, and assessment of semen quality. However, even among bulls that produce semen with identical physical characteristics, there is a wide range of variation in actual fertility when those bulls are used for natural service or artificial insemination. For the latest data available from USDA (USDA:APHIS:VS, 1998) a BSE was performed on less than 35% of breeding-eligible bulls. That proportion has declined since 1992/93. Clearly, there is room for improvement based on those statistics.

As part of a BSE, a semen sample is evaluated for sperm volume, concentration (density), motility, and morphologically abnormal cells. Of those semen traits, there is a 90% correlation between the percentage of abnormal cells and fertility of cows bred by artificial insemination (Saacke and White, 1972). At King Ranch, fertility of bulls in 1980 and 1981 was assessed comparing a random sample (gate cut) to BSE-qualified bulls with documented percentages of normal sperm cells exceeding 70 or 80% (Table 21.1). Fertility improved 5 to 6% using bulls that produced semen with fewer than 30% abnormal cells in their ejaculates when contrasted to the random sample of herdmates (Table 21.1).

In 1978, personnel at King Ranch, particularly Norm Parrish cooperating with Dr. Jim Wiltbank from the Texas A&M Research Station at Beeville, TX, started selecting bulls for scrotal circumference at 14 to 16 months of age. Scrotal circumference relates to sperm output and can be compared to a warehouse. Selecting for a large scrotal circumference will also decrease the age of pregnancy of daughters (Evans et al., 1999).

Table 21.2 illustrates the progress made at King Ranch from 1978 until 1990 in terms of distributions of bulls on the basis of scrotal circumference. From the data

TABLE 21.1
Fertility of Bulls at King Ranch Based upon Percentage of Normal Sperm in an Ejaculate

	1980		1981		
	Random Group	≥80% Normal Sperm	Random Group	≥70% Normal Sperm	≥80% Normal Sperm
No. bulls[a]	23	26	47	31	21
No. cows[b]	571	656	1179	769	522
No. pregnant	497	610	1002	700	470
% Pregnant	87%	93%	85%	91%	90%
Increase over random		6%		6%	5%

[a] Bulls were used in actual service for a 60-day breeding season.
[b] Cows were kept at a constant ratio of 25 cows per bull.

TABLE 21.2
Scrotal Circumference of 14- to 16-Month-Old Santa Gertrudis Bulls at King Ranch

Scrotal Circumference (cm)	% of Bulls[a]	
	1978 to 1979	1990
<29	21	2.2
30 to 32	37	9.7
33 to 35	34	34.5
36 to 38	7	35.3
>38	1	18.3

[a] Over 1500 bulls were evaluated in each sampling period.

TABLE 21.3
Percentage of Calves Sired per Bull in Multiple-Sire Breeding Pastures Based upon DNA-Based Parentage Test. Each Pasture Had 50 to 100 Calves Produced

	Bulls per Pasture		
	3	4	5
Dominant bull	66%	50%	40%
Subordinate bull A	28%	31%	29%
Subordinate bull B	7%	13%	19%
Subordinate bull C		6%	8%
Subordinate bull D			3%
Replicate pastures	2	7	7

presented in Table 21.2, it is apparent that the scrotal circumference recommendation of 34 cm or greater at that age would result in the majority of bulls not being qualified in 1978 to 1979. In 1978, a 33-cm or larger scrotal circumference was measured in only 42% of the bulls. By 1990, 88% exceeded 33 cm. Today, for a bull to qualify as a breeding sire, he needs to attain a 35-cm scrotal circumference when the BSE is performed at King Ranch.

NOT ALL BULLS ARE CREATED EQUAL

A common practice in the beef industry is to utilize multiple-sire breeding pastures. With the advent of DNA-based parentage tests, it becomes possible to ascertain the sire of individual calves born following those matings. Ordinarily, we assume that the distribution of calves born would be equally partitioned between the bulls used together in a multiple-sire pasture. Research performed years ago with competitive fertilization experiments proved that a hierarchy will exist among multiple sires, and that seldom will two or more males be equally fertile (Saacke et al., 1980). Data in Table 21.3 confirm those earlier observations. In general, 50% of bulls will sire 80% of the calves.

We still debate the issue of libido or serving capacity of individual bulls. If libido was highly heritable, low-libido bulls would have been self-eliminated generations ago. From a practical standpoint, serving capacity determinations will not be performed on individual ranches prior to a new breeding season. This issue will be revisited later in this chapter.

FINDING THE NEEDLE IN THE HAYSTACK

Since the early 1980s, research in Dr. R. L. Ax's lab has centered on studying cellular changes in sperm during fertilization (Miller and Ax, 1990). Most of those studies utilized laboratory tests and employed *in vitro* fertilization (IVF) as the end point. The major source of variation in IVF experimental outcomes was traced back to the semen donor. Bulls with identical physical semen characteristics differed in their ability to successfully fertilize bovine eggs during IVF. Therefore, the hunt was initiated to identify some feature in sperm that might be used to segregate high-fertility bulls from their lower-fertility herdmates.

In 1992, white mice were vaccinated with proteins extracted from sperm from high-fertility bulls. In response to that challenge of foreign proteins, the mice produced antibodies, similar to humans producing antibodies in response to a measles or tetanus immunization. The mouse that had the greatest response was sacrificed, the spleen was removed, and each spleen cell was fused to a cancer cell. Those fused cells become immortal, and each cell stock is cloned to produce thousands of identical cells. The antibody resulting from each individual cell line is referred to as a monoclonal antibody.

Using standard biochemical reagents to generate color-based detection systems, monoclonal antibodies can then be used to measure trace amounts of specific substances on tissues, in tissue extracts, or in solution. One antibody was named M1, and it can be used to identify the presence of a particular protein that is extracted from sperm surfaces. That specific protein has been coined fertility-associated antigen (FAA).

FERTILITY-ASSOCIATED ANTIGEN (FAA)

One protein found in trace amounts in semen is FAA. This protein is produced in the seminal vesicles, prostate, and Cowper's gland (accessory glands), and at the time of ejaculation, binds to sperm as they traverse the male reproductive tract. FAA is a heparin-binding protein, possesses a molecular weight of 31,000, and is nonglycosylated (McCauley et al., 1999). It represents approximately 0.5% of the protein content in seminal fluid. Sperm from the epididymis lack FAA, but if those epididymal sperm are incubated with seminal fluid from a vasectomized bull for 20 minutes, the M1 monoclonal antibody indicates that FAA is clearly distributed across distinct regions of sperm cells (McCauley et al., 1996).

To date, FAA content of sperm has been measured from 3600 bulls representing every region of the world. From those samples, 85% of bulls were classified as FAA-positive, and the difference (15%) was categorized as being FAA-negative. When FAA status is known, breeding trials can be conducted to compare breeding

TABLE 21.4
Fertility of Range Beef Bulls Segregated on the Presence or Absence of Spermatozoal Fertility-Associated Antigen (FAA)

	FAA in Sperm Membranes		
	Present	Absent	Total Numbers
No. bulls	242	192	434
No. cows bred	5317	3881	9198
No. cows pregnant	4497	2572	7069
Pregnant %	85	66	77

Adapted from Bellin et al. (1994, 1996, 1998).

efficiencies of herdmates that produce sperm with or without detectable FAA on their surfaces.

FIELD TRIALS: IS FAA STATUS IMPORTANT?

With a detection system for sperm-associated FAA in hand, fertility of bulls was compared to FAA status (Table 21.4). Prior to availability of the M1 antibody, FAA was referred to as a 30,000-molecular-weight, heparin-binding protein (HBP-30) that was one protein component of a complex of peptides that displayed the highest affinity to bind to a heparin affinity column. Therefore, data in Table 21.4 were compiled from three published reports (Bellin et al., 1994, 1996, 1998), as well as some additional field trials.

Over a 6-year span of breeding trials at King Ranch, FAA-positive bulls were 19% more fertile than their herdmates lacking FAA. In all of those field trials, bulls were used for natural matings for a 60-day breeding season at a ratio of 1 bull per 25 cows. There were approximately 12 bulls per pasture that had all been prescreened for similar protein patterns on sperm.

SERVING CAPACITY AND FAA STATUS

Data presented earlier in Table 21.3 illustrated that there are dominant bulls in every multiple-sire pasture situation. In a collaboration between the University of Arizona and Texas A & M University, bulls were evaluated for serving capacity potential, and semen samples obtained by electroejaculation were assayed for presence or absence of FAA. Bulls were then grouped by serving capacity and FAA status for a 60-day breeding season. Results are presented in Table 21.5.

From the data collectively in Tables 21.3 and 21.5, it is obvious that serving capacity plays an important role in terms of pregnancy successes and proportions of calves that will be sired by individual bulls. Coupled with that is the consideration that FAA still serves as a biochemical determinant for fertility potential (Table 21.5). Efforts are in progress to fill in the missing box in Table 21.5. How inferior will bulls be that lack sperm-associated FAA and score as having low serving capacity?

TABLE 21.5
Fertility of Santa Gertrudis Bulls Grouped According to FAA and Serving Capacity Profiles During the Breeding Season

Bull Profiles				Cows Pregnant, %				
				Breeding Season, days[e]				
FAA[a]	Serving Capacity[b]	No. Cows[c]	BCS[d]	1 to 20	21 to 40	41 to 60	Fertility,%	Differences in Fertility[f]
Positive	High	270	4.2	50	19	18	87[g]	—
Negative	High	143	4.8	45	13	20	78[h]	9
Positive	Low	238	4.3	29	13	27	69[i]	19

[a] All bulls were 2-year-old Santa Gertrudis. Bulls were grouped according to presence of fertility-associated antigen (FAA) on sperm membranes and serving capacity.
[b] Serving capacity was determined by observing and counting the number of times a bull mated (defined as intromission and/or apparent ejaculations) estrous-synchronized heifers during a 20-min period. Bulls that mated with two or more estrous-synchronized heifers were classified as high serving capacity; bulls that did not mount any heifers were classified as low serving capacity. Serving capacity tests were performed a few weeks before the breeding season.
[c] Bulls were bred to 3-year-old crossbred cows ($\frac{1}{2}$ Simmental, $\frac{1}{4}$ Hereford, $\frac{1}{4}$ Brahman). Cows were being bred to produce their third calf.
[d] Average body condition scores (BCS) for cows were estimated.
[e] The day of pregnancy was estimated by approximating the age of the fetus when cows were checked for pregnancy.
[f] Differences in fertility were calculated by subtracting the fertility of each group from the fertility of the group with FAA present and high serving capacity.
[g,h] Values differed ($P < 0.05$).
[h,i] Values differed ($P < 0.01$).

From Bellin et al. (1998). With permission.

ARTIFICIAL INSEMINATION

All of the data presented thus far in this manuscript were gleaned from field trials where natural service matings occurred (Tables 21.1 to 21.5). As stated in this text, serving capacity does contribute to variations in frequencies of calves sired by individual bulls (Tables 21.3 and 21.5). With artificial insemination (AI), serving capacity is not a factor because any cow or heifer in heat can be bred to the bull of choice once his semen sample is thawed from a nitrogen tank. Use of AI also permits us to measure actual fertility for any individual bull. From that, a range in fertility between bulls classified as FAA-positive or FAA-negative can be compared.

Table 21.6 provides a summary of data which was recently published (Sprott et al., 2000). Beef bulls with detectable FAA in sperm were 16% more fertile than herdmates with nondetectable FAA. Projected pregnancies after three AI breedings, which would correspond to a 60-day breeding season, were 88% and 73% for bulls classified as FAA-positive and FAA-negative, respectively (Table 21.6). That 15% difference calculated as projected pregnancies in Table 21.6 is virtually the same magnitude difference depicted in Table 21.4 from actual 60-day breeding seasons.

TABLE 21.6
Number of Females Pregnant to First AI Service and Projected Number Pregnant after Three Services to Semen from FAA Negative (–) or FAA Positive (+) Sperm

No. Females	No. Bulls	Bull FAA Status	No. Pregnant to 1st Service	Projected No. Pregnant by 3 Services
386	7	Negative	192 (49.7%)	283 (73.3%)
764	18	Positive	501 (65.6%)	673 (88.1%)
Total 1150	25			

TABLE 21.7
Number of FAA-Positive and FAA-Negative Bulls within Each Fertility Percentile for Three Herds

		Herd A					Herd B, C			
Fertility, %	0	10	20	30	40	50	60	70	80	90
FAA positive		2				1	4	8	1	2
FAA negative	1		1				5			

The range of actual fertility for individual bulls summarized in Table 21.6 is partitioned in Table 21.7. Differences between herds are obvious, especially for herd A. All 7 FAA-negative bulls were in the 60th percentile or lower for actual fertility. Within the group of 18 bulls determined to be FAA-positive, 11/18 animals (61%) displayed actual fertility values greater than 70%. Overall, prequalifying AI sires in terms of FAA status should be encouraged based upon data shown in Table 21.7.

IS FAA HERITABLE?

This is a commonly asked question by ranchers. Since FAA is a single protein, and every protein is coded for by a unique gene. It should be fairly safe to say, yes, it is heritable provided that fertility differences were attributable to variant FAAs, which were determined by differences in its gene. Having applied selection pressure to assure that all cows in the cow herd are only bred to bulls known to be FAA-positive, that result also indirectly supports the answer that FAA is heritable.

Table 21.8 illustrates the outcome from selecting for only FAA. Prior to the selection, 43% of the cows in the King Ranch nucleus herd calved in the first 20 days of the calving season in 1991. Selection for FAA-positive bulls exclusively as breeding bulls started in the 1992 breeding season and continued for each successive breeding season for 6 years. In 1998, 61% of the calves were born in the first 20 days of the calving season (Table 21.8). Though not presented here, in 1998, 83% of the calves were born in the first 30 days of the calving season, and 92% were born within 45 days. Therefore, a 45-day breeding season had actually occurred in the fall of 1997 at King Ranch even though the bulls were in pastures with cows for 60 days.

TABLE 21.8
Distribution of Calving Season in the Nucleus Herd at King Ranch. Cows were Bred to FAA-Positive Bulls and Their Retained Daughters Only Bred to FAA-Positive Bulls

	Prior to FAA Testing					
	1991 223 Head		1995 262 Head		1998 489 Head	
Days of Calving Season	%	Average Wean Wt (lb)	%	Average Wean Wt (lb)	%	Average Wean Wt (lb)
1 to 20	43.0	590	51.5	586	61.2	569
21 to 40	35.4	542	31.3	539	25.3	535
41 to 60	16.1	476	10.3	471	11.1	490
	94.5		93.1		97.6	

The value of testing for FAA on sperm can be justified in terms of the additional pounds of weaned beef that will result with 20 days of additional gain on 20% of the calf crop. In a 100-cow herd, at gains of 1.75 lb/day, 20 calves with 20 days earlier birth dates would translate into 700 lb of additional gain. At 50¢/lb, that would return $350 to the rancher. To breed 100 cows at a ratio of 25 cows/bull, we would have to screen four bulls at a cost of $40/bull. Therefore, a $160 investment returns $350 profit, which is slightly higher than doubling our return. Obviously, at higher prices paid for beef, the return escalates. What has also not been considered in this scenario is the higher fertility obtained from using only FAA-positive bulls, resulting in some extra calves born as well.

IS THE FAA TEST COMMERCIALLY AVAILABLE?

This technology was licensed by the University of Arizona to a company called ReproTec, Inc. The test is protected by a patent, so this is the only company that can offer the test commercially. The phone number for ReproTec is (520)-888-0401. The FAX number is (520)-888-0297. They are located at 4439 N. Highway Dr. #2, Tucson, AZ 85707-1909.

WHAT LOOMS ON THE HORIZON FOR FAA?

ReproTec, Inc., in cooperation with the Department of Animal Sciences at the University of Arizona is developing a new monoclonal antibody with the intent of being able to commercialize a chute-side fertility test. This would reduce the cost of the test, and semen samples could be screened on-site rather than having to be mailed to a lab for testing. Even in FAA-positive bulls, one should be able to quantify whether there are low, moderate, or high amounts of FAA in individual samples.

Most of the gene for FAA has been sequenced, and a mutation was found in the FAA gene from a sterile bull. The FAA gene from human prostate tissue was sequenced, and we found it to be 92% identical to bovine FAA. To date, four mutations have been identified in FAA from different human prostates (Zhang et al., 2000). Collectively, these findings point to the likelihood that DNA-based diagnostic tests will evolve and enable the screening for fertility potential at an early age. With DNA testing, heifers/cows could be screened as well, and those tests might hold promise in other livestock species.

Recently developed is a recombinant FAA. This recombinant FAA is about two thirds the size of the natural FAA (molecular weight 22,000) and enables large quantities to be produced and evaluated for therapeutic value as a semen additive. Preliminary lab experiments have shown that addition of recombinant FAA potentiates heparin-induced capacitation of bull sperm. Additionally, the product improved the percentage of intact acrosomes in bovine semen samples that were processed in the normal manner for artificial insemination and subjected to a 3-hour post-thaw stress test to evaluate acrosomal integrity (Lenz et al., 2000). There is a 90% correlation between percentage of intact acrosomes and nonreturn rates of bulls when their semen is used for AI (Saacke and White, 1972). Therefore, if recombinant FAA improves processed semen, it is reasonable to hypothesize that it will also improve pregnancy rates. A field trial to test that hypothesis is under way.

Finally, there are other proteins on sperm that could serve as markers for fertility. Researchers at Pennsylvania State University have identified two proteins as osteopontin and lipocalin-type prostaglandin D-synthase, which are incorporated into sperm membranes in the epididymis (Cancel et al., 1997; Gerena et al., 1998). No doubt other proteins will be identified in the future which will serve as useful biochemical determinants for fertility. As scientists sequence more genes for proteins like FAA, we will be able to develop precise genetic tools to improve fertility, and, ultimately, profitability of ranches that utilize this tool to practice selection.

REFERENCES

Bellin, M.E., H.E. Hawkins, and R.L. Ax. 1994. Fertility of range beef bulls grouped according to presence or absence of heparin-binding proteins in sperm membranes and seminal fluid. *J. Anim. Sci.* 72:2441.

Bellin, M.E., H.E. Hawkins, J. Oyarzo, R.J. Vanderboom, and R.L. Ax. 1996. Monoclonal antibody detection of a heparin binding protein on bull sperm corresponds to fertility. *J. Anim. Sci.* 74:173.

Bellin, M.E., J.N. Oyarzo, H.E. Hawkins, H. Zhang, R.G. Smith, D.W. Forrest, L.R. Sprott, and R.L. Ax. 1998. Fertility associated antigen (FAA) on bull sperm indicates fertility potential. *J. Anim. Sci.* 76:2032.

Cancel, A.M., D.A. Chapman, and G.J. Killian, 1997. Osteopontin is the 55-kilodalton fertility-associated protein in Holstein bulls seminal plasma. *Biol. Reprod.* 57:1293.

Chenoweth, P.J., J.C. Spitzer, and F.M. Hopkins. 1992. A new bull breeding soundness evaluation form. *Proc. Ann. Mt. Soc. Theriogenology.* San Antonio, TX.

Evans, J.L., B.L. Golden, R.M. Bourdon, and K.L. Long. 1999. Additive genetic relationships between heifer pregnancy and scrotal circumference in Hereford cattle. *J. Anim. Sci.* 77: 2621.

Gerena, R.L., D. Irikura, Y. Urade, N. Eguchi, D.A. Chapman, and G.J. Killian. 1998. Identification of a fertility-associated protein in bull seminal plasma as lipocalin-type prostaglandin D synthase. *Biol. Reprod.* 58:826.

Lenz R.W., H.M. Zhang, J.N. Oyarzo, M.E. Bellin, and R.L. Ax. 2000. Bovine fertility-associated antigen (FAA) and a recombinant segment of FFA improves sperm function. *Biol. Reprod.* 62 (Suppl. 1):137.

McCauley, T.C., M.E. Bellin, and R.L. Ax. 1996. Localization of a heparin-binding protein to distinct regions of bovine sperm. *J. Anim. Sci.* 74:429.

McCauley, T.C., H. Zhang, M.E. Bellin, and R.L. Ax. 1999. Purification and characterization of fertility-associated antigen (FAA) in bovine seminal fluid. *Mol. Reprod. Dev.* 54:145.

Miller, D.J. and R.L. Ax. 1990. Carbohydrates and fertilization in animals. *Mol. Reprod. Dev.* 26:184.

Saacke, R.G. and J.M. White. 1972. Semen quality tests and their relationship to fertility. *Proc. Fourth Tech. Conf. AI. Reprod.*, Natl. Assoc. Anim. Breed. Columbia, MO. pp. 2.

Saacke, R.G., W.E. Vinson, M.C. O'Connor, J.E. Chandler, J. Mullins, R.P. Amann, C.E. Marshall, R.A. Wallace, W.N. Vincel, and H.C. Kellgren. 1980. The relationship of semen quality and fertility: a heterospermic study. *Proc. Eighth Tech. Conf. AI. Reprod.*, Nat'l. Assoc. Anim. Breed., Columbia, MO. pp. 71.

Sprott, L.R., J. Young, D.W. Forrest, H.M. Zhang, J.N. Oyarzo, M.E. Bellin, and R.L. Ax. 2000. Artificial insemination outcomes in beef females using bovine sperm with a detectable fertility-associated antigen. *J. Anim. Sci.* 78:795.

Taylor, J.F., R. W. Everett, and B. Bean. 1985. Systematic environmental, direct, and service sire effects on conception rate in artificially inseminated Holstein cows. *J. Dairy Sci.* 68:3004.

USDA:APHIS:VS. Part IV. Changes in the U.S. beef cow-calf industry, 1993-1997. Cent. Epidemiology Anim. Health. Fort Collins, CO. #N238.398. May 1998.

Zhang, H.M., M.E. Bellin, J.N. Oyarzo, and R.L. Ax. 2000. Isolation and characterization of a human gene homologous to the gene of bovine fertility-associated antigen. *Human Genome Meet.* Vancouver, Canada (Abstr.).

Index

F

N

O

P

Q

R

S

T

U

V